CHARACTER RECOGNITION SYSTEMS

BICENTENNIAL
1807
WILEY
2007
BICENTENNIAL

THE WILEY BICENTENNIAL–KNOWLEDGE FOR GENERATIONS

*E*ach generation has its unique needs and aspirations. When Charles Wiley first opened his small printing shop in lower Manhattan in 1807, it was a generation of boundless potential searching for an identity. And we were there, helping to define a new American literary tradition. Over half a century later, in the midst of the Second Industrial Revolution, it was a generation focused on building the future. Once again, we were there, supplying the critical scientific, technical, and engineering knowledge that helped frame the world. Throughout the 20th Century, and into the new millennium, nations began to reach out beyond their own borders and a new international community was born. Wiley was there, expanding its operations around the world to enable a global exchange of ideas, opinions, and know-how.

For 200 years, Wiley has been an integral part of each generation's journey, enabling the flow of information and understanding necessary to meet their needs and fulfill their aspirations. Today, bold new technologies are changing the way we live and learn. Wiley will be there, providing you the must-have knowledge you need to imagine new worlds, new possibilities, and new opportunities.

Generations come and go, but you can always count on Wiley to provide you the knowledge you need, when and where you need it!

WILLIAM J. PESCE
PRESIDENT AND CHIEF EXECUTIVE OFFICER

PETER BOOTH WILEY
CHAIRMAN OF THE BOARD

CHARACTER RECOGNITION SYSTEMS
A Guide for Students and Practioners

MOHAMED CHERIET
École de Technologie Supérieure/University of Quebec, Montreal

NAWWAF KHARMA
Concordia University, Montreal

CHENG-LIN LIU
Institute of Automation/Chinese Academy of Sciences, Beijing

CHING Y. SUEN
Concordia University, Montreal

WILEY-
INTERSCIENCE

A JOHN WILEY & SONS, INC., PUBLICATION

Published by John Wiley & Sons, Inc., Hoboken, New Jersey
Published simultaneously in Canada

For general information on our other products and services or for technical support, please contact our Customer Care Department within the United States at (800) 762-2974, outside the United States at (317) 572-3993 or fax (317) 572-4002.

Wiley also publishes its books in a variety of electronic formats. Some content that appears in print may not be available in electronic formats. For more information about Wiley products, visit our web site at http://www.wiley.com.

Wiley Bicentennial Logo: Richard J. Pacifico

Library of Congress Cataloging-in-Publication Data

Character recognition systems : a guide for students and practioners / Mohamed Cheriet ... [et al.].
        p. cm.
Includes bibliographical references and index.
ISBN 978-0-471-41570-1 (cloth)
1.   Optical character recognition devices. I. Cheriet, M. (Mohamed)
TA1640.C47 2007
006.4'24–dc22
                                        2007011314

10 9 8 7 6 5 4 3 2 1

To my parents,
my wife Farida,
my daugters Imane,
Islem and Ihcene
(Mohamed Cheriet)

To my parents Nayef and Fauzieh,
to the children of Palestine
(Nawwaf Kharma)

To my parents,
my wife Linlin,
my son Jiayi
(Cheng-Lin Liu)

To my parents,
my wife Ling,
my sons Karwa and Karnon
(Ching Y. Suen)

CONTENTS

PREFACE

In modern society, we rely heavily on computers to process huge volumes of data. Related to this and for economic reasons or business requirements, there is a great demand for quickly inputting the mountains of printed and handwritten information into the computer. Often these data exist on paper and they have to be typed into the computer by human operators, for example, billions of letters in the mail, checks, payment slips, income tax forms, and many other business forms and documents. Such time-consuming and error-prone processes have been lightened by the invention of OCR (optical character recognition) systems that read written data by recognizing them at high speeds, one character at a time. However, the capabilities of the current OCR systems are still quite limited and only a small fraction of the above data are entered into the computer by them. Hence, much effort is still needed to enable them to read printed and handwritten characters more accurately. This book looks at some of the problems involved and describes in depth, some possible solutions, and the latest techniques achieved by researchers in the character recognition field.

Although several OCR books have been published before,[1] none of them are up-to-date now. In view of this, we have gathered together to write this book. It is our objective to provide the readers with the fundamental principles and state-of-the-art computational methods of reading printed texts and handwritten materials. We have also highlighted the key elements of character recognition[2] with extensive illustrations and references so that the readers can pursue further research and studies. It is hoped that this book will help graduate students, professors, researchers, and practitioners to master the theory and latest methodologies in character recognition, and to apply them in their research and practice.

---

[1]See references [5] to [7] at the end of Chapter 1.
[2]See references [4] to [7] at the end of Chapter 1.

This book consists of six chapters, each preceded by an introduction and followed by a comprehensive list of references for further reading and research. These chapters are summarized below:

Chapter 1 introduces the subject of character recognition, some perspectives on the history of OCR, its applications, evolution, and a synopsis of the other chapters.

Chapter 2 describes the most widely used preprocessing techniques. An intelligent form processing system is used to illustrate systematically: (a) a scale-based approach of enhancing the signal-to-noise ratio and (b) an efficient method of extracting the items-of-interest in a digital text image. This chapter also introduces the essential smoothing and noise removal techniques and key methods of normalizing the digitized characters, including the detection and correction of slants and skews, and the normalization of character sizes and positions. As well, methods of extracting character contours and skeletons are presented with appropriate illustrations.

All characters and words possess their own salient features. Chapter 3 shows various ways of extracting these features and evaluating them. Section 1 covers a large number of features commonly used for character recognition. Both structural (such as discrete features and geometric shapes) and statistical (such as moments and Fourier descriptors) features are described. Section 2 examines the relation between feature selection and classification, and the criteria for evaluating different feature subsets. Finally, a new concept of feature creation is introduced, and some pointers in the search for novel features as well as avenues for discovering some unknown ones are provided.

Chapter 4 presents some modern classification methods that have proven to be successful in character recognition. They include statistical methods, artificial neural networks (ANN), support vector machines (SVM), structural methods, and multi-classifier methods. ANNs and SVMs make use of a feature vector of fixed dimensionality mapped from the input patterns. Structural methods recognize patterns via elastic matching of strings, graphs, or other structural descriptors. Multiclassifier methods integrate the decisions of several classifiers. This chapter describes, in depth, the concepts and mathematical formulations of all these methods, and their relations to Bayesian decision theory, and to parametric and nonparametric methods. It also discusses their properties and learning procedures. Furthermore, this chapter provides comprehensive guidelines in implementing practical classifiers and in applying them to recognize characters. Examples of experimental designs, training procedures, and testing of real-life databases are also provided.

In real-world applications, documents consist of words or character strings rather than elementary units of isolated characters. However, word and character string recognition processes involve character segmentation, character recognition, and contextual analyses. Chapter 5 gives an overview of word and string recognition methods, and techniques related to segmentation and recognition. Both explicit and implicit segmentation methods are introduced, as well as holistic approaches that make use of more global features and word modeling. Because hidden Markov models (HMMs) have frequently been used in the segmentation process, their theoretical background is treated in a separate section. Implementation issues are also discussed in this chapter.

To provide further details about the practical aspects, three case studies have been included in Chapter 6. The first one illustrates an automatic method of generating pattern recognizers with evolutionary computation. The second one presents a real-world Chinese character recognition system. The third shows an automatic segmentation and recognition system for handwritten dates on Canadian bank cheques. Each study includes a description of the method and theory behind with substantial experimental results.

M. Cheriet, N. Kharma,
C.-L. Liu and C. Y. Suen
Montreal, Quebec
December 2006

ACKNOWLEDGMENTS

This textbook would have not been possible without the significant contributions from its authors as well as supporting contributors, who helped with the process of writing this book. The authors wish to thank first the following people for their deep involvement in writing parts of the book, namely

- Mr. Javad Sadri, who wrote sections 2.4, 5.2.2 and 5.5. He also spent appreciable time for proof reading and creating the index of this book. He contributed in converting the draft of the book to LaTeX format, as well. He is currently, a postdoctoral research fellow at the Center for Pattern Recognition and Machine Intelligence (CENPARMI) in Computer Science and Software Engineering Department of Concordia University, Montreal, Quebec, Canada. E-mail: j_sadri@cse.concordia.ca.
- Mr. Taras Kowaliw should be acknowledged for his contributions to the case studies chapter. Ms. Jie YAO should be acknowledged for her contributions to the feature selection section of the feature selection and creation chapter.
- Mr. Vincent Doré, Mr. Ayoub Alsarhan and Mrs. Safeya Mamish: doctoral students under Prof. Cheriet supervision, should be acknowledged for their valuable help in generating the index of this book.
- Dr. Xiangyun Ye, a former postdoctoral fellow at CENPARMI, who contributed to Sections 2.1 and 2.2.
  E-mail: xyye@ieee.org
- Dr. Mounim El-Yacoubi, a former postdoctoral fellow at CENPARMI, who contributed Section 5.4. He is currently a senior researcher at Parascript, LLC, Boulder, Colorado, USA.
  E-mail: mounim.el-yacoubi@parascript.com

- Dr. Louisa Lam, a member of CENPARMI and former research associate, who contributed to Section 6.3, is currently an associate vice-president at the Institute of Education in Hong Kong.
  E-mail: llam@ied.edu.hk

- Dr. Joanna (Qizhi) Xu, a former PhD candidate and postdoctoral fellow at CENPARMI, who contributed to Section 6.3. She is currently a research scientist at Smart Technologies Inc., Calgary, Alberta, Canada.
  E-mail: joannaxu@smarttech.com

- Ms. Jie Yao and Mr. Taras Kowaliw, PhD candidates, who contributed to Section 6.1. Both are completing their doctoral studies in the Electrical and Computer Engineering Department, Concordia University, Canada.

In addition, we would like to thank Shira Katz for proofreading numerous sections of this book.

We are also very grateful to the following organizations for funding our research programs:

- NSERC: Natural Sciences and Engineering Research Council of Canada
- FCAR: Fonds pour la Formation de Chercheurs et l'Aide a la Recherche
- CFI: Canadian Foundation for Innovations
- NSFC: National Natural Science Foundation of China
- CAS: Chinese Academy of Sciences
- Industrial organizations: A2IA Corp. (Paris, France), Bell Canada, Canada Post (Ottawa, Canada), DocImage Inc. (Montreal, Canada), Hitachi Ltd. (Tokyo, Japan), La Poste (French Post, France)

Furthermore, Mohamed Cheriet is particularly grateful to his PhD student Vincent Doré for carefully reviewing Chapter 2 and for providing considerable assistance by typesetting parts of a text in LaTeX.

Cheng-Lin Liu is grateful to his former advisors and colleagues for their help, collaboration and discussions, particularly to Prof. Ruwei Dai (CAS, China), Prof. Jin H. Kim (KAIST, Korea), Prof. Masaki Nakagawa (TUAT, Japan), Dr. Hiromichi Fujisawa and his group in Hitachi Ltd. (Japan), and Prof. George Nagy (RPI, USA).

LIST OF FIGURES

LIST OF TABLES

ACRONYMS

1-NN	Nearest Neighbor
2D	Two-Dimensional
3D	Three-Dimensional
AC	Auto-Complexity
ADF	Automatically Defined Functions
ANN	Artificial Neural Network
ANNIGMA	Artificial Neural Net Input Gain Measurement Approximation
ANSI	American National Standards Institute
ARG	Attributed Relational Graph
ASCII	American Standard Code for Information Interchange
APR	Autonomous Pattern Recognizer
ARAN	Aspect Ratio Adaptive Normalization
BAB	Branch-And-Bound
BKS	Behavior Knowledge Space
BMN	Bi-Moment Normalization
BP	Back Propagation
CC	Connected Component
CCA	Connected Component Analysis
CD	Compound Diversity
CE	Cross Entropy
CI	Class Independent
CL	Competitive Learning
CRS	Character Recognition System
DAG	Directed Acyclic Graph

DAS	Document Analysis System
DAWG	Directed Acyclic Word Graph
DCS	Dynamic Classifier Selection
DLQDF	Discriminative Learning Quadratic Discriminant Function
DP	Dynamic Programming
DPI	Dot Per Inch
DTW	Dynamic Time Warping
ECMA	European Computer Manufacturers Association
EM	Expectation Maximization
FD	Fourier Descriptor
FDA	Fisher Discriminant Analysis
FDL	Form Description Language
FIFO	First In First Out
GA	Genetic Algorithm
GLVQ	Generalized Learning Vector Quantization
GP	Genetic Programming
GPD	Generalized Probabilistic Descent
HCCR	Handwritten Chinese Character Recognition
HMM	Hidden Markov Model
HONN	Higher-Order Neural Network
HRBTA	Heuristic-Rule-Based-Tracing-Algorithm
HT	Hough Transform
h.u.p.	Heisenberg Uncertainty Principle
IAC	Indian Auto-Complexity
ICA	Independent Component Analysis
ICDAR	International Conference on Document Analysis and Recognition
ICFHR	International Conference on Frontiers in Handwriting Recognition
ICPR	International Conference on Pattern Recognition
IP	Intersection Point
ISO	International Standards Organization
IWFHR	International Workshop on Frontiers in Handwriting Recognition
KCS	Kernel with Compact Support
KKT	Karush-Kuhn-Tucker
k-NN	k-Nearest Neighbor
LAC	Latin Auto-Complexity
LDA	Linear Discriminant Analysis
LDF	Linear Discriminant Function
LN	Linear Normalization
LoG	Laplacian of Gaussian
LOO	Leave-One-Out
LVQ	Learning Vector Quantization
MAP	Maximum A Posteriori
MAT	Medial Axis Transform
MCBA	Modified Centriod-Boundary Alignment
MCE	Minimum Classification Error

MCM	Machine Condition Monitoring
ME	Merger-Enabled
ML	Maximum Likelihood
MLP	Multi-Layer Perceptron
MMI	Maximum Mutual Information
MN	Moment Normalization
MQDF	Modified Quadratic Discriminant Function
MSE	Mean Squared Error
NCFE	Normalization-Cooperated Feature Extraction
NLN	Nonlinear Normalization
NMean	Nearest Mean
NN	Neural Network
OCR	Optical Character Recognition
P2D	Pseudo-Two-Dimensional
PA	Projection Analysis
PCA	Principal Component Analysis
PCCR	Printed Chinese Character Recognition
PDA	Personal Digital Aide
PDC	Peripheral Direction Contributivity
PDF	Probability Density Function
PNC	Polynomial Network Classifier
PV	Plurality Vote
QDF	Quadratic Discriminant Function
QP	Quadratic Programming
RBF	Radial Basis Function
RDA	Regularized Discriminant Analysis
RPN	Ridge Polynomial Network
SA	Simple Average
SAT	Symmetry Axis Transform
SCL	Soft Competitive Learning
SGA	Standard Genetic Algorithm
SLNN	Single-Layer Neural Network
SMO	Sequential Minimal Optimization
SOM	Self-Organizing Map
SOR	Successive Overrelaxation
SPR	Statistical Pattern Recognition
SRM	Structural Risk Minimization
SVM	Support Vector Machine
VQ	Vector Quantization
WA	Weighted Average
WTA	Winner-Take-All

CHAPTER 1

INTRODUCTION: CHARACTER RECOGNITION, EVOLUTION, AND DEVELOPMENT

This chapter presents an overview of the problems associated with character recognition. It includes a brief description of the history of OCR (optical character recognition), the extensive efforts involved to make it work, the recent international activities that have stimulated the growth of research in handwriting recognition and document analysis, a short summary of the topics to be discussed in the other chapters, and a list of relevant references.

1.1 GENERATION AND RECOGNITION OF CHARACTERS

Most people learn to read and write during their first few years of education. By the time they have grown out of childhood, they have already acquired very good reading and writing skills, including the ability to read most texts, whether they are printed in different fonts and styles, or handwritten neatly or sloppily. Most people have no problem in reading the following: light prints or heavy prints; upside down prints; advertisements in fancy font styles; characters with flowery ornaments and missing parts; and even characters with funny decorations, stray marks, broken, or fragmented parts; misspelled words; and artistic and figurative designs. At times, the characters and words may appear rather distorted and yet, by experience and by context, most people can still figure them out. On the contrary, despite more than five decades of intensive research, the reading skill of the computer is still way behind that of human beings. Most OCR systems still cannot read degraded documents and handwritten characters/words.

1.2 HISTORY OF OCR

To understand the phenomena described in the above section, we have to look at the history of OCR [3, 4, 6], its development, recognition methods, computer technologies, and the differences between humans and machines [1, 2, 5, 7, 8].

It is always fascinating to be able to find ways of enabling a computer to mimic human functions, like the ability to read, to write, to see things, and so on. OCR research and development can be traced back to the early 1950s, when scientists tried to capture the images of characters and texts, first by mechanical and optical means of rotating disks and photomultiplier, flying spot scanner with a cathode ray tube lens, followed by photocells and arrays of them. At first, the scanning operation was slow and one line of characters could be digitized at a time by moving the scanner or the paper medium. Subsequently, the inventions of drum and flatbed scanners arrived, which extended scanning to the full page. Then, advances in digital-integrated circuits brought photoarrays with higher density, faster transports for documents, and higher speed in scanning and digital conversions. These important improvements greatly accelerated the speed of character recognition and reduced the cost, and opened up the possibilities of processing a great variety of forms and documents. Throughout the 1960s and 1970s, new OCR applications sprang up in retail businesses, banks, hospitals, post offices; insurance, railroad, and aircraft companies; newspaper publishers, and many other industries [3, 4].

In parallel with these advances in hardware development, intensive research on character recognition was taking place in the research laboratories of both academic and industrial sectors [6, 7]. Although both recognition techniques and computers were not that powerful in the early days (1960s), OCR machines tended to make lots of errors when the print quality was poor, caused either by wide variations in typefonts and roughness of the surface of the paper or by the cotton ribbons of the typewriters [5]. To make OCR work efficiently and economically, there was a big push from OCR manufacturers and suppliers toward the standardization of print fonts, paper, and ink qualities for OCR applications. New fonts such as OCRA and OCRB were designed in the 1970s by the American National Standards Institute (ANSI) and the European Computer Manufacturers Association (ECMA), respectively. These special fonts were quickly adopted by the International Standards Organization (ISO) to facilitate the recognition process [3, 4, 6, 7]. As a result, very high recognition rates became achievable at high speed and at reasonable costs. Such accomplishments also brought better printing qualities of data and paper for practical applications. Actually, they completely revolutionized the data input industry [6] and eliminated the jobs of thousands of keypunch operators who were doing the really mundane work of keying data into the computer.

1.3   DEVELOPMENT OF NEW TECHNIQUES

As OCR research and development advanced, demands on handwriting recognition also increased because a lot of data (such as addresses written on envelopes; amounts written on checks; names, addresses, identity numbers, and dollar values written on invoices and forms) were written by hand and they had to be entered into the computer for processing. But early OCR techniques were based mostly on template matching, simple line and geometric features, stroke detection, and the extraction of their derivatives. Such techniques were not sophisticated enough for practical recognition of data handwritten on forms or documents. To cope with this, the Standards Committees in the United States, Canada, Japan, and some countries in Europe designed some handprint models in the 1970s and 1980s for people to write them in boxes [7]. Hence, characters written in such specified shapes did not vary too much in styles, and they could be recognized more easily by OCR machines, especially when the data were entered by controlled groups of people, for example, employees of the same company were asked to write their data like the advocated models. Sometimes writers were asked to follow certain additional instructions to enhance the quality of their samples, for example, write big, close the loops, use simple shapes, do not link characters, and so on. With such constraints, OCR recognition of handprints was able to flourish for a number of years.

1.4   RECENT TRENDS AND MOVEMENTS

As the years of intensive research and development went by, and with the birth of several new conferences and workshops such as IWFHR (International Workshop on Frontiers in Handwriting Recognition),[1] ICDAR (International Conference on Document Analysis and Recognition),[2] and others [8], recognition techniques advanced rapidly. Moreover, computers became much more powerful than before. People could write the way they normally did, and characters need not have to be written like specified models, and the subject of unconstrained handwriting recognition gained considerable momentum and grew quickly. As of now, many new algorithms and techniques in preprocessing, feature extraction, and powerful classification methods have been developed [8, 9]. Further details can be found in the following chapters.

1.5   ORGANIZATION OF THE REMAINING CHAPTERS

Nowadays, in OCR, once a printed or handwritten text has been captured optically by a scanner or some other optical means, the digital image goes through the following stages of a computer recognition system:

---

[1] Note that IWFHR has been promoted to an international conference, namely, the International Conference on Frontiers on Handwriting Recognition (ICFHR), starting in 2008 in Montreal, where it was born in 1990.
[2] IWFHR and ICDAR series were founded by Dr. Ching Y. Suen, coauthor of this book, in 1990 and 1991, respectively.

1. The preprocessing stage that enhances the quality of the input image and locates the data of interest.
2. The feature extraction stage that captures the distinctive characteristics of the digitized characters for recognition.
3. The classification stage that processes the feature vectors to identify the characters and words.

Hence, this book is organized according to the above sequences.

REFERENCES

1. H. Bunke and P. S. P. Wang. *Handbook of Character Recognition and Document Image Analysis*. World Scientific Publishing, Singapore, 1997.
2. S. Mori, H. Nishida, and H. Yamada. *Optical Character Recognition*, Wiley Interscience, New Jersey, 1999.
3. *Optical Character Recognition and the Years Ahead*. The Business Press, Elmhurst, IL, 1969.
4. Pas d'auteur. *Auerbach on Optical Character Recognition*. Auerbach Publishers, Inc., Princeton, 1971.
5. S. V. Rice, G. Nagy, and T. A. Nartker. *Optical Character Recognition: An Illustrated Guide to the Frontier*. Kluwer Academic Publishers, Boston, 1999.
6. H. F. Schantz. *The History of OCR*. Recognition Technologies Users Association, Boston, 1982.
7. C. Y. Suen. Character recognition by computer and applications. In T. Y. Young and K. S. Fu, editors, *Handbook of Pattern Recognition and Image Processing*. Academic Press, Inc., Orlando, FL, 1986, pp. 569–586.
8. Proceedings of the following international workshops and conferences:

   - *ICPR—International Conference on Pattern Recognition*
   - *ICDAR—International Conference on Document Analysis and Recognition*
   - *DAS—Document Analysis Systems*
   - *IWFHR—International Workshop on Frontiers in Handwriting Recognition.*

9. Journals, in particular:

   - *Pattern Recognition*
   - *Pattern Recognition Letters*
   - *Pattern Analysis and Applications*
   - *International Journal on Document Analysis and Recognition*
   - *International Journal of Pattern Recognition and Artificial Intelligence.*

CHAPTER 2

TOOLS FOR IMAGE PREPROCESSING

To extract symbolic information from millions of pixels in document images, each component in the character recognition system is designed to reduce the amount of data. As the first important step, image and data preprocessing serve the purpose of extracting regions of interest, enhancing and cleaning up the images, so that they can be directly and efficiently processed by the feature extraction component. This chapter covers the most widely used preprocessing techniques and is organized in a global-to-local fashion. It starts with an intelligent form-processing system that is capable of extracting items of interest, then explains how to separate the characters from backgrounds. Finally, it describes how to correct defects such as skewed text lines and slanted characters, how to normalize character images, and how to extract contours and skeletons of characters that can be used for more efficient feature extraction.

2.1 GENERIC FORM-PROCESSING SYSTEM

When you file the annual income tax form, writing a check to pay the utility bill, or fill in a credit card application form, have you ever wondered how these documents are processed and stored? Millions of such documents need to be processed as an essential operation in many business and government organizations including telecommunications, health care, finance, insurance, and government and public utilities. Driven by the great need of reducing both the huge amount of human labor involved in reading and entering the data on to electronic databases and the associated human error, many intelligent automated form-processing systems were designed and put to prac-

Character Recognition Systems: A Guide for Students and Practitioner, by M. Cheriet, N. Kharma, C.-L. Liu and C. Y. Suen   Copyright © 2007 John Wiley & Sons, Inc.

tice. This section will introduce the general concept of an intelligent form-processing system that can serve as the kernel of all form-processing applications.

The input of a typical form-processing system consists of color or gray level images in which every several hundred pixels corresponds to a linear inch in the paper medium. The output, on the contrary, consists of ASCII strings that usually can be represented by just a few hundred bytes. To achieve such a high level of data abstraction, a form-processing system depends on the following main components for information extraction and processing:

- Image acquisition: acquire the image of a form in color, gray level, or binary format.
- Binarization: convert the acquired form images to binary format, in which the foreground contains logo, the form frame lines, the preprinted entities, and the filled in data.
- Form identification: for an input form image, identify the most suitable form model from a database.
- Layout analysis: understand the structure of forms and the semantic meaning of the information in tables or forms.
- Data extraction: extract pertinent data from respective fields and preprocess the data to remove noise and enhance the data.
- Character recognition: convert the gray or binary images that contain textual information to electronic representation of characters that facilitate postprocessing including data validation and syntax analysis.

Each of the above steps reduces the amount of the information to be processed by a later step. Conventional approaches pass information through these components in a cascade manner and seek the best solution for each step. The rigid system architecture and constant information flowing direction limit the performance of each component to local maximum. The observation from data in real-life applications and the way humans read distorted data lead us to construct a knowledge-based, intelligent form-processing system. Instead of simply concatenating general-purpose modules with little consideration of the characteristics of the input document, the performance of each module is improved by utilizing as much knowledge as possible, and the global performance is optimized by interacting with the knowledge database at run-time.

An overview of an intelligent form-processing system with some typical inputs and outputs is illustrated in Figure 2.1. As the kernel of this intelligent form-processing system, the knowledge database is composed of short-term and long-term memories along with a set of generic and specific tools for document processing. The short-term memory stores the knowledge gathered in run-time, such as the statistics of a batch of input documents. The long-term memory stores the knowledge gathered in the training phase, such as the references for recognizing characters and logos, and identifying form types. Different types of form images can be characterized by their specific layout. Instead of keeping the original images for model forms, only the extracted signatures are needed in building the database for known form types. Here,

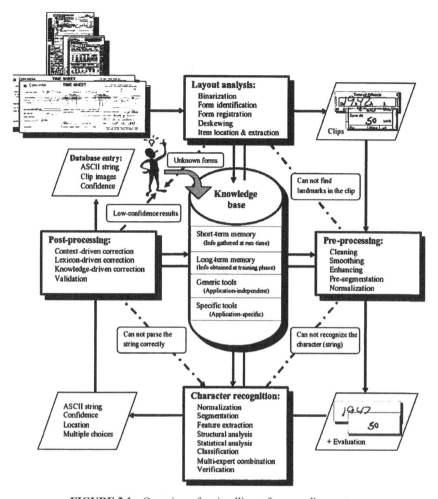

FIGURE 2.1 Overview of an intelligent form-reading system.

the signature of a form image is defined as a set of statistical and structural features extracted from the input form image. The generic tools include various binarization methods, binary smoothing methods, normalization methods, and so on. The specific tools include cleaning, recognition, or verification methods for certain types of input images. When an input document passes the major components of the systems, the knowledge in the database helps the components to select the proper parameters or references to process. At the same time, each component gathers respective information and sends back to the knowledge base. Therefore, the knowledge base will adapt itself dynamically at run-time. At run-time, for each unknown document image, features are extracted and compared with signatures of the prototypes exposed to the system in the learning phase.

As of understanding the form structures and extracting the areas of interest, there are two types of approaches in form-reading techniques: The first type of form readers relies upon a library of form models that describe the structure of forms being processed. A form identification stage matches the extracted features (e.g., line crossings) from an incoming form against those extracted from each modeled design in order to select the appropriate model for subsequent system processes. The second type of form readers do not need an explicit model for every design that may be encountered, but instead rely upon a model that describes only design-invariant features associated with a form class. For example, in financial document processing, instead of storing blank forms of every possible checks or bills that may be encountered, the system may record rules that govern how the foreground components in a check may be placed relative to its background baselines. Such a generic financial document processing system based on staff line and a form description language (FDL) is described [28]. In either type of form readers, the system extracts the signature, such as a preprinted logo or a character string, of each training form during the training phase, and stores the information in the long-term memory of the knowledge base. In the working phase, the signature of an input form is extracted and compared statistically and syntactically to the knowledge in the database. By defining a similarity between signatures from the two form images, we will be able to identify the format of the input form. If the input form is of known type, according to the form registration information, the pertinent items can be extracted directly from approximate positions. Otherwise, the signature of this unknown form can be registered through human intervention.

Through form analysis and understanding, we are able to extract the form frames that describe the structure of a form document, and thus extract the user filled data image from the original form image. Instead of keeping the original form image, only extracted user filled data images need to be stored for future reference. This will help to minimize storage and facilitates accessibility. Meanwhile, we can use the "signature" of form images to index and retrieve desired documents in a large database. As the kernel of this *up* to this moment, the useful information in a form is reduced from a document image that usually contains millions of gray level or color pixels to only the subimages that contain the items of interest.

2.2   A STROKE MODEL FOR COMPLEX BACKGROUND ELIMINATION

Before the items of interest can be automatically recognized and thus converted to simple ASCII strings, a crucial step is to extract only the pixels that belong to the characters to be recognized. For documents that are printed in "dropout" colors that are brighter than the black or blue ink with which people fill forms, the backgrounds can be dropped by applying color image filters either during or after the image acquisition step. This will be very useful in separating the user-entered data from machine-preprinted texts. Especially, when the user-entered data touch or cross the preprinted entities, color may be the only information that can distinguish these two counterparts sharing similar spatial features. Color image processing is time and space consuming. Moreover, a majority of the forms are printed without dropout ink for the sake of low cost and convenience. For throughput and cost concerns, most document processing systems resort to gray level, or even binary images.

In gray level documents, the data extraction procedure often requires binarizing the images, which discards most of the noise and replaces the pixels in the characters and the pixels in the backgrounds with binary 0s and 1s, respectively. The accuracy of data extraction is critical to the performance of a character recognition system, in that the feature extraction and the recognition performance of the characters largely depend on the quality of the data extracted. The variability of the background and structure of document images, together with the intrinsic complexity of the character recognition problem, makes the development of general algorithms and strategies for automatic character extraction difficult.

Modeling character objects is thus an attractive topic in researches of character extraction methods. Different methods taking advantage of various aspects of the definition of a character stroke evolved and flourished in different applications. Consider a document image with character data appearing darker than the background; the following facts are often used to derive different character extraction techniques:

- The characters are composed of strokes.
- A stroke is an elongated connected component that is darker than the pixels in its local neighborhood.
- A stroke has a nominally constant width that is defined by the distance between two close gray level transitions of the opposite directions.
- The local background of a stroke has a much smaller variation compared to that of a stroke.

The simplest property that the pixels in a character can share is intensity. Therefore a natural way to extract the characters is through thresholding, a technique that separates a pixel to object or background by comparing its gray level or other feature to a reference value. Binarization techniques based on gray levels can be divided into two classes: global and local thresholdings. Global thresholding algorithms use a single threshold for the entire image. A pixel having a gray level lower than the threshold value is labeled as print (black), otherwise background (white). Locally, adaptive binarization methods compute a designated threshold for each pixel based on a neighborhood of the pixel. Some of the methods calculate a smooth threshold surface over the entire image. If a pixel in the original image has a gray level higher than the threshold surface value at the corresponding location, it is labeled as background, otherwise it is labeled as print. Other locally adaptive methods do not use explicit thresholds, but search for object pixels in a feature space such as gradient magnitude among others.

2.2.1 Global Gray Level Thresholding

For document images in which all character strokes have a constant gray level that is unique among all background objects, the global thresholding methods are the most efficient methods. The histogram of these document images are bimodal so that the characters and the backgrounds contribute to distinct modes, and a global threshold can be deliberately selected to classify best the pixels in the original image as object or background. The classification of each pixel simply depends on their gray level in the global view regardless of local features. Among many global thresholding techniques, the Otsu threshold selection [25] is ranked as the best and the fastest global thresholding method [29, 30]. Based on discriminant analysis, Otsu thresholding method is also being referred to as "optimal thresholding" and has been used in a wide range of machine vision-related applications that involve a two-class classification problem. The thresholding operation corresponds to partitioning the pixels of an image into two classes, C_0 and C_1, at a threshold t. The Otsu method solves the problem of finding the optimal threshold t^* by minimizing the error of classifying a background pixel as a foreground one or vice versa.

Without losing generality, we define objects as dark characters against lighter backgrounds. For an image with gray level ranges within $G = \{0, 1, \ldots, L-1\}$, the object and background can be represented by two classes, as $C_0 = \{0, 1, \ldots, t\}$ and $C_1 = \{t+1, t+2, \ldots, L-1\}$. As elegant and simple criteria for class separability, the within and between class scatter matrices are widely used in discriminant analysis of statistics [12]. The within-class variance, between-class variance, and total-variance reach the maximum at equivalent threshold t. Using σ_W^2, σ_B^2, and σ_T^2 to represent them, respectively, the optimal threshold t^* can be determined by maximizing one of the following functions against the threshold:

$$\lambda = \frac{\sigma_B^2}{\sigma_W^2}, \ \eta = \frac{\sigma_B^2}{\sigma_T^2}, \ \kappa = \frac{\sigma_T^2}{\sigma_W^2}. \tag{2.1}$$

Taking η, for example, the optimal threshold t^* is determined as

$$t^* = \arg \max_{t \in [0, L-1]} (\eta), \tag{2.2}$$

in which

$$P_i = n_i/n, \ w_0 = \sum_{i=0}^{t} P_i, \ w_1 = w - w_0, \tag{2.3}$$

$$\mu_T = \sum_{i=0}^{L-1} i P_i, \ \mu_t = \sum_{i=0}^{t} i P_i, \ \mu_0 = \frac{\mu_t}{w_0}, \ \mu_1 = \frac{\mu_T - \mu_t}{1 - \mu_0}, \tag{2.4}$$

$$\sigma_T^2 = \sum_{i=0}^{L-1} (i - \mu_T)^2 P_i, \ \sigma_B^2 = w_0 w_1 (\mu_1 - \mu_0)^2. \tag{2.5}$$

Here n_i is the ith element of the histogram, that is, the number of pixels at gray level i; $n = \sum_{i=0}^{L-1} n_i$ is the total number of pixels in the image; $P_i = n_i/n$ is the probability of occurrence at gray level i. For a selected threshold t^* of a given image, the class probabilities w_o and w_1 represent the portions of areas occupied by object and background classes, respectively. The maximal value of η, denoted by η^*, can serve as a measurement of class separability between the two classes, or the bimodality of the histogram. The class separability η lies in the range $[0, 1]$; the lower bound is reached when the image has a uniform gray level, and the upper bound is achieved when the image consists of only two gray levels.

2.2.2 Local Gray Level Thresholding

Real-life documents are sometimes designed deliberately with stylistic, colorful, and complex backgrounds, causing difficulties in character extraction methods. While global thresholding techniques can extract objects from simple, uniform backgrounds at high speed, local thresholding methods can eliminate varying backgrounds at a price of long processing time.

A comparative study of binarization methods for document images has been given by Trier et al. [29, 30]. Eleven most promising locally adaptive algorithms have been studied: *Bernsen's* method, *Chow* and *Kaneko's* method, *Eikvil's* method, *Mardia* and *Hainsworth's* method, *Niblack's* method, *Taxt's* method, *Yanowitz* and *Bruckstein's* method, White and Rohrer's *Dynamic Threshold* Algorithm, White and Rhorer's *Integrated Function* Algorithm, *Parker's* method, and *Trier* and *Taxt's* method (corresponding references can be found in [29, 30]). Among them the first eight methods use explicit thresholds or threshold surfaces, whereas the latter three methods search for printed pixels after having located the edges. Trier and Jain concluded that Niblack's [24] and Bernsen's [5] methods along with postprocessing step proposed by Yannowitz and Bruckstein [37] were the fastest and best ones concerning a set of subjective and goal-directed criteria. The postprocessing step here is used to improve the binary image by removing "ghost" objects. The average gradient value at the edge of each printed object is calculated, and objects having an average gradient below a threshold are labeled as misclassified and therefore removed.

The Bernsen's local thresholding method computes the local minimum and maximum for a neighborhood around each pixel $f(x, y) \in [0, L - 1]$, and uses the median of the two as the threshold for the pixel in consideration:

$$g(x, y) = (F_{\max}(x, y) + F_{\min}(x, y))/2, \tag{2.6}$$

$$b(x, y) = \begin{cases} 1 & \text{if } f(x, y) < g(x, y), \\ 0 & \text{otherwise.} \end{cases} \tag{2.7}$$

$F_{\max}(x, y)$ and $F_{\min}(x, y)$ are the maximal and minimal values in a local neighborhood centered at pixel (x, y).

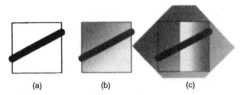

(a)   (b)   (c)

FIGURE 2.2 Character strokes on different types of background; (a) Simple background, easy to binarize by global thresholding; (b) Slowly varying background, which can be removed by local thresholding; (c) Fast varying background, which need to be removed by feature thresholding.

To avoid "ghost" phenomena, the local variance $c(x, y)$ can be computed as

$$c(x, y) = F_{\max}(x, y) - F_{\min}(x, y). \tag{2.8}$$

The classification of a foreground pixel $f(x, y)$ can be verified by examining the local variance being higher than a threshold. Each pixel is classified as object or background pixel according to the following condition:

$$b(x, y) = \begin{cases} 1 & \text{if } (f(x, y) < g(x, y) \text{ and } c(x, y) > c^*) \text{ or} \\ & \quad (f(x, y) < f^* \text{ and } c(x, y) \le c^*), \\ 0 & \text{otherwise.} \end{cases} \tag{2.9}$$

The thresholds c^* and f^* are determined by applying Otsu's method to the histogram of $c(x, y)$ and $g(x, y)$, respectively. The characters are extracted and stored as $B = \{b(x, y) | (x, y) \in F\}$.

This local thresholding method is sufficient for document images with slowly varying nonuniform backgrounds. However, when the background tends to vary at sharp edges, the edge of background can be taken as objects, causing great difficulty to the ensuing recognition stage. Figure 2.2 shows some examples of character strokes on different types of backgrounds, each posing a greater difficulty in extracting the strokes. Figure 2.2(a) is an ideal case for simple background that can be removed by a global threshold; Figure 2.2(b) illustrates a slow-varying background that can be removed by the above mentioned local thresholding method; Figure 2.2(c) illustrates a background image with a sharp edge in the middle, which will produce an artifact if the local thresholding method is applied. The solution to this problem requires a more precise and robust feature to describe the objects of our interest.

2.2.3   Local Feature Thresholding—Stroke-Based Model

When the characters in a document image cannot distinguish themselves from the backgrounds in the gray level space, the character extraction can be done in a feature space in which the characters can be distinctively described.

Edge-based image analysis is a widely used method in computer vision. Marr [22] describes any type of significant scene structures in an image as discontinuities in the intensity, generally called edge. From the point of view of extracting scene structures, the objective is not only finding significant intensity discontinuities but also suppressing unnecessary details and noise while preserving positional accuracy. Bergholm [4] proposed that in any image, there are only four elementary structures: the "*step edge*," the "*double edge*," the "corner edge" (L-junction), and the "*edge box*" (blob):

$$step\ edge\ f(x, y) = \begin{cases} 1 & \text{if } x > a, \\ 0 & \text{elsewhere} \end{cases}$$

$$double\ edge\ f(x, y) = \begin{cases} 1 & \text{if } a < x < b, \\ 0 & \text{elsewhere} \end{cases}$$

$$corner\ edge\ f(x, y) = \begin{cases} 1 & \text{if } x > a \text{ and } y > b, \\ 0 & \text{elsewhere} \end{cases}$$

$$edge\ box\ f(x, y) = \begin{cases} 1 & \text{if } a < x < b \text{ and } c < y < d. \\ 0 & \text{elsewhere} \end{cases}$$

The *double-edge* model, delimited by a positive edge and a negative one nearby, is a good description of a stroke in a local region. An *edge box* can also be considered as *double edges* in two orthogonal directions. Based on this observation, Palumbo [26] proposed a second derivative method to extract the characters; White and Rohrer [32] proposed an integrated function technique to extract the sequence of $+$, $-$, and 0 in the differential image and take the pixels between sequences "$+-$" and "$+$" as objects; Liu et al. [21] extracted baselines from bank checks by searching for a pair of opposite edges in a predefined distance. One common problem for these methods is the sensitivity to noise because the depths of the strokes have not been taken into account. Based on this observation, many approaches were proposed to describe character strokes in different ways, with the common features behind are the thickness and darkness of a stroke. In this sense, the local threshold techniques based on stroke width restriction can be unified into searching for *double-edge* features occurring within a certain distance. In one-dimensional profile, the intensity of a stroke with thickness W can be estimated as a *double edge*, whose intensity is

$$DE^+(x) = \max_{i \in [1, W-1]} \{\min(f(x - i), f(x + W - i))\} - f(x) \qquad (2.10)$$

for dark-on-light strokes, or

$$DE^-(x) = f(x) - \min_{i \in [1, W-1]} \{\max(f(x - i), f(x + W - i))\} \qquad (2.11)$$

for light-on-light strokes.

For a given pixel p on the stroke, we assume that d_1 and d_2 are the distances between p and the nearest opposite boundaries p_1 and p_2. Therefore the *double-*

edge feature value at pixel p is positive if and only if $(d_1 + d_2) < W$. Taking all possible directions of a stroke into consideration, we can use $d = 0, 1, 2, 3$ to refer to four directions $\{0, \pi/4, \pi/2, 3\pi/4\}$ that approximate most possible edge directions around a pixel. Therefore the intensity or darkness of a two-dimensional stroke can be approximated as

$$DE_w(p) = \max_{d \in [0,3]} \{DE_{Wd}\}, \tag{2.12}$$

whose positive values indicate the existence of a dark stroke written on a lighter background. The depth or intensity of a dark-on-light stroke is measured by the magnitude of the positive *double-edge* features being extracted. Similarly, the *double-edge* features can also be used to extract light-on-dark strokes that compose characters in scene images, such as car license plates, or road signs, and so on. The problem of extracting character strokes from a gray level image is thus converted to binarizing an image whose pixel values correspond to the magnitude of the *double-edge* feature in the original image. Although a global thresholding of the *double-edge* feature image can usually generate a satisfactory result, a hysteresis thresholding may further remove noise and tolerate the variation of gray scales within a stroke. Such a step can be implemented using region growing or conditional dilation techniques.

$$B_i = (B_{i-1} \oplus D) \bigcap |DE_W|_{T_1}, \tag{2.13}$$

$$i = 1, 2, \ldots, \quad B_0 = |DE_W|_{T_0}, \quad |F|_T = \{f_T(x, y)|(x, y) \in F\},$$

$$f_T(x, y) = \begin{cases} \begin{cases} 1 & \text{if } f(x, y) < T \\ 0 & \text{otherwise} \end{cases} & \text{if } T \text{ is a constant} \\ \begin{cases} 1 & \text{if } f(x, y) < T \\ 0 & \text{otherwise} \end{cases} & \text{if } T \text{ is an image} \end{cases},$$

Here, $|F|_T$ is a thresholding operation with either a constant threshold T or a threshold surface defined by an image, and $0 < T_1 < T_0 < L$ are two thresholds for the intensity of strokes, and D is a 3×3 square structuring element. Equation 2.13 is repeated until there is no change from B_{i-1} to B_i. The characters are thus extracted and stored as B_n, where $B_n = B_{n-1}$ holds. One practical choice of the thresholds T_0, T_1, as 1.2 and 0.8 times the Ostu optimal threshold for the *double-edge* feature image. A low threshold of T_0 tends to expand the extracted strokes to background regions, whereas a high T_1 excludes isolated light handwritten strokes from the final results.

Figure 2.3 shows a close-up of character extraction results from web images. The hysteresis thresholding of the *double-edge* feature images ensures that the complex backgrounds around the texts are removed. More examples are shown in Figure 2.4 and 2.5. Further postprocessing that utilizes topological information of the binarized

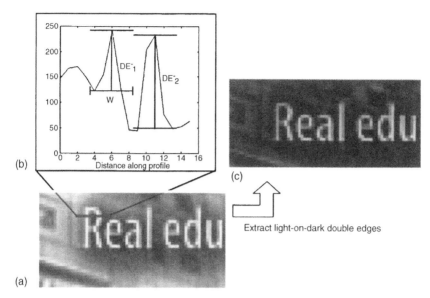

FIGURE 2.3   (a) A subimage of a web page image with light-on-dark text; (b) A one-dimensional profile taken from the subimage, at the position marked with blue line; two positive *double-edge* peaks are detected, with intensity of DE-1 and DE-2, respectively. The width of the *double-edge* detector is W (5 pixels); (c) The light-on-dark *double-edges* extracted from (a).

foregrounds can remove the isolated blobs that do not belong to a text line. Such techniques are very useful for content-based image indexing, including searching for web images that contain certain textual information.

2.2.4   Choosing the Most Efficient Character Extraction Method

As stated in the previous section, the selection of an efficient character extraction method depends largely on the complexity of the images. Real-life documents are sometimes designed deliberately with stylistic, colorful, and complex backgrounds, causing difficulties in choosing appropriate character extraction methods. Although global thresholding techniques can extract objects from simple or uniform backgrounds at high speed, local thresholding methods can eliminate varying backgrounds at a price of processing time. Document images such as checks or maps may contain complex backgrounds consisting of pictures and designs printed in various colors and intensities, and are often contaminated by varying backgrounds, low contrast, and stochastic noise. When designing a character recognition system for these documents, the locally adaptive algorithms are apparently better choices. For the thresholding techniques discussed in this section, regardless of the space from which the thresholds are computed, the computational complexity of a sequential implementation of the thresholding is measured by the number of operations required to compute the threshold for each pixel.

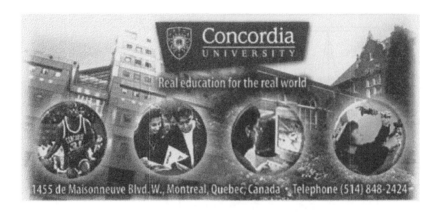

FIGURE 2.4   Examples of character extraction from web page images.

- Local gray level thresholding (enhanced Bernsen's [5] method). The local gray level threshold at size W is calculated by Eqs. 2.6–2.9. The major part of the computations involved in this method is the calculation of local maximum and local minimum described in formula 2.6. Because $N - 1$ comparisons are

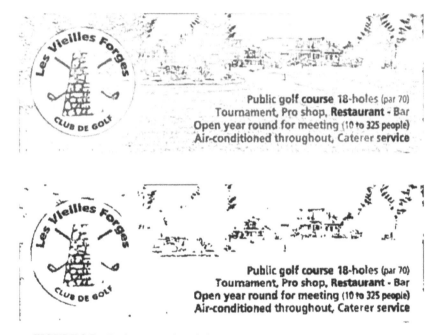

FIGURE 2.5   Further examples of character extraction from web page images.

required to extract the maximum or minimum value of N elements, $(W^2 - 1) \times 2$ comparisons are required for each pixel in the worst case. Regardless of the size of the structuring elements, erosion and dilation by linear structuring elements can be implemented with only three comparisons for each pixel and with six comparisons by rectangular structuring elements. Therefore, the total number of comparisons required in calculating the local maximum and minimum is

$$C_1 = 6 \times 2 = 12. \tag{2.14}$$

• Double-edge-based feature thresholding. The intensity of a *double edge* with thickness W is calculated by Eqs. 2.10 and 2.11. Similarly, $[2 + (W + 1)]$

comparisons are required for each $DE_d(p)(d = 0, 1, 2, 3)$, and the total number of comparisons required in calculating *double edge* for each pixel is

$$C_2 = [2 + (W + 1)] \times 4 + 3 = 4W + 7. \tag{2.15}$$

When W becomes larger than three, the fast algorithm independent of kernel size can be used to calculate the local maxima. In this case, the total number of comparisons required for calculating *double edge* at each pixel in the raw image becomes

$$C_3 = \begin{cases} 4W + 7 & \text{if } W \le 3, \\ (3 + 1) \times 4 + 3 = 19 & \text{otherwise.} \end{cases} \tag{2.16}$$

Because each method improves its robustness at the price of reducing the processing speed, when designing a high-throughput application system that needs to handle document images with a spectrum of complexity in the backgrounds, it is very important to have a balanced solution to process large quantities of documents at both high speed and high reliability. The first step of a balanced solution is to differentiate complex images from simple ones, so that various types of images can be processed with the appropriate methods. Meanwhile, the classification of document images should involve as little computation as possible, so that no extra computation is needed for processing simple images.

An image complexity measurement such as the class separability in the Otsu's method can help to make the decision of choosing the appropriate character extraction approach. The maximal ratio of the between-class variance and total-variance corresponds to the optimal threshold η^* and can serve as a measurement of the separability between the two classes and the bimodality of the histogram. Because the larger the η is, the simpler the background, a document image can be classified according to its class separabilities and treated with different thresholding techniques. For an ideal case when characters and background are composed of only two distinct gray levels, the class separability reaches the maximum, η^*.

By choosing one or several proper thresholds for class separability, we are able to classify document images as relatively "simple" or "complex," and thus apply corresponding binarization methods to optimize the batch processing speed and meanwhile preserve recognition rate of the overall system. The thresholds for class separability are established from a training set of real-life document images, depending on the application. The training set should be composed of document images at different complexity levels, and the distribution of the complexity should conform to that of the real inputs to the system.

Although there is no explicit function of image complexity versus class separability, it is clearly a nonincreasing function. We may choose two thresholds η_0 and η_1, $0 \le \eta_0 \le \eta_1 \le 1$, to classify the document images into three classes: "simple" images with $\eta_1 < \eta \le 1$, "medium" images with $\eta_0 < \eta \le \eta_1$, or "complex" images with $0 \le \eta \le \eta_0$ and assign a specific thresholding technique to individual ranges. Suppose the average processing time of one image by global gray level, local gray level, and feature thresholding methods are t_0, t_1, and t_2, respectively; the average processing time can be expected as

$$\bar{t} = \sum_{i=0}^{2} p_i.t_i, \tag{2.17}$$

in which p_i represents the probability of different types of images in a testing set, estimated by the total number of images in the testing set and the number of images belonging to various classes. For the simplest case, when $p_i = 1/3$ ($i = 0, 1, 2$), the expectation of the processing time is the algebraic average of $t_i (i = 0, 1, 2)$, which is a substantial improvement over a sole feature thresholding technique when large quantities of images are processed.

2.2.5   Cleaning Up Form Items Using Stroke-Based Model

The stroke model is useful not only for extracting characters from gray level documents images but also for cleaning up the form entries from binary images. As a prerequisite step of a form-processing system, the major task in form dropout is to separate and remove the preprinted entities while preserving the user entered data by means of image-processing techniques. Many form-processing systems have been found successful when the filled-in items are machine-printed characters. However, as discussed in [39] and [1], a number of these approaches perform the extraction of filled-in items without paying attention to field overlap problem. This happens when the filled-in items touch or cross form frames or preprinted texts. For the approaches that can drop out form frames or straight lines, the preprinted texts remain an unsolved problem. When the filled-in items are unconstrained handwritings, this problem is more pronounced and can prevent the whole processing system from functioning properly.

A subimage obtained from the item location and extraction module of a form registration system usually consists of three components:

- Form frames, including black lines, usually called baselines, and blocks;
- Preprinted data such as logos and machine preprinted characters;
- User filled-in data (including machine-typed and/or handwritten characters and some check marks) located in predefined areas, called filled-in data areas, which are bounded by baselines and preprinted texts.

These three components actually carry two types of information: preprinted entities, which give instructions to the users of the form; and the filled-in data. In most applications where rigid form structures are used, the preprinted entities appear at the same expected positions. In an ideal case, the filled-in items can be extracted by a simple subtraction of a registered form model from the input image. With distortion, skewing, scaling, and noise introduced by the scanning procedure, it is almost impossible to find an exact match between the input image and the model form. One way of dealing with this degradation is to define a region in which the preprinted entities are mostly likely to appear and remove them by the stroke width difference between the filled data and the preprinted entities. Because the strokes of the filled-in characters can be either attached to or located across the form frames and preprinted texts, the problem of item cleaning involves the following steps:

- Estimating the positions of preprinted entities;
- Separating characters from form frames or baselines;
- Reconstructing strokes broken during baseline removal;
- Separating characters from preprinted texts.

One of the most important characteristics of character objects is the stroke width. For each foreground pixel in a binary image, the stroke width can be approximated as $SW(x, y) = \min(SW_H, SW_V)$, in which SW_H and SW_V are the distances between the two closest background pixels in horizontal and vertical directions. We have observed that in most real-life applications, the form frames and the instructions are printed in relatively small fonts. When the users fill in the forms with ball or ink pen, the stroke width of the handwritings is usually larger than that of the preprinted entities. The histograms of the handwritings and the preprinted entities in 10 form samples scanned at 300 DPI are shown in Figure 2.6, which clearly shows the different distributions. This observation helps us to distinguish the preprinted frames and texts from handwritings by eliminating the pixels whose corresponding stroke width is less than a threshold. The stroke width of the handwriting (t_{hw}) and the preprinted entities (t_{pp}) can be estimated at run-time by collecting histograms of stroke widths:

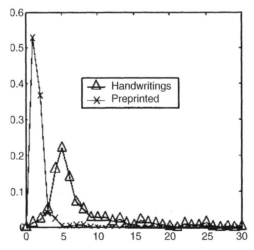

FIGURE 2.6   Histograms of the stroke widths of handwritings and preprinted entities obtained from 10 sample form images scanned at 300 DPI.

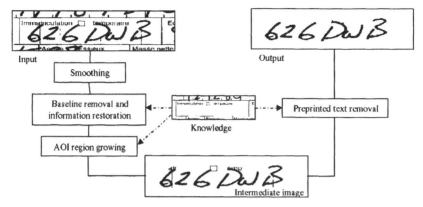

FIGURE 2.7   Removal of preprinted texts.

$$t_{hw} = \arg \max \{\text{hist}[SW(x, y)|(x, y) \in \text{fill in region}]\}, \qquad (2.18)$$

$$t_{pp} = \arg \max \{\text{hist}[SW(x, y)|(x, y) \in \text{preprinted region}]\}. \qquad (2.19)$$

Therefore the threshold can be chosen as $(t_{hw} + t_{pp})/2$. Following a unified scheme of baseline removal and information restoration, a set of binary morphological operators at different sizes can be used to extract the stroke width at each foreground pixel [38], and thus remove the connected preprinted texts from the handwriting. An example of the cleaning procedures and the corresponding intermediate results are illustrated in Figure 2.7.

2.3 A SCALE-SPACE APPROACH FOR VISUAL DATA EXTRACTION

When we design a recognition engine, we may assume to have access to clearly written material (printed or handwritten) on a homogenous background. This is not the case in most real applications. The challenge increases when the image space is in gray level and the material is handwritten, in the presence of a noisy and complex background. In this section, we focus on extracting handwritten material from gray level form images with slow-varying backgrounds. Having this in mind, our aim is to extract handwritten data that contains characters of different sizes, with variable distances and where their corresponding image intensity changes over a wide range of scales. Usually these data are decorated with uniformly distributed graphical symbols in their backgrounds. Examples can be found in mail pieces, bank checks, business forms, and so on. When such a gray-scale image is processed, it is not simple for a computer to distinguish the pertinent data from the background symbols. There is a great amount of information loss, resulting in characters with the presence of noise and unnecessary information. Visual shapes of handwriting characters are very important in improving recognition [10, 27]; results will depend mostly on the quality of the extracted characters. Therefore, it is necessary to perform a preprocessing procedure to extract the pertinent information before applying the recognition algorithm. For this purpose, various mathematical tools have been developed; multiscale representation has received considerable interest in this field because of its efficiency in describing real-world structures [17]. It has been shown that the concept of scale is crucial when describing the structure of images, which is needed for our application. It is therefore important to focus on image regularization, as images need to respect the Nyquist criterion when going through a smaller scale to larger one. Thus, the purpose of this study is thus to introduce a new low-pass filter that is an alternative to the approximation of the sample Gaussian filter.

2.3.1 Image Regularization

2.3.1.1 Linear Filters Degradations in document images are principally due to the acquisition system and other complex phenomena that will be highlighted below. Hence, effective tools are necessary for their automatic treatment, such as the compression, the segmentation, or, well again, the recognition of visual shapes. To grasp the issues well, we propose to return to the fundamental definition of an image, which may be explained as an intensity distribution of focused light on a plane. From a numerical standpoint, this is a matrix of numbers. These digital images are made of very irregular values for several reasons:

- Having visual objects at different scales, and their fine details come to perturb the objects of large sizes;
- The sensors record noisy data (electronic noise);
- The images are quantified (noise of quantification);
- The compression of certain images adds undesirable effects that are more or less visible;
- Information can be lost at the time of an eventual transmission;
- To these degradations, we add the one caused by the optical system (diffraction inducing fuzziness, deformation, and chromatic aberration), and by the weak movement of the acquisition system during the image acquisition.

For all of these reasons, the images are complex objects, and we are far from being able to completely analyze the shapes that they contain completely in an automatic manner. For better addressing this challenging problem, we rely on human perception to contribute to sturdy solutions. We consider the observed image as composed of the real signal and a white noise with unknown variance:

$$u_{\mathrm{obs}}(x, y) = u_{\mathrm{real}}(x, y) + \eta_\sigma(x, y).$$

To use an approach based on linear filtering, it is necessary to suppose that the signal is stationary. Intuitively, we look for a low-pass filter to eliminate the variations in very high frequencies. These filters are only mediocre for processing an image because the latter is not stationary; they considerably degrade the object contours and return a fuzzy image. But used within effective multiscale processes to offset this deficiency, an accurate contour extraction is possible. One of the two linear filters that are used the most is the Gaussian filter, as it has been proven to be the unique multiscale kernel [16] and its resultant reduced complexity is due to the kernel separability.

2.3.1.2 The Wiener Filtering Approach The Wiener filtering approach [33] consists of looking for a filter h that minimizes a cost function, which describes the discrepancy between the original signal and the restored signal. For images, the cost function should measure the visual degradation that is often difficult to model. The Euclidean distance does not model the visual degradations perfectly, but it remains mathematically correct. A cost function that is equal to the square of the Euclidean distance is calculated as the risk in average:

$$r = \mathbb{E}\{\|u_{\mathrm{obs}} \times h - u_{\mathrm{obs}}\|^2\}.$$

The transfer function of this filter is

$$H(u, v) = \left[1 + \frac{p_{\mathrm{n}}(u, v)}{p_{\mathrm{f}}(u, v)}\right]^{-1},$$

where $p_{\mathrm{n}}(u, v)$ and $p_{\mathrm{f}}(u, v)$ are, respectively, the spectrum of the power of the noise and the image. Unfortunately, the difficulty of estimating the parameters of the noise and the signal are limited to the application of this filter.

2.3.1.3 Multiscale Representation The multiscale representation [7, 34] furnishes a regularizing transformation from a fine scale to a coarse scale. This transformation can be seen as a simplification of the image. The fine structures disappear in a monotonous manner with the increase of the scale. It is also important that the structures not belonging to the fine scales are not created in the coarse ones (the kernel is said causal). Our goal in this chapter is to extract visual handwritten data from noisy backgrounds. Considering the large variation of stroke width and intensity in the handwritten data, a multiscale approach that preserves the data seems therefore to be very appropriate.

2.3.1.4 Choice of the Regularizing Kernel Meer et al. [23] suggested that the regularizing filter should be the closest to an ideal low-pass filter. They proposed to select a filter whose Fourier transform is close to an ideal low pass. An alternate approach is to use a positive and unimodal kernel that has a bell shape in both spatial and frequency domains. Thus, the high frequency components of the signal are allocated weights that are inferior to the low frequency components. The kernel we are seeking must be causal, isotropic (all the spatial points must be treated in the same manner), and homogenous (the spatial dimensions and the scale dimension must be treated in the same manner). From these kernel conditions, it was shown that

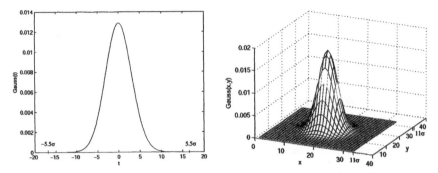

FIGURE 2.8 One-dimensional and two-dimensional profiles of a Gaussian kernel with variance σ equal to 9 and a mask support of size 11σ.

multiscale representation of a 2D signal satisfies the heat equation:

$$
\begin{cases}
\dfrac{\partial u}{\partial t} = \Delta u = div(\nabla u) \\[2ex]
\displaystyle\int \|u(.,t) - u_0\| \underset{t \to 0}{\longrightarrow} 0,
\end{cases}
$$

of which the solution is the signal convolved with the Gaussian kernel [16]. The Gaussian kernel possesses two very interesting properties:

- It has a semigroup property (the same properties of the group one except for the nonexistence of the inverse);
- It minimizes the Heisenberg uncertainty criterion, that is, all time-frequency atoms verify that $\sigma_t\sigma_w \geq 1/2$ where σ_t is the temporal variance and σ_w is the frequency variance. The one-dimensional and two-dimensional profiles of a Gaussian kernel are shown in Figure 2.8. The Gaussian kernel is known to preserve best the energy in the time-frequency resolution.

2.3.2 Data Extraction

2.3.2.1 An Algorithm Using the Gaussian Kernel   The Gaussian kernel can be used to detect edges [22, 4, 11], which usually define the boundaries of objects, and thus can be used for object segmentation in the presence of noise. Although the intensity changes may occur over a large range of spatial scales in noisy gray level images, a multiscale approach is used to achieve good quality data. The decision criterion in segmenting data [9, 10] is the detection of convex parts of the smoothed image I_i, at each scale level of resolution (ith step). A part of the smoothed image is convex if the $LoG \times I_{i-1}$ is positive. The scale-space parameter in this approach is the standard deviation σ; the mask (kernel) size, Dim, is therefore deduced at each step. The detailed algorithm is as follows.

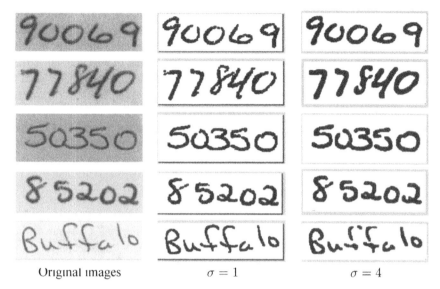

Original images	$\sigma = 1$	$\sigma = 4$

FIGURE 2.9   Original images and their segmentation results using Gaussian kernels.

Procedure Multiresolution ( )

Step 0: Let I^0 be a working space image; turn all its pixels on.

Step 1: Apply the $L \circ G$ operator to the input image I with a scale parameter $\sigma_1 = \sigma_{max}$ that is sufficiently large to extract all the information.

Step 2: Threshold the output image by turning off pixels in I^0 for which $L \circ G_{\sigma_1} \times I$ is negative to obtain I^1. Let i be the iteration index; set i to 1.

Step 3: Turn on the neighbors of pixels in I^i (for continuity).

Step 4: Decrease the value of σ by a step $\Delta\sigma$.

Step 5: Apply the $L \circ G$ operator to input image I with scale parameter σ, only on the pixels being turned on in the previous step in I^i.

Step 6: Threshold the output image as in Step 2, to obtain I^{i+1}. The image I^{i+1} constitutes the desired result at scale σ. Set i to $i + 1$.

Step 7: Repeat the process from Step 3 to Step 7 until $\sigma_n = \sigma_{min}$.

Step 8: Obtain the desired fine image $I^f = I^n$.

Some examples are illustrated in Figure 2.9. These figures show the influence of the variance σ, and thus the mask size parameter (which is set to 11σ) on the shape of the segmented images.

2.3.2.2 Practical Limitations of the Gaussian Kernel   The accuracy of computation using Gaussian kernels depends on the mask size; wide masks give precise computation at the cost of processing time, whereas small mask sizes decrease the processing time, but the accuracy is sometimes severely compromised and induces information loss, and this creates high frequencies due to the support truncate of the kernel. The inaccuracy caused by kernel truncation and the prohibitive processing time are the two fundamental practical limitations of the Gaussian kernel. Some solutions are proposed in the literature to overcome these problems such as the approximation of the Gaussian by recursive filters [14], or the use of truncated exponential functions instead of the Gaussian kernel [17]. These solutions reduce the processing time by approximating the Gaussian kernel, but the information loss still persists and sometimes it gets worse. In order to reduce the mask size and to prevent the information loss, Remaki and Cheriet proposed [27] a new family of kernels with compact supports. They are called KCS (kernel with compact support), and they can be used to generate scale-spaces in multiscale representations.

2.3.2.3 The KCS Filter   The KCS ρ_γ is derived from the Gaussian kernel, by composing the Gaussian $G_{1/2\gamma}$ variance $1/2\gamma$ and a C^1-diffeomorphism h that transforms the half plane to a unit ball. Let us define a function h as the following:

$$[0, 1] \rightarrow \mathbb{R}^+$$
$$r \rightarrow h(r) = \sqrt{\frac{1}{1-r^2} - 1},$$

we note :

$$\rho_\gamma(x, y) = G_{1/2\gamma} \circ h\left(\sqrt{x^2 + y^2}\right) = \begin{cases} \dfrac{1}{C_\gamma} e^{\left(\frac{\gamma}{x^2+y^2-1}+\gamma\right)} & \text{if } x^2 + y^2 < 1, \\ 0 & \text{otherwise,} \end{cases}$$

where C_γ is a normalizing constant. The multiscale family of KCS is defined as

$$\rho_{\sigma,\gamma}(x, y) = \frac{1}{\sigma^2}\rho_{\sigma,\gamma}\left(\frac{x}{\sigma}, \frac{y}{\sigma}\right) = \begin{cases} \dfrac{1}{C_\gamma \sigma^2} e^{\left(\frac{\gamma\sigma^2}{x^2+y^2-1}+\gamma\right)} & \text{if } x^2 + y^2 < \sigma^2, \\ 0 & \text{otherwise.} \end{cases}$$

In contrast to the Gaussian, the KCS possesses two scale parameters: σ controls the support size and γ controls the width of the kernel mode. The support of the kernel is 2σ, which is significantly smaller than the support of Gaussian, 11σ. The parameter γ controls the distance between the zero-crossings of the KCS and the origin of the axes. The selection of parameter γ does not affect the nature of the functions $\rho_{\sigma,\gamma}$; they remain kernels with compact support. Furthermore, it is worth noting that if $\gamma \geq 2$ then the filter is causal (there is no creation of new structures when decreasing the value of the σ parameter). Mostly, we set $\gamma = 3$ so that it only has to fit a single parameter. Figure 2.10 shows the one-dimensional and two-dimensional profiles of a KCS kernel.

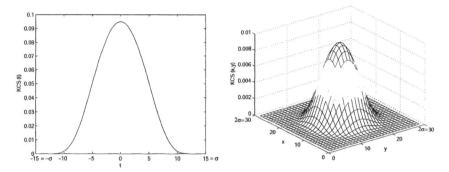

FIGURE 2.10 One-dimensional and two-dimensional profiles of a KCS kernel with parameters $\sigma = 15$ and $\gamma = 5$.

When building a smoothing kernel, it is important to check that it is a low-pass filter as well; so that the low frequencies are retained whereas the high ones are attenuated. When plotting the magnitude response of the truncated Gaussian (Fig. 2.11), we first observe that it is Gaussian-like, with side lobes. Actually, in the time space, the truncated Gaussian is a Gaussian multiplied by an indicator function, which corresponds to the convolution of a Gaussian with a cardinal sine, in the frequency space. The ripples introduced with the sine mean that the kernel is not decreasing while the frequencies increase. Thus, in the convolution, a frequency can have a weight larger than a lower frequency, which is not desired. As we are looking for a compact support kernel in the space, we will always have side lobes in the response. It is therefore wished to use a kernel with low side lobes. When comparing responses of the KCS and truncated Gaussian, we noticed that the KCS possesses a strong reduction of the secondary lobe in comparison to the size of its temporal support (Fig. 2.12). The secondary lobe of a truncated Gaussian kernel is located at 12DB, whereas one of the KCS is at 25DB. The KCS is therefore a better low-pass filter. The width of the principal lobe can be tailored based on the parameter γ.

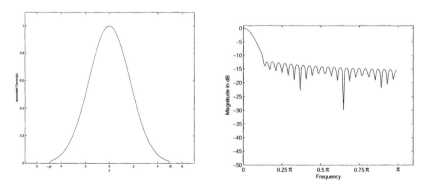

FIGURE 2.11 Truncated Gaussian kernel with the same variance and the same mask support as the KCS kernel shown below and the logarithm of its magnitude response.

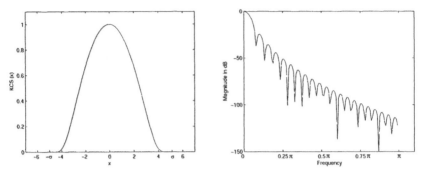

FIGURE 2.12   KCS kernel and the logarithm of its magnitude response.

2.3.2.4 *Features of the KCS*

The main purpose of constructing the KCS family is to give an effective response to the two practical limitations of using the Gaussian kernel. These limitations include the information loss and the high frequency creation caused by the truncation, and the prohibitive processing time due to the wide mask size (we do not use the separability of the Gaussian for the convolution with the truncated kernel, as by construction it is not any more separable). The authors have shown in [3rivieres] that the KCS kernel keeps the most important properties relative to the image segmentation process of the Gaussian kernel, but obviously not all of the properties that make the Gaussian kernel unique.

The properties of a signal convolved with the KCS are

- recovery of the initial signal when the scale parameter σ approaches zero. This is a necessary condition to construct the scale space;
- continuity in comparison with the scale parameter;
- strong regularization;
- decreasing number of zero-crossings following each scale;
- for large values of γ, the uncertainty principle of Heisenberg is minimized. Actually, we noticed that for $\gamma = 3$ the hup (Heisenberg uncertainty principal) is not so far from its minimum value. By setting this parameter to 3, the filter use is easier because only one parameter remains to be fitted, depending on the application. This property may be the most interesting; we showed that even if the support of the KCS is compact, it is still well located in the frequencies. This property is independent from the scale parameter γ.

2.3.2.5 *Data Extraction Using KCS*

The performance of the KCS is also judged by a visual inspection using the image shape criteria, such as thickness, intensity, and connectivity of strokes of the character shapes, with a weight given to each. Some sample images and their segmented images are shown in Figure 2.13

As we can see, the KCS operators give respectable results with a high visual quality. Additionally, the processing time required for the LoKCS (Laplacian of KCS) operator

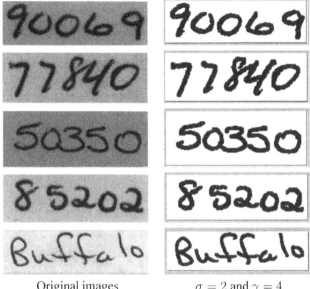

Original images $\sigma = 2$ and $\gamma = 4$

FIGURE 2.13   Original form images and their segmentation results using a KCS.

is drastically less than that required by the LoG (Laplacian of Gaussian) operator. Thanks to their compact support property, the KCS kernels give an effective solution to the two practical limitation problems when using the Gaussian kernel, namely the information loss caused by the truncation and the prohibitive processing time due to the wide mask size. A practical comparison between the LoG and the LoKCS operators has been widely developed by the authors in Ref. [2].

2.3.3   Concluding Remarks

In the previous section we have shown the derivation of a new family of kernels (KCS) to generate scale space in multiscale representation. We have appreciated the nice features of the KCS and its effectiveness in preventing information loss and reducing the processing time, as compared to the Gaussian kernel. The question one may ask now is, how can we further increase the performance of the KCS? With this aim in mind, we propose a new separable version of the KCS, denoted as SKCS (separable kernel with compact support) in order to perform the KCS and to further minimize the processing time. We leave it up to the reader to consult our publication [2, 3] for more information. In order to create the SKCS, we had to resort to some properties of the Gaussian (semigroup and separability). Other kernels can thus be created by giving a greater place to other properties, but the Heisenberg criterion must still be kept in mind.

2.4 DATA PREPROCESSING

The conversion of paper-based documents to electronic image format is an important process in computer systems for automated document delivery, document preservation, and other applications. The process of document conversion includes scanning, displaying, quality assurance, image processing, and text recognition. After document scanning, a sequence of data preprocessing operations are normally applied to the images of the documents in order to put them in a suitable format ready for feature extraction. In this section, all possible required preprocessing operations are described. Conventional preprocessing steps include noise removal/smoothing [13], document skew detection/correction [15, 41], connected component analysis, normalization [34, 17–19], slant detection/correction [36], thinning [40, 42, 43], and contour analysis [44].

2.4.1 Smoothing and Noise Removal

Smoothing operations in gray level document images are used for blurring and for noise reduction. Blurring is used in preprocessing steps such as removal of small details from an image. In binary (black and white) document images, smoothing operations are used to reduce the noise or to straighten the edges of the characters, for example, to fill the small gaps or to remove the small bumps in the edges (contours) of the characters. Smoothing and noise removal can be done by filtering. Filtering is a neighborhood operation, in which the value of any given pixel in the output image is determined by applying some algorithm to the values of the pixels in the neighborhood of the corresponding input pixel. There are two types of filtering approaches: linear and nonlinear. Here we look at some simple linear filtering approaches in which the value of an output pixel is a linear combination of the values of the pixels in the input pixel's neighborhood. A pixel's neighborhood is a set of pixels, defined by their locations relative to that pixel. In Figure 2.14(a) and (b), two examples of averaging mask filters are shown.

Each of these filters can remove the small pieces of the noise (salt and pepper noise) in the gray level images; also they can blur the images in order to remove the unwanted details. Normally, these averaging filter masks must be applied to the image a predefined number of times, otherwise all the important features (such as details and

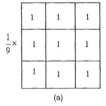
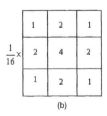

(a)   (b)

FIGURE 2.14   Two 3 by 3 smoothing (averaging) filter masks. The sum of all the weights (coefficients) in each mask is equal to one.

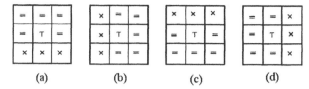

(a)   (b)   (c)   (d)

FIGURE 2.15 Examples of filter masks (3 by 3) that are used for smoothing of binary document images, (b), (c), and (d) are rotated versions of the filter in (a) by 90°.

edges, etc.) in the image will be removed completely. For smoothing binary document images, filters shown in Figure 2.15, can be used to smooth the edges and remove the small pieces of noise.

These masks are passed over the entire image to smooth it, and this process can be repeated until there is no change in the image. These masks begin scanning in the lower right corner of the image and process each row moving upward row by row. The pixel in the center of the mask is the target. Pixels overlaid by any square marked "X" are ignored. If the pixels overlaid by the squares marked "=" all have the same value, that is, all zeros, or all ones, then the target pixel is forced to match them to have the same value, otherwise it is not changed. These masks can fill or remove single pixel indentation in all the edges, or single bumps. Also all the single pixel noises (salt and pepper), or lines that are one pixel wide will be completely eroded. See Figure 2.16 for some examples of binary smoothed digits.

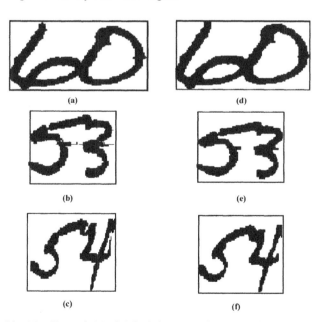

FIGURE 2.16 (a), (b), and (c) Original images; (d), (e), and (f) Smoothed images, respectively.

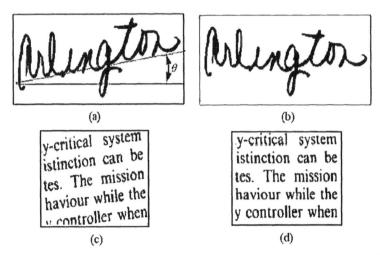

FIGURE 2.17 Deviation of the baseline of the text from horizontal direction is called skew. (a) A skewed handwritten word; (b) skew corrected handwritten word in (a); (c) a skewed typewritten text; (d) skew corrected image in (c).

2.4.2 Skew Detection and Correction

During the document scanning process, the whole document or a portion of it can be fed through the loose-leaf page scanner. Some pages may not be fed straight into the scanner, however, causing skewing of the bitmapped images of these pages. So, document skew often occurs during document scanning or copying. This effect visually appears as a slope of the text lines with respect to the x-axis, and it mainly concerns the orientation of the text lines; see some examples of document skew in Figure 2.17

Without skew, lines of the document are horizontal or vertical, depending on the language. Skew can even be intentionally designed to emphasize important details in a document. However, this effect is unintentional in many real cases, and it should be eliminated because it dramatically reduces the accuracy of the subsequent processes, such as page segmentation/classification and optical character recognition (OCR). Therefore, skew detection is one of the primary tasks to be solved in document image analysis systems, and it is necessary for aligning a document image before further processing. Normally subsequent operations show better performance if the text lines are aligned to the coordinate axes. Actually the algorithms for layout analysis and character recognition are generally very sensitive to page skew, so skew detection and correction in document images are the critical steps before layout analysis. In an attempt to partially automate the document processing systems as well as to improve the text recognition process, document skew angle detection and correction algorithms can be used.

2.4.2.1 Skew Detection There are several commonly used methods for detecting skew in a page. Some methods rely on detecting connected components (connected

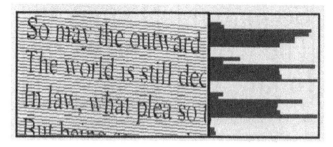

FIGURE 2.18   Sample image and its projection parallel to the text lines.

components or CCs are roughly equivalent to characters) and finding the average angles connecting their centroids, others use projection profile analysis. Methods based
on projection profiles are straightforward. In these methods, a document page is projected at several angles, and the variances in the number of black pixels per projected
line, are determined. The projection parallel to the correct alignment of the lines will
likely have the maximum variance, as in parallel projection, each given ray projected
through the image will hit either almost no black pixels (as it passes between text
lines) or many black pixels (while passing through many characters in sequence). An
example of parallel projection is shown in Figure 2.18

Oblique projections that are not parallel to the correct alignment of the text lines,
will normally pass through lines of the text, and spaces between the lines. Thus those
variance in the number of pixels hit by the individual rays will be smaller than in the
parallel case. An example is shown in Figure 2.19

2.4.2.2   *Skew Correction*   After the skew angle of the page has been detected,
the page must be rotated in order to correct this skew. A rotation algorithm has to be
both fairly fast and fairly accurate. A coordinate rotation transformation in Eq. 2.20
can be used in order to correct the skew of the document image. Given a point, its
new coordinates after rotating the entire image around its origin by angle (which is

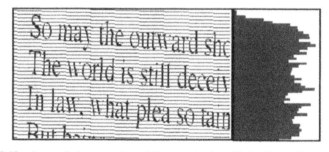

FIGURE 2.19   A sample image and an oblique projection. Notice the uniform nature of the
distribution.

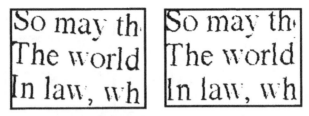

FIGURE 2.20 A rotation transformation can correct the skew of the document image.

the detected skew angle of the document image) can be satisfied by Eq. 2.20:

$$\begin{pmatrix} x' \\ y' \end{pmatrix} = \begin{pmatrix} \cos(\theta) & \sin(\theta) \\ -\sin(\theta) & \cos(\theta) \end{pmatrix} \begin{pmatrix} x \\ y \end{pmatrix}. \tag{2.20}$$

An example of skew correction is shown in Figure 2.20

2.4.3 Slant Correction

The character inclination that is normally found in cursive writing is called slant. Figure 2.21 shows some samples of slanted handwritten numeral string. Slant correction is an important step in the preprocessing stage of both handwritten words and numeral strings recognition. The general purpose of slant correction is to reduce the variation of the script and specifically to improve the quality of the segmentation candidates of the words or numerals in a string, which in turn can yield a higher recognition accuracy. Here, a simple and efficient method of slant correction of isolated handwritten characters or numerals is presented.

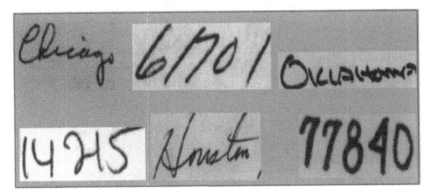

FIGURE 2.21 Images of handwritten words and handwritten numeral strings, selected from USPS-CEDAR CDROM1 database.

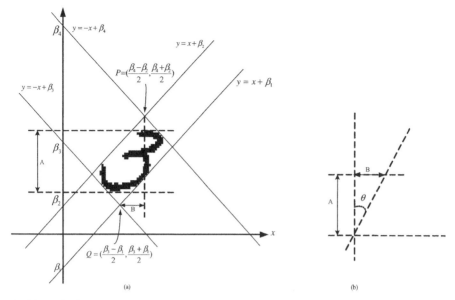

FIGURE 2.22   (a) A digit circumscribed by a tilted rectangle; (b) slant angle estimation for the digit in (a).

2.4.3.1   *Slant Angle Estimation*   A character or numeral image is circumscribed by a tilted rectangle, as shown in Figure 2.22 In other words, each character or digit is surrounded by the following four lines:

$$y = x + \beta_1, \tag{2.21}$$

$$y = x + \beta_2, \tag{2.22}$$

$$y = -x + \beta_3, \tag{2.23}$$

$$y = -x + \beta_4, \tag{2.24}$$

where $\beta_2 > \beta_1$ **and** $\beta_4 > \beta_3$, respectively.

As seen in Figure 2.22(b), the slant angle can be calculated as $\theta = \arctan(B/A)$. These parameters are shown in Figure 2.22(a), where A is the height of the character/digit and $B = (\beta_4 + \beta_1 - \beta_3 - \beta_2)/2$. Corresponding to the orientation of the slant to the left or right, θ can have positive or negative values, respectively.

2.4.3.2   *Slant Correction*   After estimation of the slant angle (θ), a horizontal shear transform is applied to all the pixels of the image of the character/digit in order to shift them to the left or to the right (depending on the sign of the θ). The transformation

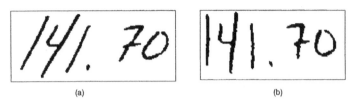

(a) (b)

FIGURE 2.23 (a) Original digits; (b) slant corrected digits.

expressions are given below:

$$x' = x - y \cdot \tan(\theta),$$
$$y' = y. \tag{2.25}$$

In these equations, x and y are original horizontal/vertical coordinates of the pixels in the image; x' and y' are corresponding transformed coordinates, respectively. An example of slant corrected digits is shown in Figure 2.23

2.4.4 Character Normalization

Character normalization is considered to be the most important preprocessing operation for character recognition. Normally, the character image is mapped onto a standard plane (with predefined size) so as to give a representation of fixed dimensionality for classification. The goal for character normalization is to reduce the within-class variation of the shapes of the characters/digits in order to facilitate feature extraction process and also improve their classification accuracy. Basically, there are two different approaches for character normalization: linear methods and nonlinear methods. In this section, we introduce some representative linear and nonlinear normalization methods, which are very effective in character recognition.

As seen in Figure 2.24, we denote the width and height of the original character by W_1 and H_1, the width and height of the normalized character by W_2 and H_2, and the size of the standard (normalized) plane by L. The standard plane is usually considered as a square and its size is typically 32×32 or 64×64, among others. We define the aspect ratios of the original character (R_1) and the normalized character (R_2) as

$$R_1 = \frac{\min(W_1, H_1)}{\max(W_1, H_1)}, \tag{2.26}$$

and

$$R_2 = \frac{\min(W_2, H_2)}{\max(W_2, H_2)}, \tag{2.27}$$

which are always considered in the range of [0, 1].

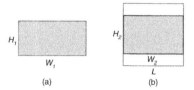

(a)   (b)

FIGURE 2.24   (a) Original character; (b) normalized character filled in standard (normalized) plane.

In the so-called aspect ratio adaptive normalization (ARAN) strategy [19], the aspect ratio of the normalized character (R_2) is adaptively calculated based on that of the original character (R_1) using a mapping function such as those in Table 2.1 In the implementation of ARAN, the normalized character image is filled into a plane of flexible size (W_2, H_2), and then this flexible plane is shifted to overlap the standard plane by aligning the center (shown in Fig. 2.24). If the normalized image fills one dimension of the standard plane, that is, the dimension of $\arg\max(W_1, H_1)$, then L is considered as equal to $\max(W_2, H_2)$ and the other dimension is centered in the standard plane. Having R_2 and $L (= \max(W_2, H_2))$, we can compute $\min(W_2, H_2)$ from Eq. (2.27). Thus, the size of the normalized character (W_2, H_2) can be obtained.

The transformation of the coordinates from the original character plane to normalized plane can be accomplished by forward mapping or backward mapping. If we denote the original image and the normalized image as $f(x, y)$ and $g(x', y')$, respectively, the normalized image is generated by $g(x', y') = f(x, y)$ based on coordinate mapping. The forward mapping and backward mapping are given by

$$\begin{aligned} x' &= x'(x, y), \\ y' &= y'(x, y), \end{aligned} \tag{2.28}$$

and

$$\begin{aligned} x &= x(x', y'), \\ y &= y(x', y'), \end{aligned} \tag{2.29}$$

respectively.

TABLE 2.1   Functions for aspect ratio mapping.

Method	Function
Fixed aspect ratio	$R_2 = 1$
Aspect ratio preserved	$R_2 = R_1$
Square root of aspect ratio	$R_2 = \sqrt{R_1}$
Cubic root of aspect ratio	$R_2 = \sqrt[3]{R_1}$
Sine of aspect ratio	$R_2 = \sqrt{\sin(\frac{\pi}{2} R_1)}$

TABLE 2.2   Coordinate mapping functions of normalization methods.

Method	Forward mapping	Backward mapping
Linear	$x' = \alpha x$ $y' = \beta y$	$x = x'/\alpha$ $y = y'/\beta$
Moment	$x' = \alpha(x - x_c) + x_c'$ $y' = \beta(y - y_c) + y_c'$	$x = (x' - x_c')/\alpha + x_c$ $y = (y' - y_c')/\beta + y_c$
Nonlinear	$x' = W_2 h_x(x)$ $y' = H_2 h_y(y)$	$x = h_x^{-1}(x'/W_2)$ $y = h_y^{-1}(y'/H_2)$

In forward mapping, x and y are discrete, but $x'(x, y)$ and $y'(x, y)$ are not necessarily discrete; whereas in backward mapping, x' and y' are discrete, but $x(x', y')$ and $y(x', y')$ are not necessarily discrete. Further, in forward mapping, the mapped coordinates (x', y') do not necessarily fill all pixels in the normalized plane. Therefore, coordinate discretization or pixel interpolation is needed in the implementation of normalization. By discretization, the mapped coordinates (x', y') or (x, y) are approximated by the closest integer numbers, $([x'], [y'])$ or $([x], [y])$. In the discretization of forward mapping, the discrete coordinates (x, y) scan the pixels of the original image and the pixel value $f(x, y)$ is assigned to all the pixels ranged from $([x'(x, y)], [y'(x, y)])$ to $([x'(x + 1, y + 1)], [y'(x + 1, y + 1)])$.

Backward mapping is often adopted because the discretization of mapped coordinates (x, y) is trivial. However, for some normalization methods, the inverse coordinate functions $x(x', y')$ and $y(x', y')$ cannot be expressed explicitly.

In the following, we give the coordinate functions of three popular normalization methods: linear normalization, moment normalization [8], and nonlinear normalization based on line density equalization [31, 35]. The forward and backward mapping functions are tabulated in Table 2.2 In this table, α and β denote the ratios of scaling, given by

$$\alpha = W_2/W_1,$$
$$\beta = H_2/H_1. \tag{2.30}$$

For linear normalization and nonlinear normalization, W_1 and H_1 are the horizontal span and vertical span of strokes of the original character (size of minimum bounding box). For moment normalization, the character is rebounded according to the centroid and second-order moments.

2.4.4.1 *Moment-Based Normalization*   Moment normalization cannot only align the centroid of character image but also correct the slant or rotation of character [8]. In the following, we first address moment-based size normalization (referred to as moment normalization in Table 2.2) and then turn to moment-based slant correction.

In moment normalization, the centroid and size of the normalized image are determined by moments. Let (x_c, y_c) denotes the center of gravity (centroid) of the original character, given by

$$x_c = m_{10}/m_{00},$$
$$y_c = m_{01}/m_{00}, \tag{2.31}$$

where m_{pq} denotes the geometric moments:

$$m_{pq} = \sum_x \sum_y x^p y^q f(x, y). \tag{2.32}$$

(x_c', y_c') denotes the geometric center of the normalized plane, given by

$$x_c' = W_2/2,$$
$$y_c' = H_2/2. \tag{2.33}$$

In moment normalization, the original image is viewed to be centered at the centroid and its boundaries are reset to $[x_c - \delta_x/2, x_c + \delta_x/2]$ and $[y_c - \delta_y/2, y_c + \delta_y/2]$. The reset dimensions $W_1 = \delta_x$ and $H_1 = \delta_y$ are calculated from moments:

$$\delta_x = a\sqrt{\mu_{20}/m_{00}},$$
$$\delta_y = a\sqrt{\mu_{02}/m_{00}}, \tag{2.34}$$

where μ_{pq} denotes the central moments:

$$\mu_{pq} = \sum_x \sum_y (x - x_c)^p (y - y_c)^q f(x, y), \tag{2.35}$$

and the coefficient a can be empirically set to be around 4. The original image plane is expanded or trimmed so as to fit the reset boundaries. The aspect ratio R_1 of the original image is then calculated from W_1 and H_1, and by using one of the formulas in Table 2.1 the aspect ratio R_2 of the normalized character is calculated. In turn, the width W_2 and height H_2 of normalized character are determined from R_2 and L.

On obtaining the centroid of original character, the width and height of original character and normalized image, the normalization of character image is performed by pixel mapping according to forward or backward coordinate mapping functions as in Table 2.2 After pixel mapping, the centroid of normalized character is aligned with the center of the normalized plane.

Size normalization can be preceded by a slant correction procedure as described in Section 2.4.3. The slant angle can also be estimated from moments. If preserving the centroid of character image in slant correction, the coordinate functions are given by

$$x' = x - (y - y_c)\tan\theta,$$
$$y' = y. \tag{2.36}$$

The aim is to eliminate the covariance of pixels (second-order central moment) after slant correction:

$$\mu'_{11} = \sum_x \sum_y [x - x_c - (y - y_c) \tan \theta](y - y_c) f(x, y) = 0, \tag{2.37}$$

which results in

$$\tan \theta = -\mu_{11}/\mu_{02}. \tag{2.38}$$

2.4.4.2 Nonlinear Normalization Nonlinear normalization is efficient to correct the nonuniform stroke density of character images and has shown superiority in handwritten Japanese and Chinese character recognition. The basic idea of nonlinear normalization is to equalize the projections of horizontal and vertical line density functions. Denote the horizontal and vertical density functions of character image $f(x, y)$ by $d_x(x, y)$ and $d_y(x, y)$, respectively, the (normalized to unit sum) projections on horizontal axis and vertical axis, $p_x(x)$ and $p_y(y)$, are calculated by

$$p_x(x) = \frac{\sum_y d_x(x, y)}{\sum_x \sum_y d_x(x, y)}, \tag{2.39}$$

$$p_y(y) = \frac{\sum_x d_y(x, y)}{\sum_x \sum_y d_y(x, y)}. \tag{2.40}$$

The normalized projections (also viewed as histograms) are accumulated to generate normalized coordinate functions (ranged in [0,1]):

$$h_x(x) = \sum_{u=0}^{x} p_x(u),$$
$$h_y(y) = \sum_{v=0}^{y} p_y(v). \tag{2.41}$$

The coordinates of normalized character image are then obtained by multiplying the normalized coordinate functions with W_2 and H_2, respectively, as shown in Table 2.2 After this coordinate transformation, the projections of line density of normalized image are approximately uniform.

The local line density $d_x(x, y)$ and $d_y(x, y)$ can be calculated in various ways, among which are the popular ones of Tsukumo and Tanaka [31] and Yamada et al. [35]. Both methods perform superiorly but the one of Tsukumo and Tanaka is simpler in implementation. According to Tsukumo and Tanaka, $d_x(x, y)$ is taken as the reciprocal of horizontal run-length of background. If the pixel (x, y) is in stroke region, the density is assumed to be a (empirical) constant. For background runs at left/right margins or blank rows, the density can be empirically set to $1/(rl + W_1)$, where rl is the run-length within the minimum bounding box of character. $d_y(x, y)$ is similarly calculated from vertical runs. On the contrary, Yamada et al. define the density in all regions of character rigorously and force $d_x(x, y)$ and $d_y(x, y)$ to be equal.

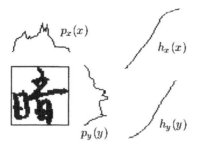

FIGURE 2.25 Character image, horizontal and vertical density projections, and coordinate functions.

Figure 2.25 shows an example of nonlinear normalization by line density equalization. The horizontal and vertical density projection profiles are shown to the above and the right of the character, and the normalized coordinate functions are shown to the right of the density projection profiles.

The forward coordinate mapping functions of line-density-based nonlinear normalization are not smooth, and their inverse functions cannot be expressed explicitly. Fitting smooth coordinate functions with intensity projection profiles can yield comparable recognition performance as line-density-based nonlinear normalization [18]. Further, the above one-dimensional normalization methods, with coordinate mapping functions like $x'(x)$ and $y'(y)$, can be extended to pseudo-two-dimensional for higher recognition performance [20].

An example of normalization is shown in Figure 2.26 where a digit "9" is transformed by linear normalization, line-density-based nonlinear normalization, and moment normalization, with and without slant correction.

2.4.5   Contour Tracing/Analysis

Contour tracing is also known as border following or boundary following. Contour tracing is a technique that is applied to digital images in order to extract the boundary of an object or a pattern such as a character. The boundary of a given pattern P is

FIGURE 2.26 From left: original image, linear normalization, nonlinear normalization, moment normalization; upper/lower: without/with slant correction.

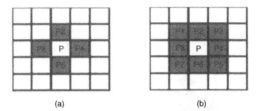

FIGURE 2.27   (a) 4-neighbors (4-connected); (b) 8-neighbors (8-connected).

defined as the set of border pixels of P. In digital images, each pixel of the image is considered as a square. As a result, there are two kinds of border (boundary) pixels for each pattern: 4-border pixels and 8-border pixels. Before describing these two types of borders in patterns, we define two types of connectivity or neighborhoods in digital images: 4-connectivity and 8-connectivity.

2.4.5.1   *Defining Connectivity*   A pixel, Q, is a 4-neighbor or 4-connected of a given pixel, P, if both Q and P share an edge. The 4-neighbors of pixel P (namely pixels P_2, P_4, P_6, and P_8) are shown in Figure 2.27(a) below. A pixel, Q, is an 8-neighbor or 8-connected of a given pixel, P, if both Q and P share a vertex or edge with that pixel. These pixels are namely P_1, P_2, P_3, P_4, P_5, P_6, P_7, and P_8 shown in Figure 2.27(b).

2.4.5.2   *Defining 4-Connected Component*   A set of black pixels, P, is a 4-connected component object if for every pair of pixels P_i and P_j in P, there exists a sequence of pixels $P_i, \ldots, P_j$ such that

(a) All pixels in the sequence are in the set P and
(b) Every 2 pixels of the sequence that are adjacent in the sequence are 4-neighbors or 4-connected.

Two examples of 4-connected patterns are shown in Figure 2.28

2.4.5.3   *Defining 8-Connected Component*   A set of black pixels denoted by P, is an 8-connected component (or simply a 8-connected component) if for every

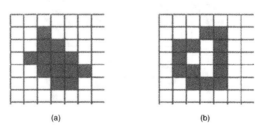

FIGURE 2.28   Two examples of 4-connected component objects.

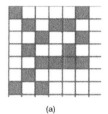 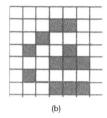

(a)   (b)

FIGURE 2.29 (a) An 8-connected component pattern; (b) a pattern that is not 4- or 8-connected.

pair of pixels P_i and P_j in P, there exists a sequence of pixels $P_i, \ldots, P_j$ such that

(a) All pixels in the sequence are in the set P and

(b) Every 2 pixels of the sequence that are adjacent are 8-connected.

One example of 8-connected pattern is shown in Figure 2.29(a). The pattern in Figure 2.29(b) is neither 4-connected nor 8-connected.

Now we can define 4-connected and 8-connected borders (contour) of patterns. A black pixel is considered as a 4-border pixel if it shares an edge with at least one white pixel. On the contrary, a black pixel is considered an 8-border pixel if it shares an edge or a vertex with at least one white pixel. (Note a 4-border pixel is also an 8-border pixel, however an 8-border pixel may or may not be a 4-border pixel).

2.4.5.4 Contour-Tracing Algorithms

Here, we describe a simple contour-tracing algorithm, namely square-tracing algorithm, which is easy to implement and, therefore, it is used frequently to trace the contour of characters. This algorithm will ignore any "holes" present in the pattern. For example, if a pattern is given like that of Figure 2.30(a), the contour traced by the algorithms will be similar to the one shown in Figure 2.30(b) (the black pixels represent the contour). This could be acceptable in some applications but not in all applications. Sometimes we want to trace the interior of the pattern as well in order to capture any holes that identify a certain character. Figure 2.30(c) shows the "complete" contour of the pattern. A "hole-

 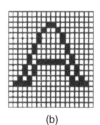 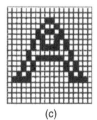

(a)   (b)   (c)

FIGURE 2.30 (a) Original pattern in black and white format; (b) outer contour of the pattern; (c) both outer and inner contours of the pattern.

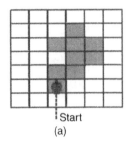 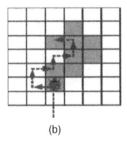

Start

(a)   (b)

FIGURE 2.31   (a) Starting point for contour tracing; (b) contour-tracing procedure.

searching" algorithm should be used to first extract the holes in a given pattern and then apply a contour-tracing algorithm on each hole in order to extract the complete contour. We can also reverse the original pattern (black pixels change to white, and white pixels change to black) and apply the contour-tracing algorithm to find the contour of the inner holes.

2.4.5.5 Square-Tracing Algorithm   The idea behind the square-tracing algorithm is very simple. It can be described as follows: given a digital pattern, that is, a group of black pixels on a white background, such as the one shown in Figure 2.31, one can locate a black pixel on the image and declare it as the "starting" pixel. This can be done in a number of ways; for example, one can start at the bottom left corner of the grid and scan each column of pixels from the bottom going upward starting from the leftmost column and proceeding to the right until a black pixel is encountered. One will declare that pixel as "starting" pixel. Now, imagine that you are a bug standing on the starting pixel as in Figure 2.31 below. In order to extract the contour of the pattern, you have to do the following: every time you find yourself hitting on a black pixel, turn left, and every time you find yourself standing on a white pixel, turn right, until you encounter the starting pixel again. The black pixels you walked over will be the contour of the pattern.

The important thing in the square-tracing algorithm is the "sense of direction." The left and right turns you make are with respect to your current positioning, which depends on your direction of entry to the pixel you are standing on. Therefore, it is important to keep track of your current orientation in order to make the right moves. The following is a formal description of the square-tracing algorithm:

Input: An image I containing a connected component P of black pixels.
Output: A sequence $B(b_1, b_2, \ldots, b_k)$ of boundary pixels, that is, the outer contour.

Begin

- Set B to be empty;
- From bottom to top and left to right scan the cells of I until a black pixel, S, of P is found;

- Insert S in B;
- Set the current pixel, P, to be the starting pixel, S;
- Turn left, that is, visit the left adjacent pixel of P;
- Update P, that is, set it to be the current pixel;
- While P not equal to S do
  - If the current pixel P is black;
    - Insert P in B and turn left (visit the left adjacent pixel of P),
    - Update P, that is, set it to be the current pixel,
  - Else
    - Turn right (visit the right adjacent pixel of P),
    - Update P, that is, set it to be the current pixel.

End

2.4.5.6 *Contours for Feature Extraction*   Contour tracing is one of many pre-processing techniques performed on character images in order to extract important information about their general shape. Once the contour of a given character or pattern is extracted, its different characteristics will be examined and used as features that will be used later on in pattern classification. Therefore, correct extraction of the contour will produce more accurate features that will increase the chances of correctly classifying a given character or pattern. But one might ask why first extract the contour of a pattern and then collect its features? Why not collect features directly from the pattern? One answer is, the contour pixels are generally a small subset of the total number of pixels representing a pattern. Therefore, the amount of computation is greatly reduced when we run feature extracting algorithms on the contour instead of the whole pattern. Because the contour shares a lot of features with the original pattern, but has fewer pixels, the feature extraction process becomes much more efficient when performed on the contour rather on the original pattern. In conclusion, contour tracing is often a major contributor to the efficiency of the feature extraction process, which is an essential process in pattern recognition.

2.4.6 Thinning

The notion of skeleton was introduced by Blum as a result of the medial axis transform (MAT) or symmetry axis transform (SAT). The MAT determines the closest boundary point(s) for each point in an object. An inner point belongs to the skeleton if it has at least two closest boundary points. A very illustrative definition of the skeleton is given by the prairie-fire analogy: the boundary of an object is set on fire and the skeleton is the loci where the fire fronts meet and quench each other. The third approach provides a formal definition: the skeleton is the locus of the center of all maximal inscribed hyperspheres (i.e., discs and balls in 2D and 3D, respectively). An inscribed hypersphere is maximal if it is not covered by any other inscribed hypersphere. See Figure 2.32 for an example of skeletonization.

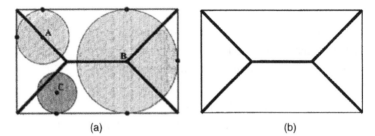

(a)   (b)

FIGURE 2.32 Finding the skeleton of a rectangular object: (a) The centers of all the maximal circles form the skeleton of the object, points A, B are on the skeleton, but C does not belong to the skeleton; (b) the resulting skeleton of the rectangular object.

There are many approaches to find the skeleton of an object. One approach is thinning. Thinning is the process of peeling off a pattern as many pixels as possible without affecting the general shape of the pattern. In other words, after pixels have been peeled off, the pattern can still be recognized. Hence, the skeleton obtained must have the following properties:

- Must be as thin as possible;
- Connected;
- Centered.

When these properties are satisfied, the algorithm must stop. As two examples, in Figures 2.33 and 2.34, a character pattern and a text line and their skeletons are shown, respectively.

Thinning Algorithms: Many algorithms have been designed for skeletonization of digital patterns. Here we describe two of those algorithms that can be used for skeletonization of binary patterns or regions in digital images: first, Hilditch's, and second, Zhang-Suen thinning algorithms. These algorithms are very easy to implement. In applying these algorithms, we assume that background region pixels (white pixels) have the value "0," and the object regions (black pixels) have the value "1."

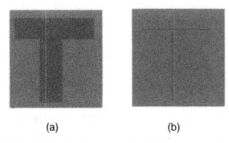

(a)   (b)

FIGURE 2.33 (a) Original pattern; (b) skeleton as a result of thinning.

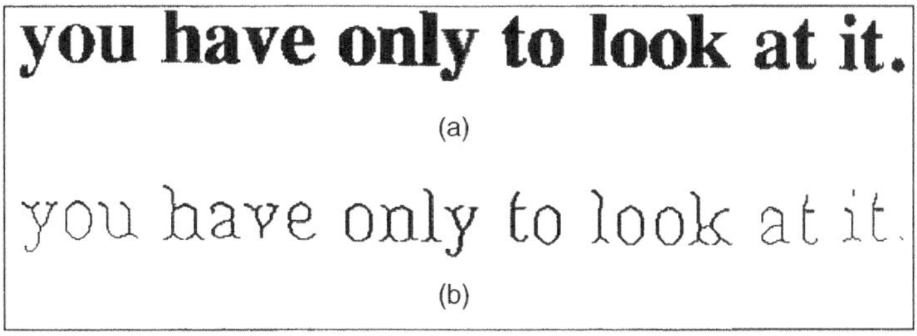

you have only to look at it.

(a)

you have only to look at it

(b)

FIGURE 2.34 (a) Original text; (b) skeleton of the text line in (a).

2.4.6.1   Hilditch's Thinning Algorithms

Consider the following 8-neighborhood of a pixel P_1, in Figure 2.35 To determine whether to peel off P_1 or keep it as part of the resulting skeleton, we arrange the eight neighbors of P_1 in a clockwise order and define the two functions $A(P_1)$ and $B(P_1)$ as follows:

- $A(P_1)$ = number of 0, 1 patterns (transitions from 0 to 1) in the ordered sequence of $P_2, P_3, P_4, P_5, P_6, P_7, P_8, P_9, P_2$.
- $B(P_1) = P_2 + P_3 + P_4 + P_5 + P_6 + P_7 + P_8 + P_9$ (number of black or 1 pixel, neighbors of P_1).

As mentioned above, in the definition of these functions black is considered equal to "1," and white is considered equal to "0," two examples of computations of these functions are shown in Figure 2.36

Hilditch's algorithm consists of performing multiple passes on the pattern, and on each pass the algorithm checks all the pixels and decides to change a pixel from black to white if it satisfies the following four conditions:

$$
\begin{aligned}
&2 <= B(P_1) <= 6, \\
&A(P_1) = 1, \\
&P_2 \cdot P_4 \cdot P_8 = 0 \text{ or } A(P_2) \neq 1, \\
&P_2 \cdot P_4 \cdot P_6 = 0 \text{ or } A(P_4) \neq 1.
\end{aligned}
\tag{2.42}
$$

The algorithm stops when changes stop (no more black pixels can be removed). Now we review each of the above conditions in detail.

P9	P2	P3
P8	P1	P4
P7	P6	P5

FIGURE 2.35   A 3 by 3 window that shows the pixels around P_1 and its neighboring pixels.

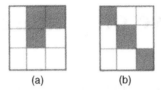

FIGURE 2.36   Two examples of computations of functions of $A(P_1)$ and $B(P_1)$: in (a), $A(P_1) = 1$, $B(P_1) = 2$, and in (b) $A(P_1) = 2$, $B(P_1) = 2$.

- *Condition 1:* $2 <= B(P_1) <= 6$. This condition combines two subconditions, first that the number of nonzero neighbors of P_1 is greater than or equal to 2 and second that it should be less than or equal to 6. The first condition ensures that no endpoint pixel and no isolated one be deleted (any pixel with 1 black neighbor is an endpoint pixel), the second condition ensures that the pixel is a boundary pixel.
- *Condition 2:* $A(P_1) = 1$. This is a connectivity test. For example, if you consider Figure 2.37, where $A(P_1) > 1$, you can see that by changing P_1 to white the pattern will become disconnected.
- *Condition 3:* $P_2 \cdot P_4 \cdot P_8 = 0$ or $A(P_2) \neq 1$. This condition ensures that 2-pixel wide vertical lines do not get completely eroded by the algorithm. An example for these conditions is shown in Figure 2.38
- *Condition 4:* $P_2 \cdot P_4 \cdot P_6 = 0$ or $A(P_4) \neq 1$. This condition ensures that 2-pixel wide horizontal lines do not get completely eroded by the algorithm. An example for these conditions is shown in Figure 2.39

2.4.6.2   Zhang-Suen Thinning Algorithm   This algorithm has two steps, which will be successively applied to the image. In each step contour points of the region that can be deleted are identified. Contour points are defined as points that have value "1," and they have at least one 8-neighbor pixel value equal to "0." Step 1 of the algorithm flags a contour point P_1 for deletion if the following conditions are satisfied:

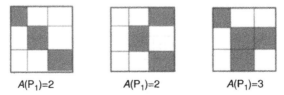

$A(P_1)=2$ $A(P_1)=2$ $A(P_1)=3$

FIGURE 2.37   In examples shown in this figure, condition $A(P_1) = 1$ is violated, so by removing P_1 the corresponding patterns will become disconnected.

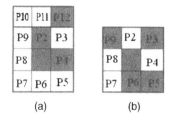

FIGURE 2.38   (a) An example where $A(P_2) \neq 1$; (b) an example where $P_2 \cdot P_4 \cdot P_8 = 0$; (c) an example where $P_2 \cdot P_4 \cdot P_8 \neq 0$ and $A(P_2) = 1$.

Condition 1: $2 <= B(P_1) <= 6$;
Condition 2: $A(P_1) = 1$;
Condition 3: $P_2 \cdot P_4 \cdot P_6 = 0$;
Condition 4: $P_4 \cdot P_6 \cdot P_8 = 0$.

Step 2 of the algorithm flags a contour point P_1 for deletion if the following conditions are satisfied:

Condition 1: $2 <= B(P_1) <= 6$ (the same condition as step 1);
Condition 2: $A(P_1) = 1$ (the same condition as step 1);
Condition 3: $P_2 \cdot P_4 \cdot P_8 = 0$;
Condition 4: $P_2 \cdot P_6 \cdot P_8 = 0$.

In this algorithm, Step 1 is applied to all the pixels of the image. If all the conditions of this step are satisfied, the point will be flagged to be removed. However, the deletion of the flagged points will be delayed until all the pixels of the image have been visited. This delay avoids changing of pixels of the image during each step. After Step 1 has been applied to all pixels of the image, the flagged pixels will be removed from the image. Then Step 2 is applied to the image exactly the same as Step 1. Applying Steps 1 and 2, one after the other is continued iteratively, until no further changes occur in the image. The resulting image will be the skeleton. In Figure 2.40 some results of skeletonization of the characters by these two algorithms are displayed. As these figure show there are minor differences between the results of these two algorithms.

FIGURE 2.39   (a) An example where $A(P_4) \neq 1$; (b) an example where $P_2 \cdot P_4 \cdot P_6 = 0$; (c) an example where $P_2 \cdot P_4 \cdot P_6 \neq 0$ and $A(P_4) = 1$.

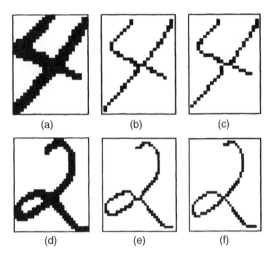

FIGURE 2.40 (a) and (d) Original images; (b) and (e) skeletons by Hilditch algorithm; (c) and (f) skeletons by Zhang-Suen algorithm.

In general, all thinning algorithms can be summarized as follows:

Repeat

    Change *removable* black pixels to white.

Until no pixels are removed.

2.4.6.3 Feature Extraction from Skeleton A skeleton shows the general shape of a pattern, and some important features can be extracted from it, such as intersection or branching points, number of strokes, their relative position, or the situation of their connections. Skeleton is also useful when we are interested in the relative positions of the strokes in the pattern but not in the size of the pattern itself.

2.5 CHAPTER SUMMARY

The purpose of the first important step, image, and data preprocessing, of a character recognition system is to prepare subimages with optimal quality so that the feature extraction step can work correctly and efficiently. This chapter covers most widely used preprocessing techniques and provides a global-to-local view of how information with different perspectives should be used in extracting the most pertinent features from characters to be recognized.

An intelligent form-processing system is introduced for extracting items of interest. Depending on the complexity of the background images, data extraction is achieved by thresholding based on global or local thresholding, on gray level or topological feature. A time critical application may use a balanced combination of different approaches. When the scale of the data to be extracted is unknown, a scale-space-based approach can enhance the signal-to-noise ratio and extract information with most efficiency. After the characters are separated from its background, defects such as skewed text lines and slanted characters are detected and corrected. Finally, preprocessing techniques such as contour tracking and skeleton extraction can substantially reduce the redundancy in the information to be processed by the ensuing feature extraction steps. At this stage, the information to be processed further has been reduced from millions of gray or color pixels to just small subimages of binarized characters, or even simply their contours or skeletons.

REFERENCES

1. H. Arai, and K. Odaka. Form reading based on background region analysis. In *Proceedings of the 4th International Conference on Document Analysis and Recognition.* Ulm, Germany, 1997, pp. 164–169.

2. E. Ben Braiek, M. Cheriet, and V. Doré. SKCS—a separable kernel family with compact support to improve visual segmentation of handwritten data. *Electronic Letters on Computer Vision and Image Analysis.* **5**(1), 14–29, 2004.

3. E. Ben Braiek, M. Cheriet, and V. Doré. SKCS—une nouvelle famille de noyaux à support compact appliquée à la segmentation des données visuelles manuscrites. *Traitement du Signal.* **23**(1), 27–40, 2006.

4. F. Bergholm. Edge focusing. *IEEE Transactions on Pattern Analysis and Machine Intelligence.* **9**(6), 726–741, 1987.

5. J. Bernsen. Dynamic thresholding of gray level images. In *Proceedings of the 8th International Conference on Pattern Recognition.* Paris, France, 1986, Vol. 2, pp.1251–1255.

6. D. S. Britto JR., R. Sabourin, E. Lethelier, F. Bortolozzi, and C. Y. Suen. Improvement in Handwritten Numeral String Recognition by Slant Correction and Contextual Information. In Proceedings of International Workshop on Frontiers in Handwriting Recognition (IWFHR), pages 323–332, Amsterdam, Septmber 2000.

7. P. J. Burt. Fast filter transform for image processing. *Computer Vision, Graphics, and Image Processing.* **16**, 20–51, 1981.

8. R.G. Casey. Moment normalization of handprinted character. *IBM Journal of Research and Development.* **14**, 548–557, 1970.

9. M. Cheriet, R. Thibault, and R. Sabourin. A multiresolution based approach for handwriting segmentation in gray-scale images. In *Proceedings of the 1st International Conference on Image Processing.* Austin, Texas, USA, 1994, pp. 159–168.

10. M. Cheriet. Extraction of handwritten data from noisy gray level images using multi-scale approach. *International Journal of Pattern Recognition and Artificial Intellignence.* **13**(5), 665–685, 1999.

11. J. L. Crowley. *A representation for visual information. Ph.D. dissertation.* Carnegie Mellon University, Pittsburgh, PA, 1981.

12. K. Fukunaga. *Introduction to Statistical Pattern Recognition.* 2nd edition, Academic Press, New York, 1990.

13. R. C. Gonzalez and R. E. Woods. *Digital Image Processing.* 2nd edition, Addison Wesley, 2001.

14. R. Horaud and O. Monga. *Vision par Ordinateur, Outils Fondamentaux.* Hermes, Paris, France, 1995, p.425.

15. J. J. Hull. Document image skew detection: Survey and annotated bibliography. In J. J. Hull and S. L. Taylor, editors, *Document Analysis Systems II*, World Scientific, Singapore, 1998, pp. 40–64.

16. J. J. Koenderink. The structure of images. *Biological Cybernetics.* **50**, 363–370, 1984.

17. T. Lindeberg. *Scale-Space Theory in Computer Vision.* Kluwer Academic Publishers, 1994.

18. C.-L. Liu, H. Sako, and H. Fujisawa. Handwritten Chinese character recognition: Alternatives to nonlinear normalization. In *Proceedings of the 7th International Conference on Document Analysis and Recognition.* Edinburgh, Scotland, 2003, pp. 524–528.

19. C.-L. Liu, K. Nakashima, H. Sako, and H. Fujisawa. Handwritten digit recognition: Investigation of normalization and feature extraction techniques. *Pattern Recognition.* **37**(2), 265–279, 2004.

20. C.-L. Liu and K. Marukawa. Pseudo two-dimensional shape normalization methods for handwritten Chinese character recognition. *Pattern Recognition.* **38**(12), 2242–2255, 2005.

21. K. Liu, C. Y. Suen, M. Cheriet, J. N. Said, C. Nadal, and Y. Y. Tang. Automatic extraction of baselines and data from check images. *International Journal of Pattern Recognition and Artificial Intelligence.* **11**(4), 675–697, 1997.

22. D. Marr and E. Hildreth. Theory of edge detection. *Proceedings of the Royal Society of London.* **B-207**, 187–217, 1980.

23. P. Meer, E. S. Baugher, and A. Rosenfeld. Frequency domain analysis and synthesis of image pyramid generating kernels. *IEEE Transactions on Pattern Analysis and Machine Intelligence.* **9**, 512–522, 1987.

24. W. Niblack. *An Introduction to Digital Image Processing.* Prentice Hall, 1986.

25. N. Otsu. A threshold selection method from gray-scale histogram. *IEEE Transactions on System, Man, and Cybernetics.* **9**, 62–66, 1979.

26. P. W. Palumbo, P. Swaminathan, and S. N. Srihari. Document image binarization: Evaluation of algorithms. In *Proceedings of the SPIE Symposium on Applications of Digital Image Processing IX.* 1986, Vol. 697, pp. 278–285.

27. L. Remaki and M. Cheriet. KCS—a new kernel family with compact support in scale space: Formulation and impact. *IEEE Transactions on Image Processing.* **9**(6), 970–981, 2000.

28. Y. Y. Tang, C. Y. Suen, C. D. Yan, and M. Cheriet. Financial document processing based on staff line and description language. *IEEE Transactions on System, Man, and Cybernetics.* **25**(5), 738–754, 1995.

29. O. D. Trier and A. K. Jain. Goal-directed evaluation of binarization methods. *IEEE Transactions on Pattern Analysis and Machine Intelligence.* **17**, 1191–1201, 1995.

30. O. D. Trier and T. Taxt. Evaluation of binarization methods for document images. *IEEE Transactions on Pattern Analysis and Machine Intelligence.* **17**, 312–314, 1995.

31. J. Tsukumo and H. Tanaka. Classification of handprinted Chinese characters using nonlinear normalization and correlation methods. In *Proceedings of the 9th International Conference on Pattern Recognition.* Rome. Italy, 1988, pp. 168–171.

32. J. M. White and G. D. Rohorer. Image thresholding for optical character recognition and other applications requiring character image extraction. *IBM Journal of Research and Development*. **27**(4), 400–411, 1983.

33. N. Wiener. *Extrapolation, Interpolation, and Smoothing of Stationary Time Series*. MIT Press, Cambridge, MA, 1942.

34. A. P. Witkin. Scale-space filtering. In *Proceedings of the International Joint Conference on Artificial Intelligence*. Karlsruhe, Germany, 1983, pp. 1019–1022.

35. H. Yamada, K. Yamamoto, and T. Saito. A nonlinear normalization method for hanprinted Kanji character recognition—line density equalization. *Pattern Recognition*. **23**(9), 1023–1029, 1990.

36. T. Yamaguchi, Y. Nakano, M. Maruyama, H. Miyao, and T. Hananoi. Digit classification on signboards for telephone number recognition. In *Proceedings of the 7th International Conference on Document Analysis and Recognition*. Edinburgh, Scotland, 2003, pp. 359–363.

37. S. D. Yanowitz and A. M. Bruckstein. A new method for image segmentation. *Computer Vision, Graphics and Image Processing*. **46**(1), 82–95, 1989.

38. X. Ye, M. Cheriet, and C. Y. Suen. Stroke-model-based character extraction from gray level document images. *IEEE Transactions on Image Processing*. **10**(8), 1152–1161, 2001.

39. B. Yu and A. K. Jain. A generic system for form dropout. *IEEE Transactions on Pattern Analysis and Machine Intelligence*. **18**(11), 1127–1132, 1996.

40. T. Y. Zhang and C.Y. Suen. A fast parallel algorithm for thinning digital patterns. *Communication of the ACM*. **27**(3), 236–239, 1984.

41. http://www.cs.berkeley.edu/~fateman/kathey/skew.html

42. http://www.inf.u-szeged.hu/~palagyi/skel/skel.html

43. http://cgm.cs.mcgill.ca/~godfried/teaching/projects97/azar/skeleton.html

44. http://www.cs.mcgill.ca/~aghnei/alg.html

CHAPTER 3

FEATURE EXTRACTION, SELECTION, AND CREATION

"The only way of finding the limits of the possible is by going beyond them into the impossible."

—Arthur C. Clarke

The purpose of *feature extraction* is the measurement of those attributes of patterns that are most pertinent to a given classification task. The task of the human expert is to select or invent features that allow effective and efficient recognition of patterns. Many features have been discovered and used in pattern recognition. A partial collection of such features occupies the first part of this chapter. Given a large set of features of possible relevance to a recognition task, the aim of *feature selection* is to find a subset of features that will maximize the effectiveness of recognition, or maximize the efficiency of the process (by minimizing the number of features), or both, done with or without the involvement of a classifier. The second part of this chapter is dedicated to feature selection. The final part is dedicated to the promising discipline of *feature creation*, which aims to delegate the task of making or discovering the best features for a given classification task to a machine.

3.1 FEATURE EXTRACTION

This section describes a collection of popular feature extraction methods, which can be categorized into geometric features, structural features, and feature space transformations methods. The geometric features include moments, histograms, and direction

Character Recognition Systems: A Guide for Students and Practitioner, by M. Cheriet, N. Kharma, C.-L. Liu and C. Y. Suen   Copyright © 2007 John Wiley & Sons, Inc.

features. The structural features include registration, line element features, Fourier descriptors, topological features, and so on. The transformation methods include principal component analysis (PCA), linear discriminant analysis (LDA), kernel PCA, and so on.

3.1.1   Moments

Moments and functions of moments have been employed as pattern features in numerous applications to recognize two-dimensional image patterns. These pattern features extract global properties of the image such as the shape area, the center of the mass, the moment of inertia, and so on. The general definition of moment functions m_{pq} of order $(p + q)$ for an $X \times Y$ continuous image intensity function $f(x, y)$ is as follows:

$$m_{pq} = \int_y \int_x \psi_{pq}(x, y) f(x, y) dx \, dy, \tag{3.1}$$

where p, q are integers between $[0, \infty)$, x and y are the x- and y-coordinate of an image point, and $\psi_{pq}(x, y)$ is the basis function. Valuable properties, such as orthogonality, of the basis functions are inherited by the moment functions.

Similarly, the general definition for an $X \times Y$ digital image can be obtained by replacing the integrals with summations

$$m_{pq} = \sum_y \sum_x \psi_{pq}(x, y) f(x, y), \tag{3.2}$$

where p, q are integers between $(0, \infty)$ and represent the order, x and y are the x- and y-pixel of the digital image, and $\psi_{pq}(x, y)$ is the basis function.

A desirable property for any pattern feature is that it must preserve information under image translation, scaling, and rotation. Hu was the first to develop a number of nonlinear functions based on geometric moments (discussed in the next section) that were invariant under image transformation [44]. The sequel presents the definitions of various types of moments and their properties.

3.1.1.1   *Geometric Moments*   Geometric moments or regular moments are by far the most popular types of moments. The geometric moment of order $(p + q)$ of $f(x, y)$ is defined as [51]

$$m_{pq} = \int_{-\infty}^{+\infty} \int_{-\infty}^{+\infty} x^p y^q f(x, y) dx \, dy, \tag{3.3}$$

where p, q are integers between $(0, \infty)$, x and y are the x- and y-coordinate of the image, and $x^p y^q$ is the basis function.

For a digital image, the definition is given by

$$m_{pq} = \sum_{-\infty}^{+\infty} \sum_{-\infty}^{+\infty} x^p y^q f(x, y), \tag{3.4}$$

where p, q are integers between $(0, \infty)$ and represent the order, and x and y are the coordinates of a pixel of the digital image.

As may be observed from the definition, the basis function for the geometric moment is the monomial product $x^p y^q$, which is not orthogonal. As a result, recovering an image from geometric moments is difficult and computationally expensive.

3.1.1.2 Zernike and Pseudo-Zernike Moments

Zernike defined a complete orthogonal set $\{V_{nm}(x, y)\}$ of complex polynomials over the polar coordinate space inside a unit circle (i.e., $x^2 + y^2 = 1$) as follows [51]:

$$V_{nm}(x, y) = V_{nm}(\rho, \theta) = R_{nm}(\rho)e^{jm\theta}, \tag{3.5}$$

where $j = \sqrt{-1}$, $n \geq 0$, m is a positive or negative integer, $|m| \leq n$, $n - |m|$ is even, ρ is the shortest distance from the origin to (x, y) pixel, θ is the angle between vector ρ and x-axis in counterclockwise direction, and $R_{nm}(\rho)$ is the orthogonal radial polynomial given by

$$R_{nm}(\rho) = \sum_{s=0}^{n-|m|/2} (-1)^s \cdot \frac{(n-s)!}{s!(\frac{n+|m|}{2} - s)!(\frac{n-|m|}{2} - s)!} \rho^{n-2s}. \tag{3.6}$$

Note that $R_{n-m}(\rho) = R_{nm}(\rho)$. These polynomials are orthogonal and satisfy the following condition:

$$\int\int_{x^2+y^2 \leq 1} [V_{nm}(x, y)]^* V_{pq}(x, y)dx\,dy = \frac{\pi}{n+1}\delta_{np}\delta_{mq}, \tag{3.7}$$

where

$$\delta_{ab} = \begin{cases} 1; & \text{if } a = b, \\ 0; & \text{otherwise.} \end{cases} \tag{3.8}$$

Zernike moments are the projection of the image intensity function $f(x, y)$ onto the complex conjugate of the previously defined Zernike polynomial $V_{nm}(\rho, \theta)$, which is defined only over the unit circle

$$A_{nm} = \frac{n+1}{\pi} \int\int_{x^2+y_2 \leq 1} f(x, y)V_{nm}^*(\rho, \theta)dx\,dy. \tag{3.9}$$

For a digital image, Zernike moments are given by

$$A_{nm} = \frac{n+1}{\pi} \sum_x \sum_y f(x, y) V_{nm}^*(\rho, \theta), \quad x^2 + y^2 \leq 1. \tag{3.10}$$

To evaluate the Zernike moments, a given image is mapped to a unit circle using polar coordinates, where the center of the image is the origin of the unit circle. Pixels falling outside the unit circle are not taken into consideration.

Bhatia and Wolf derived pseudo-Zernike moments from Zernike moments; therefore, both these moments have analogous properties [108]. The difference lies in the definition of the radial polynomial, which is given by

$$R_{nm}(\rho) = \sum_{s=0}^{n-|m|} (-1)^s \cdot \frac{(2n+1-s)!}{s!(n-|m|-s)!(n+|m|+1-s)!} \rho^{n-s}, \tag{3.11}$$

where $n \geq 0$, m is a positive or negative integer subject to $|m| \leq n$ only [108]. Replacing the radial polynomial in Zernike moments with the radial polynomial defined above yields pseudo-Zernike moments. Pseudo-Zernike moments offer more feature vectors than Zernike moments due to the condition that $n - |m|$ is even for the latter, which reduces the polynomial by almost half. In addition, pseudo-Zernike moments perform better on noisy images.

3.1.1.3 Legendre Moments
The well-known Legendre polynomials form the basis function in the Legendre moments. Recall that the *p*th order Legendre polynomial is defined as

$$P_p(x) = \frac{1}{2^p p!} \frac{d^p}{dx^p} (x^2 - 1)^p, \quad x \in [-1, 1]. \tag{3.12}$$

The definition of the $(p + q)$ order Legendre moment is

$$L_{pq} = \frac{(2p+1)(2q+1)}{4} \int_{-1}^{1} \int_{-1}^{1} P_p(x) P_q(y) f(x, y) dx \, dy, \tag{3.13}$$

where p, q are integers between $(0, \infty)$, and x and y are the x- and y-coordinate of the image.

Similarly, the Legendre moment for a $(N \times N)$ digital image is given by

$$L_{pq} = \sum_{m=0}^{N-1} \sum_{n=0}^{N-1} P_p(m_N) P_q(n_N) f(m, n), \tag{3.14}$$

where

$$m_N = \frac{2m - N + 1}{N - 1}. \tag{3.15}$$

To compute the Legendre moments, a given image is mapped into the limit domain $[-1, 1]$ as the Legendre polynomial is only defined over this range.

3.1.1.4 Tchebichef Moments Mukundan et al. introduced the Tchebichef moments where the basis function is the discrete orthogonal Tchebichef polynomial [84]. For a given positive integer N (normally the image size), the Tchebichef polynomial is given by the following recurrence relation:

$$t_n(x) = \frac{(2n-1)t_1(x)t_{n-1}(x) - (n-1)(1 - \frac{(n-1)^2}{N^2})t_{n-2}(x)}{n}, \qquad (3.16)$$

with the initial conditions

$$\begin{aligned} t_0(x) &= 1, \\ t_1(x) &= (2x + 1 - N)/N, \end{aligned} \qquad (3.17)$$

where $n = 0, 1, \ldots, N - 1$.

The Tchebichef moment of order $(p + q)$ of an image intensity function is defined as

$$T_{nm} = \frac{1}{\rho(n, M)\rho(n, N)} \sum_{x=0}^{N-1} \sum_{y=0}^{N-1} t_m(x)t_n(y)f(x, y), \qquad (3.18)$$

where $n, m = 0, 1, \ldots, N - 1$. The Tchebichef polynomial satisfies the property of orthogonality with

$$\rho(n, N) = \frac{N(1 - \frac{1}{N^2})(1 - \frac{2^2}{N^2}) \cdots (1 - \frac{n^2}{N^2})}{2n + 1}. \qquad (3.19)$$

Note that with Tchebichef moments, the problems related to continuous orthogonal moments are purged by using a discrete orthogonal basis function (i.e., Tchebichef polynomial). In addition, no mapping is required to compute Tchebichef moments as the Tchebichef polynomials are orthogonal in the image coordinate space.

3.1.2 Histogram

Histograms are charts displaying the distribution of a set of pixels of gray-level images $f(x, y)$. This section briefly discusses the amplitude histogram.

The amplitude histogram of a gray-level image $f(x, y)$ is the table $H(k)$ that shows the number of pixels that have gray-level value k. The definition of an amplitude histogram for an $(X \times Y)$ digital image is given by

$$\begin{aligned} H(k) = \quad & \text{cardinal}\{f(x, y)| f(x, y) = k\}, \\ & k \in [0, k_{\max}], \ (x, y) \in [(0, 0), (X - 1, Y - 1)], \end{aligned} \qquad (3.20)$$

where cardinal{} is the number of pixels in a given set.

The above definition may be generalized to compute the histogram of a $(W \times H)$ image over any range B. In addition, divide the range of gray levels into bins B_n and let $k_{\min}$ be the minimum gray value. For instance, if $B = 10$, $k_{\min} = 1000$, and $B_n = 100$, then bin 0 (B_0) corresponds to all pixel values ranging from 1000 to 1010. Therefore, the generalized amplitude histogram $H_r(k)$ is given by

$$H_r(k) = \text{cardinal}\{f(x, y)|(k_{\min} + B_k) \leq f(x, y) < k_{\min} + B_{k+1}\}, \quad (3.21)$$
$$k \in [0, B_n - 1], \ (x, y) \in [(0, 0), (W - 1, H - 1)].$$

3.1.3 Direction Features

Characters comprise strokes that are oriented lines, curves, or polylines. The orientation or direction of strokes plays an important role in differentiating between various characters. For a long time, stroke orientation or direction has been taken into account in character recognition based on stroke analysis. For statistical classification based on feature vector representation, characters have also been represented as vectors of orientation/direction statistics. To do this, the stroke orientation/direction angle is partitioned into a fixed number of ranges, and the number of stroke segments in each angle range is taken as a feature value. The set of numbers of orientational/directional segments thus forms a histogram, called orientation or direction histogram. To further enhance the differentiation ability, the histogram is often calculated for local zones of the character image, giving the so-called local orientation/direction histogram. Figure 3.1 shows an example of contour orientation histogram (four orientations, 4×4 zones).

Both orientation and direction histogram features can be called direction features in general. In early stages, character recognition using direction features was called directional pattern matching [32, 119], wherein the character image is decomposed into orientation planes (each recording the pixels of a particular stroke orientation), and a distance measure is computed between the planes and the template of each class. The local stroke orientation/direction of a character can be determined in different ways: skeleton orientation [119], stroke segment [117], contour chaincode [52, 73], gradient direction [104, 74], and so on. The contour chaincode and gradient direction features are now widely adopted because they have simple implementation and are approximately invariant to stroke-width variation.

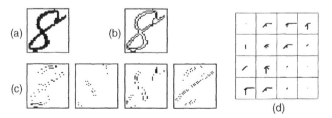

FIGURE 3.1 (a) Character image; (b) contour of character; (c) orientation planes; (d) local orientation histograms.

On decomposing the (size normalized) character image into direction planes, a simple way to obtain feature values is to partition each direction plane into a number of equal-sized zones and take the sum of pixel values in each zone as a feature value. The values of all directions in a zone form a local direction histogram as that in Figure 3.1. By this crisp partitioning, the feature values become sensitive to the variation of stroke position and stroke width. To alleviate this problem, dynamic partitioning [117] and soft partitioning have been tried. A blurring technique, akin to soft zone partitioning, is now commonly adopted. Blurring, initially proposed by Iijima in 1960s [46], is equivalent to low-pass spatial filtering. It blurs the position of strokes and the boundary between zones, thus improving the tolerance of character shape variation.

In the following section, we briefly introduce the blurring technique and then describe the directional decomposition according to chaincode and gradient.

3.1.3.1 *Blurring*   Blurring is applicable to orientation/direction planes as well as to any kind of image for reducing noise and improving the translation invariance. It is performed by low-pass spatial filtering, usually with a Gaussian filter, whose impulse response function is given by

$$h(x, y) = \frac{1}{2\pi\sigma_x^2} \exp\left(-\frac{x^2 + y^2}{2\sigma_x^2}\right),$$ (3.22)

and its frequency transfer function is

$$H(u, v) = \exp\left(-\frac{x^2 + y^2}{2\sigma_u^2}\right),$$ (3.23)

where $\sigma_u = \frac{1}{2\pi\sigma_x}$. $h(x, y)$ is usually approximated to a finite-domain 2D window, which is also called a blurring mask.

Denote the image to be filtered by $f(x, y)$; the pixel values of the filtered image are computed by convolution as follows:

$$F(x_0, y_0) = \sum_x \sum_y f(x, y)h(x - x_0, y - y_0).$$ (3.24)

This can be viewed as a weighted sum of the shifted window $h(x - x_0, y - y_0)$ with a region of image $f(x, y)$. The values of filtered image are not necessarily computed for all pixels (x_0, y_0). According to the Nyquist sampling theorem, an image of low-frequency band can be reconstructed from a down-sampled image, provided that the sampling rate is higher than twice the bandwidth. Approximating the bandwidth of Gaussian filter to $u_{max} = v_{max} = \sqrt{2}\sigma_u$ and making the sampling rate $f_x = 2u_{max}$ relates the Gaussian filter parameter σ_x to the sampling rate by

$$\sigma_x = \frac{\sqrt{2}}{\pi f_x} = \frac{\sqrt{2}t_x}{\pi},$$ (3.25)

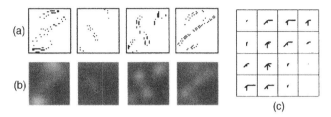

FIGURE 3.2 (a) Contour orientation planes; (b) blurred orientation planes; (c) sampled values of blurred planes.

where t_x is the sampling interval [73]. Formula (3.25) guides us to determine the Gaussian filter parameter from the sampling interval, which can be predefined empirically. For example, to sample 4×4 values from an image of 32×32 pixels, the sampling interval is 8.

Figure 3.2 shows an example of the blurring and down-sampling of the orientation planes. The orientation planes of Figure 3.1 (copied in Fig. 3.2(a)) are blurred by a Gaussian filter (filtered images shown in Figure 3.2(b)), and the 4×4 values are sampled from each filtered image. At each sampling point, the sampled values of the four orientations form a local orientation histogram, as shown in Figure 3.2(c), which is similar to that of Figure 3.1. In practice, only the values of sampling points are calculated in spatial filtering according to (3.24).

3.1.3.2 *Directional Decomposition*   We introduce here two techniques for generating direction planes from stroke edges: contour chaincode decomposition and gradient direction decomposition.

The contour pixels of a binary image are commonly encoded into chaincodes when they are traced in a certain order, say counterclockwise for outer loop and clockwise for inner loop. Each contour pixel points to its successor in one of eight directions as shown in Figure 3.3. As the contour length is approximately invariant to stroke width, and the chaincode direction reflects the local stroke direction, it is natural to assign contour pixels of the same chaincode to a direction plane and extract feature values from the direction planes.

There are basically two ways to determine the direction codes of contour pixels. In one way, the order of contour pixels is obtained by contour tracing, and then

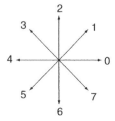

FIGURE 3.3   Eight chaincode directions.

d_3	d_2	d_1
d_4	c	d_0
d_5	d_6	d_7

FIGURE 3.4   Pixels in the neighborhood of current pixel "c."

the chaincodes are calculated. Contour tracing, however, is not trivial to implement. Another way examines a 3×3 window centered at each contour pixel (a black pixel with at least one of the 4-neighbors being white). The contour pixel is assigned to one or two directions according to the configuration of neighborhood [48]. In the following, we introduce an algorithm that efficiently determines the chaincodes without contour tracing [73].

Before the chaincode decomposition of binary image $f(x, y)$, the eight direction planes, $f_i(x, y)$, $i = 1, \ldots, 8$, are empty, that is, $f_i(x, y) = 0$. Scanning the pixels of image $f(x, y)$ in an arbitrary order and denoting the 8-connected neighbors of each pixel by d_i, $i = 0, \ldots, 7$ results in Figure 3.4. We can see that the neighboring pixels correspond to the eight chaincode directions. If the current pixel $c = f(x, y) = 1$ (black pixel), check its 4-neighbors:

    for $i = 0, 2, 4, 6$
        if $d_i = 0$,
            if $d_{i+1} = 1$, then $f_{i+1}(x, y) = f_{i+1}(x, y) + 1$,
            else if $d_{(i+2)\%8} = 1$, then $f_{(i+2)\%8}(x, y) = f_{(i+2)\%8}(x, y) + 1$,
            end if
        end if
    end for.

In this procedure, $i + 1$ or $(i + 2)\%8$ is exactly the chaincode connecting the pixel $c = (x, y)$ and the pixel d_{i+1} or the pixel $d_{(i+2)\%8}$.

For extracting orientation feature, the eight direction planes are merged into four orientation planes by $f_i(x, y) = f_i(x, y) + f_{i+4}(x, y)$, $i = 0, 1, 2, 3$.

Considering that the stroke edge direction is approximately normal to the gradient of image intensity, we can alternatively decompose the local gradient into directional planes and extract the gradient direction feature. Unlike the chaincode feature, which is applicable to binary images only, the gradient feature is applicable to gray-scale images as well and is more robust against image noise and edge direction fluctuations. The gradient can be computed by the Sobel operator, which has two masks for the gradient components in horizontal and vertical directions, as shown in Figure 3.5. Accordingly, the gradient vector $\mathbf{g} = (g_x, g_y)^T$ at a pixel (x, y) is computed by

-1	0	$+1$
-2	0	$+2$
-1	0	$+1$

$+1$	$+2$	$+1$
0	0	0
-1	-2	-1

FIGURE 3.5   Sobel masks for gradient.

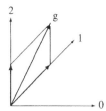

FIGURE 3.6   Gradient vector decomposition.

$$
\begin{aligned}
g_x(x, y) = \ & f(x+1, y-1) + 2f(x+1, y) + f(x+1, y+1) \\
& -f(x-1, y-1) - 2f(x-1, y) - f(x-1, y+1), \\
g_y(x, y) = \ & f(x-1, y+1) + 2f(x, y+1) + f(x+1, y+1) \\
& -f(x-1, y-1) - 2f(x, y-1) - f(x+1, y-1).
\end{aligned}
\tag{3.26}
$$

Unlike the discrete chaincodes, the direction of a gradient vector is continuous. To obtain direction histogram feature, the gradient direction can be partitioned into crisp regions of angle [104]. A gradient vector decomposition method, originally proposed in online character recognition [49], can give better recognition performance [74]. By specifying a number of standard directions (e.g., eight chaincode directions), a gradient vector of arbitrary direction is decomposed into two components coinciding with the two neighboring standard directions (Fig. 3.6). Denoting the component lengths of two standard directions by l_1 and l_2, the corresponding two direction planes are updated by $f_1(x, y) = f_1(x, y) + l_1$ and $f_2(x, y) = f_2(x, y) + l_2$, respectively. The direction planes are completed by decomposing the gradient vectors at all pixels. In order to extract the gradient orientation feature, each pair of direction planes of opposite directions is merged into an orientation plane.

Figure 3.7 shows the orientation planes of gradient on the character image of Figure 3.1 and their blurred images. We can see that both the orientation planes and the blurred images are similar to those of chaincode feature in Figure 3.2.

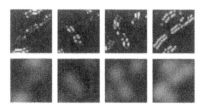

FIGURE 3.7   Orientation planes merged from eight direction planes of gradient (upper) and their blurred images (lower).

3.1.4   Image Registration

In image processing applications, image registration is a necessity as it allows us to compare and to integrate pictures taken at different times, from different sensors or from different viewpoints [14]. Combining these images provides a comprehensive and global picture.

3.1.4.1 Definition   Image registration is defined as a mapping between two or more images, both spatially and with respect to intensity [14]. Consider two two-dimensional images (2D) with intensity functions, $f_1(x, y)$ and $f_2(x, y)$. Both of these functions map to their corresponding intensity values in their respective images. Therefore, the mapping between two images can be described as follows:

$$f_2(x, y) = g(f_1(T(x, y))), \tag{3.27}$$

where $f_1(x, y)$ and $f_2(x, y)$ are the intensity functions of two 2D images, T is a 2D spatial-coordinate transformation, and g is a one-dimensional (1D) intensity transformation.

The objective of image registration is to determine the optimal spatial and intensity transformation, that is, T and g, respectively, in order to match images for (1) determining the parameters of the matching transformation or (2) exposing important differences between images [14].

3.1.4.2 Transformation   Transformations are required to eliminate the variation in images and to register them properly. The type of spatial transformation depends on the image registration technique used. The registration methodology must choose the optimal set of transformations to eliminate the variation of images caused by differences in acquisition or intrinsic changes in the scene. Popular transformations include rigid, affine, projective, perspective, and global polynomial. A brief definition of each type of transformation and its properties is given in the sequel.

Rigid (or linear) transformations consist of a mixture of a rotation, a translation, and a scale change [14]. A transformation is said to be rigid if it preserves relative distance before and after the transformation. Such transformations are linear in nature. A transformation T is *linear* if

$$\begin{aligned} T(x_1 + x_2) &= T(x_1) + T(x_2), \\ cT(x) &= T(cx), \end{aligned} \tag{3.28}$$

where c is a constant [14].

Affine transformations are a generalization of linear transformations; therefore, they are able to cope with more complex distortions. Nevertheless, they still hold some nice mathematical characteristics. A transformation is *affine* if $T(x) - T(0)$ is linear [14]. Typically, it maps a point (x_1, y_1) of the first image to a point (x_2, y_2) of the second image using the following mathematical relation:

$$\begin{pmatrix} x_2 \\ y_2 \end{pmatrix} = \begin{pmatrix} t_x \\ t_y \end{pmatrix} + s \begin{pmatrix} \cos\theta & -\sin\theta \\ \sin\theta & \cos\theta \end{pmatrix} \begin{pmatrix} x_1 \\ y_1 \end{pmatrix}, \tag{3.29}$$

or

$$\bar{p}_2 = \bar{t} + sR\bar{p}_1, \tag{3.30}$$

where $\bar{p}_1$, $\bar{p}_2$ are the coordinate vectors of the two images, $\bar{t}$ is the translation vector, s is a scalar scale factor, and R is the rotation matrix. Note that the rotation matrix R is orthogonal, therefore, the angles and lengths in the first images are conserved. The scale factor s is a scalar that changes both lengths of x and y with respect to the first image. Affine transformations are linear if the translation vector $\bar{t}$ is ignored.

Generalizing the mathematical relation of affine transformation yields

$$\begin{pmatrix} x_2 \\ y_2 \end{pmatrix} = \begin{pmatrix} a_{13} \\ a_{23} \end{pmatrix} + s \begin{pmatrix} a_{11} & a_{12} \\ a_{21} & a_{22} \end{pmatrix} \begin{pmatrix} x_1 \\ y_1 \end{pmatrix}. \tag{3.31}$$

The rotation matrix is no longer orthogonal, nor are the angles and lengths conserved; however, parallel lines remain parallel. Complex distortions, namely shears and changes in aspect ratio, may be held with the general affine transformation. Shears are distortions of pixels along one axis proportional to their location in the other axis [14]. The shear matrices, which act along the x-axis and the y-axis, are described as

$$\text{shear}_x = \begin{pmatrix} 1 & a \\ 0 & 1 \end{pmatrix}, \quad \text{shear}_y = \begin{pmatrix} 1 & 0 \\ b & 1 \end{pmatrix}, \tag{3.32}$$

respectively. Aspect ratio is the relative scale between the x and y axes. The aspect ratio is changed by scaling x-axis and y-axis separately. Mathematically, scaling is defined as

$$\begin{pmatrix} s_x & 0 \\ 0 & s_y \end{pmatrix}. \tag{3.33}$$

Combinations of rigid transformations, shears, and aspect ratio distortions can be eliminated using the general affine transformation.

Perspective transformation is required for the distortion that occurs when a three-dimensional (3D) scene is projected through an idealized optical image system [14]. For instance, a 3D object projection on a flat screen results in perspective distortion. Under perspective distortion, the image appears to be smaller, the farther it is from the camera, and more compressed, the more it is inclined away from the camera [14]. Let (x_0, y_0, z_0) be a point in the original scene; the corresponding point in the image

(x_i, y_i) is defined by

$$x_i = \frac{-fx_0}{z_0 - f}, \quad y_i = \frac{-fy_0}{z_0 - f}, \tag{3.34}$$

where f is the position of the center of the camera lens.

Projective transformation is a special case of perspective transformation. It accounts for situations when the scene is composed of a flat plane skewed with respect to the image plane. Projective transformation maps the scene plane into an image that is tilt-free and has a set scale. A point in the scene plane (x_p, y_p) is mapped into a point in the image plane (x_i, y_i) using the following relation:

$$x_i = \frac{a_{11}x_p + a_{12}y_p + a_{13}}{a_{31}x_p + a_{32}y_p + a_{33}},$$

$$y_i = \frac{a_{21}x_p + a_{22}y_p + a_{23}}{a_{31}x_p + a_{32}y_p + a_{33}}, \tag{3.35}$$

where the a terms are constants that are dependant on the equations of the scene and image plane [14].

When all the above transformations are inept to eliminate the image distortions, then polynomial transformation may be utilized to achieve a global alignment. This type of transformation is defined in the following section.

3.1.4.3 Image Variations
The first step to optimal image registration is to understand the different types of image variations. Image variations occur when there are changes in the scene due to the sensor and its position or viewpoint. They are classified into three different types: corrected distortions, uncorrected distortions, and variations of interest. *Corrected distortions* are spatial distortions due to a difference in acquisition that may be corrected using an appropriate transformation technique. *Uncorrected distortions* are volumetric distortions, caused by lighting and atmospheric conditions, which remain uncorrected after using a transformation. *Variations of interests* are distortions that enclose important information, such as object movements or growths, and must be exposed after transformation. Differentiating the types of image variations helps immensely the process of image registration [14].

3.1.4.4 Registration Methods
In the following, we outline four classes of registration methods: correlation, Fourier methods, point mapping, and elastic methods.

(1) Correlation methods
Correlation methods are fundamental techniques used in registration methods. They are founded on cross-correlation that matches a template or a pattern in a picture. In principle, cross-correlation measures the degree of similarity between an image and a template. Consider an $(X \times Y)$ template T and an $(U \times V)$ image I, where T is

smaller than I, then the 2D normalized cross-correlation function is defined as

$$C(u, v) = \frac{\sum_x \sum_y T(x, y)I(x - u, y - v)}{\sqrt{\left[\sum_x \sum_y I^2(x - u, y - v)\right]}},$$

(3.36)

for each translation, where (u, v) is a point located on the image I [14]. To find the degree of similarity between a template and an image, the cross-correlation must be computed over all possible translations. Note that as the number of translation increases, the computation cost also increases. Therefore, correlation methods are mostly used for images that are misaligned by small rigid or affine transformations.

(2) *Fourier methods*
Fourier methods are based on the fact that information in a picture can be represented in the frequency domain. In addition, geometric transformations, such as translation, rotation, and reflection, all have their counterparts in the frequency domain. Therefore, all computations are done in the frequency domain in the Fourier methods. Kuglin and Hines [14] introduced *phase correlation* to achieve alignment between two images. Before discussing phase correlation, it is necessary to define a few terms.

The Fourier transform of an intensity image function $f(x, y)$ is given by

$$F(w_x, w_y) = R(w_x, w_y) + jI(w_x, w_y),$$

(3.37)

where $R(w_x, w_y)$ and $I(w_x, w_y)$ are the real and imaginary parts at each frequency (w_x, w_y), respectively. The Fourier transform can also be expressed in the exponential form as follows:

$$F(w_x, w_y) = |F(w_x, w_y)|e^{j\phi(w_x, w_y)},$$

(3.38)

where $|F(w_x, w_y)|$ and $\phi(w_x, w_y)$ are the magnitude and the phase angle of the Fourier transform, respectively. An important property of the Fourier transform is its translation property, also known as shift theorem. Consider two images f_1 and f_2, which are related by

$$f_2(x, y) = f_1(x - d_x, y - d_y),$$

(3.39)

where d_x and d_y are constants. The corresponding Fourier transforms F_1 and F_2 are defined as

$$F_2(w_x, w_y) = e^{-j(w_x d_x + w_y d_y)} F_1(w_x, w_y).$$

(3.40)

Hence, displacement in the spatial domain is equivalent to a phase shift in the frequency domain. In the frequency domain, the phase shift or phase correlation can be computed using the cross-power spectrum of the two images:

$$\frac{F_1(w_x, w_y)F_2^*(w_x, w_y)}{|F_1(w_x, w_y)F_2^*(w_x, w_y)|} = e^{(w_x d_x + w_y d_y)}, \tag{3.41}$$

where F^* is the complex conjugate of F. Taking the inverse Fourier transform of the cross-power spectrum yields an impulse function $\delta(dx, dy)$. The impulse is located at the displacement required to match the two images and thus achieves optimal image registration. Like the correlation methods, Fourier methods are mostly employed for images that are misaligned by small, rigid, or affine transformations.

(3) *Point mapping*
Point mapping is the primary registration method employed when the misalignment between two images is not identifiable, for instance, when two images suffer from perspective distortion (see Section 3.1.3.2). In principle, the standard point-mapping technique has three stages. The first stage entails computing the features in the image. In the second stage, feature points (also known as control points) in the template or pattern are matched with feature points in the picture. In the third stage, a spatial mapping or transformation is found taking into consideration the matched feature points obtained in the second stage [14]. This spatial mapping (along with interpolation) is thereafter used to resample one image onto the other.

(4) *Elastic methods*
Elastic methods are the latest engineered methods for image registration. In these methods, the distortion in the image is modeled as an elastic body; therefore, no transformations are computed. The elastic body is bent and stretched and thus deformed. The amount of bending and stretching is described by the energy state of the elastic material. In principle, the control points are modeled as external forces that deform the body. These forces are compensated by stiffness or smoothness constraints that are usually parameterized to give the user some flexibility [14]. In this process, the registration is defined by determining the minimum energy state that corresponds to the given deformation transformation. As elastic methods imitate physical deformations, they are widely employed for problems in shape and motion reconstruction and medical imaging.

3.1.5   Hough Transform

A popular feature extraction method used in digital image processing is the Hough transform (HT). It is able to detect straight lines, curves, or any particular shape that can be defined by parametric equations. In essence, this method maps the figure points from the picture space to the parameter space and, thereafter, extracts the features. There are two types of Hough transform: standard HT and randomized HT. The sequel briefly expounds these two types of HT.

3.1.5.1 *Standard Hough Transform*   As mentioned above, the objective of the Hough transform is to detect a possible line, curve, or shape that passes through the

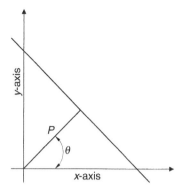

FIGURE 3.8   Polar space representation of a straight line.

figure points in a digital picture. In the standard hough transform, all figure points $\{(x_1, y_1), \ldots, (x_n, y_n)\}$ from the picture space are mapped to the parameter space [25]. For instance, a line in the picture space, $y = mx + b$ is defined in the parameter space as follows:

$$\rho = x \cos \theta + y \sin \theta, \qquad (3.42)$$

where ρ is the length of the normal from the origin to the line and θ is the angle of ρ with respect to the x-axis (see Fig. 3.8). Therefore, a line in the picture space is mapped to a unique point (e.g., (θ_0, ρ_0)) in the parameter space. The parameter space is divided into finite intervals, often called accumulator bins, which are assigned a value that is incremented every time a figure point is mapped to that given interval in the parameter space. Maxima in the accumulator array indicate the detection of a straight line or shape in the picture. Mapping the maxima in the accumulator array back to the picture space completes the detection of a feature [25].

3.1.5.2 *Randomized Hough Transform*   randomized Hough transform is an enhanced approach for detecting features in binary images. It takes into account the drawbacks of standard Hough transform, namely long computation time and large memory requirements.

It is based on the fact that a curve or a shape can be defined in the parameter space with a pair or n-tuple of points (depending on the shape to be detected) from the original binary picture. Consider a set of figure points $P = \{(x_1, y_1), \ldots, (x_n, y_n)\}$ in the binary picture. Now, let (θ, ρ) be the two parameters of the lines to be detected. A pair of points $(p_i(x, y), p_j(x, y))$ is selected randomly from the set P and mapped to the parameter space. The corresponding accumulator bins are then incremented. Unlike standard Hough transform where each figure point is mapped to the parameter space, randomized Hough transform maps a set of figure points that may or may not form a shape; therefore, a considerable decrease in the computation time is noticed. The result of continuing this process will be the appearance of maxima in the accumulator bins of the parameter space. These maxima can thereafter be used to detect the lines in the binary pictures [64].

3.1.6   Line-Based Representation

Though the strokes in character images are usually thicker than one pixel, line-based features can be extracted from the stroke edge (contour) or the skeleton.

3.1.6.1 Edge (Contour)   An edge is defined as the boundary between two regions with different gray-level properties [37]. It is possible to detect an edge because the gray-level properties of an image at an edge drastically change, which results in gray-level discontinuity.

In most edge-detection methods, the local first and second derivatives are computed. A drastic change in the magnitude of the first derivative signifies the presence of an edge in an image. Similarly, using the sign of the second derivative, it is possible to determine whether an edge pixel lies on the dark or on the light side of an edge. The second derivative is positive for a dark side and negative for a light side. Therefore, if a second derivative of a given boundary in an image is a negative spike followed by a positive spike, then there is a detection of a light-to-dark edge.

Tracing a contour involves the detection of edges. As a contour is composed of a set of boundary points, it is possible to extend any edge-detection method to detect contours.

3.1.6.2 Thinning   *Thinning* is a process during which a generally elongated pattern is reduced to a line-like representation. This line-like representation or a collection of thin arcs and curves is mostly referred to as a *skeleton*. Most thinning algorithms are iterative in that they delete successive layers of pixels on the edge of the pattern until only a skeleton is left [125]. A set of tests is performed to assess if a given black pixel p is to be deleted or to be preserved. Obviously, the deletion or preservation of p depends on its neighboring pixels' configuration. Thinning algorithms are divided into two categories: sequential and parallel.

In a sequential algorithm, the sequence in which the pixels are tested for deletion is predetermined, and each iteration depends on all the previous iterations. That is, the deletion of p in the nth iteration depends on all the operations that have been performed so far, that is, on the result of the $(n-1)th$ iteration as well as on the pixels already processed in the nth iteration [67].

In a parallel algorithm, the deletion of pixels in the nth iteration would depend only on the result that remains after the $(n-1)th$; therefore, all pixels can be examined independently in a parallel manner in each iteration [67].

3.1.6.3 Tracing   Tracing is a method designed to recover the sequence of the segments that compose handwritten English or handwritten signatures [72]. Before introducing the tracing algorithm, a few definitions are provided in order to ease its understanding.

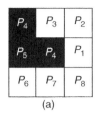

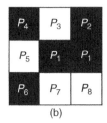

(a)   (b)

FIGURE 3.9   (a) An end point pixel P_i; (b) a junction point pixel P_i.

(1) *Definitions*

A signature is made up of multiple segments, better known as *strokes*. A stroke is a series of pixels that are placed according to the order in which they are traced. The stroke originates at a starting point and concludes at an ending point. The notation of a stroke can be given by

$$S_i^j \xrightarrow{\text{tracing}} E_i^j, \tag{3.43}$$

where i is the stroke number, j is the number of time a given stroke is traced, S_i^j is the starting point (may be traced more than once $j > 1$), $\rightarrow$ is the tracing direction, and E_i^j is the ending point [72].

A transition function $T(x)$ computes the connectivity of a given pixel P_i with its eight neighboring pixels. In other words, it returns the number of pixels that are connected to a given pixel P_i.

An end point (EP) is a pixel P_i that is connected to one immediate neighboring pixel (i.e. $T(P_i) = 1$). A junction point (JP) is a pixel P_i that is connected to three or more adjacent pixels (i.e., $T(P_i) \geq 3$). Figure 3.9 illustrates an EP and a JP with $T(P_i) = 1$ and $T(P_i) = 3$, respectively.

Another important parameter during tracing is the performance measure (PM). Performance measure is the tracing direction between a pair of successive pixels and is coded according to the eight directional chaincodes shown in Figure 3.10 (a). For instance, the pixel configuration of Figure 3.10 (b) is given a performance measure of 3, that is, PM $= 3$.

Obtaining the direction of a stroke requires another parameter, the intrastroke performance measure (intra-PM). This measures the sums and averages of all the successive PMs of the stroke. The mathematical expression for the intra-PM of the *ith* stroke (intra-PM$_i$) is as follow

$$\text{intra-PM}_i = \sum_{j=1}^{n-1} \frac{\text{PM}_{ij}}{n-1} = \frac{\text{PM}_{i1} + \text{PM}_{i2} + \cdots + \text{PM}_{i(n-1)}}{n-1}, \tag{3.44}$$

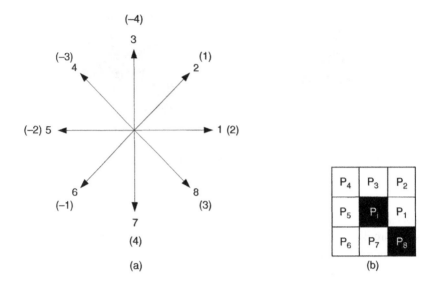

FIGURE 3.10 (a) Directional chaincodes; (b) pixel configuration with PM = 3.

where PM_{ij} is the performance measure between pixel j and $j+1$ in the ith stroke of the signature, and n is the total number of pixels of the ith stroke.

A simple test is performed to detect the direction or sequence of the stroke:

- If (intra-PM_i) > threshold (zero), the original stroke sequence is preserved.
- If (intra-PM_i) ≤ threshold (zero), the original stroke sequence is reversed to enhance the intra-PM of the ith stroke.

(2) *Tracing algorithm*
After defining key parameters used in tracing, the sequel presents a heuristic-rule-based tracing algorithm (HRBTA). Before starting the tracing process, there is some preliminary work to be done:

- An integer array with one element per black pixel is needed to track the number of times each pixel has been traced. Obviously, all elements in the array are initialized to zero.
- Two first in, first out (FIFO) queues are required to track the x and y coordinates of the EPs and JPs of the given signature.

The pseudocode of the HRBTA is as follows [72]:

Procedure HRMTA(*sig_array*, *ep_q*, *jp_q*, *traced_array*)
begin

    while there are EPs or JPs in their respective FIFO queue do
    begin

        Find the starting point of the current stroke from *ep_q* and
        jp_q. Traced from the selected starting point until
        an ending point is encountered.
        If intra-PM of the stroke ≤ 0 then
        begin

            Reverse the sequence of the traced stroke.
        end
        Append the stroke to *traced_array* and pad it with a SEPARATOR.
        Delete the starting point from its corresponding FIFO queue
    end
    call subroutine ReOdr_Stkr(*traced_array*, *sorted_array*))
end

As you may notice, the tracing algorithm has four inputs: *sig_array*, *ep_q*, *jp_p*, and *traced_array*. *sig_array* is a 448×120 array that saves the skeletal image of a given signature. *ep_q* and *jp_p* are two FIFO queues that store the EPs and JPs of a given signature. Finally, *traced_array* is an array that stores the sequence of traced strokes of a signature.

The algorithm starts by tracing from a starting point and finishes at an ending point. This stroke is then concluded. It reverses the stroke sequence if the intra-PM is negative. This stroke is then saved in the *traced_array* and followed by a SEPARATOR. The current starting point is deleted from its FIFO and a new starting point is selected to repeat the process. The process is repeated until both FIFO queues are empty.

The subroutine ReOdr_Stkr plays a crucial role in the algorithm. Occasionally, strokes are not in order when the tracing process is complete due to backtracking strokes and the minimum hand-movement behavior of a signature. Therefore, ReOdr_Stkr rearranges the stroke sequence to eliminate the remaining stroke disorders.

3.1.7   Fourier Descriptors

Nowadays in image processing, a fundamental need is to retrieve a particular shape from a given image. Numerous shape retrieval methods have been developed in the literature; however, Fourier descriptors (FD) is one of the most popular and efficient methods. In principle, it represents the shape of the object in the frequency domain. By doing so, Fourier descriptors enjoy multiple advantages over their counterparts. These advantages include strong discrimination ability, low noise sensitivity, easy normalization, and information preservation [125]. The sequel first discusses different shape signatures and thereafter expounds two types of FD: standard Fourier descriptors and elliptical Fourier descriptors.

3.1.7.1 Shapes Fourier descriptors are used for describing the shape (closed curve) of any object found on an input image. However, before computing the Fourier descriptors, the input image is digitalized and boundary information of the object is extracted [89] and normalized. During normalization, the data points of the shape boundary of the object and model are sampled to have the same number of data points. As Fourier descriptors require a 1D representation of the boundary information, shape signatures are used. A shape signature maps a 2D representation of a shape to a 1D representation. Although there are numerous shape signatures, complex coordinates, centroid distance, curvature, and cumulative angular function are the four shape signatures that will be considered hereafter.

Complex Coordinates
Consider a binary shape boundary that is made of L pixels, where each pixel has the coordinates $(x(t), y(t))$ with $t = 0, 1, \ldots, L - 1$. A boundary coordinate pair $(x(t), y(t))$ can be expressed in terms of complex coordinates using the following statistical relation:

$$z(t) = [x(t) - x_c] + j[y(t) + y_c], \tag{3.45}$$

where $j = \sqrt{-1}$ and (x_c, y_c) is the shape's centroid given by

$$x_c = \frac{1}{L} \sum_{t=0}^{L-1} x(t), \quad y_c = \frac{1}{L} \sum_{t=0}^{L-1} y(t), \tag{3.46}$$

where L is the number of pixels that form the shape boundary. Shifting the complex representation by the centroid is not necessary; however, it makes the shape representation invariant to translation [125].

Centroid Distance
The centroid distance function represents the boundary information by the distance separating the centroid and the boundary points. For a given shape, the centroid distance function is given by

$$r(t) = \sqrt{([x(t) - x_c]^2 + [y(t) - y_c]^2)}, \tag{3.47}$$

where (x_c, y_c) is the shape's centroid defined by Eq. (3.1.40).

Curvature Signature
Curvature signature can be obtained by differentiating successive boundary angles calculated in a window w. The boundary angle is defined as

$$\theta(t) = \arctan \frac{y(t) - y(t - w)}{x(t) - x(t - w)}. \tag{3.48}$$

The curvature is given by

$$K(t) = \theta(t) - \theta(t-1).$$

(3.49)

Note that this definition of curvature function has discontinuities at size 2π in the boundary. Thus, it is preferable to use the following definition:

$$K(t) = \varphi(t) - \varphi(t-1),$$

(3.50)

where $\varphi(t)$ is defined by Eq. (3.1.44).

Cumulative angular function shapes can also be defined according to the boundary angles; however, boundary angles can only take values in range of $[0, 2\pi]$ as the tangent angle function $\theta(t)$ has discontinuities at size 2π. Zahn and Roskies introduced the cumulative angular function $\varphi(t)$ to eliminate the problem of discontinuity. They defined it as the net amount of angular bend between the starting position $z(0)$ and position $z(t)$ on the shape boundary:

$$\varphi(t) = [\theta(t) - \theta(0)] \mathrm{mod}(2\pi),$$

(3.51)

where $\theta(t)$ is the boundary angle defined by Eq. (3.1.42).

A normalized cumulative angular function $\psi(t)$ is used as the shape signature (assuming shape is traced in the anticlockwise direction):

$$\psi(t) = \varphi\left(\frac{L}{2\pi}t\right) - t,$$

(3.52)

where L is the number of pixel forming the shape boundary and $t = 0, 1, \ldots, L-1$.

3.1.7.2 Standard Fourier Descriptors

The Fourier transform for a shape signature formed with L pixels, $s(t), t = 0, 1, \ldots, L$, assuming it is normalized to N points in the sampling stage, is given by

$$u_n = \frac{1}{N} \sum_{t=0}^{N-1} s(t) e^{\left(\frac{-j2\pi nt}{N}\right)}, \quad n = 0, 1, \ldots, N-1.$$

(3.53)

The coefficients u_n are called the Fourier descriptors of the sample. They may also be denoted by FD_n.

3.1.7.3 Elliptical Fourier Descriptors

Kuhl and Giardina [63] introduced another approach for shape description [109]. They approximated a closed contour formed of L pixels, $(x(t), y(t)), t = 0, 1, \ldots, L-1$ as

$$\hat{x}(t) = A_0 + \sum_{n=1}^{L} \left[a_n \cos \frac{2n\pi t}{T} + b_n \sin \frac{2n\pi t}{T} \right],$$

$$\hat{y}(t) = C_0 + \sum_{n=1}^{L} \left[c_n \cos \frac{2n\pi t}{T} + d_n \sin \frac{2n\pi t}{T} \right], \tag{3.54}$$

where $n = 1, 2, \ldots, L$, T is the total length of the contour and with $\hat{x}(t) \equiv x(t)$ and $\hat{y}(t) \equiv y(t)$ in the limit when $N \to \infty$. The coefficients are given by

$$\begin{aligned}
A_0 &= \tfrac{1}{T} \int_0^T x(t)dt , \\
C_0 &= \tfrac{1}{T} \int_0^T y(t)dt , \\
a_n &= \tfrac{2}{T} \int_0^T x(t) \cos \frac{2n\pi t}{T} dt , \\
b_n &= \tfrac{2}{T} \int_0^T x(t) \sin \frac{2n\pi t}{T} dt , \\
c_n &= \tfrac{2}{T} \int_0^T y(t) \cos \frac{2n\pi t}{T} dt , \\
d_n &= \tfrac{2}{T} \int_0^T y(t) \sin \frac{2n\pi t}{T} dt .
\end{aligned} \tag{3.55}$$

3.1.8   Shape Approximation

The contour or skeleton of a character image can be approximated into piecewise line segments, chaincodes, and smooth curves.

3.1.8.1 Polygonal Approximation   As the name implies, polygonal approximation approximates a digital contour as a 2D polygon. For a given contour, the most exact approximation is defined as the number of segments of the polygon being equal to the number of points of the contour. A polygonal approximation offers the possibility to describe the essential contour shape with the fewest possible polygonal segments. For instance, minimum perimeter polygons approximate a contour by enclosing the contour by a set of concatenated cells and regarding the contour as a rubber band contained within the inside and outside boundaries of the strip of cells [97].

3.1.8.2 Chaincodes   Freeman [31] introduced chaincodes to represent a shape boundary by a connected sequence of straight lines that have fixed length and direction. The direction of the lines is based on 4- or 8-connectivity (see Fig. 3.11(a) and 3.11(b)) and coded with the corresponding code. For instance, the 4-connectivity code 00112233 represents a line sequence that forms a perfect square having the bottom left corner as the starting point (see fig. 3.11(c)). Similarly, any shape can be coded depending on the scheme using 4- or 8-connectivity [83, 37].

3.1.8.3 Bezier Curves   Shape approximation may also be achieved using *Bezier curves*. Paul Bezier developed Bezier curves in 1960 for designing cars. In principle, the shape of the smooth Bezier curve is controlled by a set of data points $\{P_0, P_1, \ldots, P_n\}$, called control points. These control points determine the slopes of the segments that form the smooth Bezier curve and thus contribute to the overall shape of the curve. For instance, the slope of the curve at P_0 is going to be the same

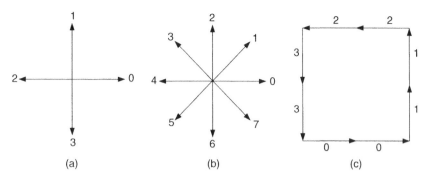

FIGURE 3.11   Chaincodes: (a) 4-connectivity; (b) 8-connectivity; (c) example.

as the slope of the line connecting $\mathbf{P}_0$ and $\mathbf{P}_1$. Therefore, the location of $\mathbf{P}_1$ relative to $\mathbf{P}_0$ controls the shape of the Bezier curve.

A Bezier curve of order n can formally be defined as a weighted sum of $n + 1$ control points $\{\mathbf{P}_0, \mathbf{P}_1, \ldots, \mathbf{P}_n\}$. The weights are the Bernstein polynomials, also called blending functions:

$$B_i^n(t) = \binom{n}{i}(1 - t)^{n-i}t^i, \tag{3.56}$$

where n is the order of the curve and $i = 0, 1, \ldots, n$.

The mathematical definition of Bezier curve is as follow:

$$\bar{P}(t) = \sum_{i=0}^{n} \binom{n}{i}(1 - t)^{n-i}t^i \bar{P}_i, \ 0 \leq t \leq 1, \tag{3.57}$$

where n is the order of the curve and $i = 0, 1, \ldots, n$. It is important to note that the curve generally does not pass through the control points except for the first and the last. Also, the Bezier curve is always enclosed within the convex hull of the control points. Figure 3.12 illustrates a common application of Bezier curve: font definition.

FIGURE 3.12   Bezier curve used in font definition.

3.1.8.4 B-Splines
B-splines are a generalization of the Bezier curves. A B-spline is defined as follows:

$$P(t) = \sum_{i=1}^{n+1} N_{i,k}(t) P_i, \quad t_{min} \leq t \leq t_{max}, \tag{3.58}$$

where $n + 1$ are the control points $(P_1, P_2, \ldots, P_{n+1})$, $N_{i,k}$ are the blending functions of order $k \in [2, n + 1]$, and degree $k - 1$ and t are the prespecified values at which the pieces of the curve join. This set of values is called a knot vector $(t_1, t_2, \ldots, t_{k+(n+1)})$. The values of t are subject to the following condition $\forall_i$, $t_i \leq t_{i+1}$. The B-spline blending functions are recursively defined as follows:

$$N_{i,k} = \begin{cases} 1 & \text{if } t_i < t < t_{i+1} \\ 0 & \text{otherwise} \end{cases} \tag{3.59}$$

for $k = 1$, and

$$N_{i,k}(t) = \frac{(t - t_i) N_{i,k-1}(t)}{t_{i+k-1} - t_i} + \frac{(t_{i+k} - t) N_{i+1,k-1}(t)}{t_{i+k} - t_{i+1}}. \tag{3.60}$$

3.1.9 Topological Features

We describe two types of topological features extracted from skeleton or traced contour: feature points and curvature.

3.1.9.1 Feature Points
In a thinned image, black pixels that have a number of black neighbors not equal to 0 or 2 are given the name of feature points. Feature points have been classified into three types:

1. End points;
2. Branch points;
3. Cross points.

The first type of point, the end point, is defined as the start or the end of a line segment. A branch point is a junction point connecting three branches. A cross point is another junction point that joins four branches [5]. Figure 3.13 visually illustrates the three types of feature points.

3.1.9.2 Curvature Approximation
Curvature of a given curve or binary contour is another essential local feature. In calculus, the curvature c at a point p on a continuous plane curve C is defined as $c = \lim_{\Delta s \to 0} \frac{\Delta \alpha}{\Delta s}$, where s is the distance to the point p along the curve and $\Delta \alpha$ is the change in the angles of the tangents to the curve at the distances s and $s + \Delta s$, respectively [24].

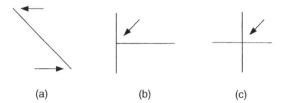

(a)   (b)   (c)

FIGURE 3.13   Types of feature points: (a) end point; (b) branch point; (c) cross point.

In a binary curve, the above definition of curve cannot be applied directly as the analytical format of the curve is not available. Nevertheless, the curvature can be obtained using a unity interval $\Delta s = 1$. The curvature is therefore given by $c = \Delta \alpha$. In order to compute the curvature, the digital curve C must be a sequence of consecutive points on the x–y plane: $C = p_1 p_2 \ldots p_N$, where $p_n = (x_n, y_n)$, the distance from p_n to p_{n+1} is 1 (since $\Delta s = 1$), and N is the number of sampled points in the trace [24].

To approximate the curvature of the given curve C, the sequences of angles from point to point must be calculated:

$$A = \alpha_1 \alpha_2 \ldots \alpha_L, \qquad (3.61)$$

where the angle α_n has the range of $[-180°, 180°]$ and is given by

$$\alpha_n = \tan^{-1} \frac{y_{n+1} + y_n}{x_{n+1} + x_n}. \qquad (3.62)$$

From the definition of A, it is easy to show that the sequence of changes in angles is given by

$$\Delta A = \Delta \alpha_1 \Delta \alpha_2 \ldots \Delta \alpha_n, \qquad (3.63)$$

where the difference in angles is defined as

$$\Delta \alpha_n = (\alpha_n - \alpha_{n-1}) \mathrm{mod} 360°. \qquad (3.64)$$

To remove the digitization and quantization noise from the above definition, the sequence ΔA should be convolved with a Gaussian filter G, that is, $\Delta A \times G = \Delta A^*$. Finally, the curvature C of curve can be computed as follows:

$$c_n = \Delta \alpha_n^*. \qquad (3.65)$$

3.1.10   Linear Transforms

Feature space transformation methods are often used in pattern recognition to change to feature representation of patterns for improving the classification performance.

Linear transforms are usually used for reducing the dimensionality of features, and some of them can also improve the classification accuracy.

3.1.10.1 *Principal Component Analysis*   Principal component analysis is a statistical technique belonging to the category of factor analysis methods. The common goal of these methods is the simplification of a set of features by the identification of a smaller subset of transformed features that contains essentially the "same" information as the original set. "Dimensionality reduction" is also a common way of describing this goal. The reason why there are many factor analysis methods is that there are many ways of finding reasonable subsets of transformed features with roughly the same information. For instance, independent component analysis (ICA) seeks independent factors, but a drawback of this strong condition of independence is that it will not work if the features are normally distributed.

PCA works regardless of the distributional characteristics of the features, as it enforces the weaker condition of noncorrelation of the transformed features. Moreover, it is a linear technique, in the sense that each new transformed feature is a linear combination of the original set of features.

Another way to consider PCA is from the concept of *regression*. Classical least squares regression is concerned with finding a simple way to explain the dependency between a dependent variable Y and an independent variable X. The simple linear relationship is the one that minimizes the sum of squares of the regression errors. One important thing to remember about this is that regression minimizes the errors considered as the *vertical* distances between Y and its approximation. In many applications, such as when one regresses a variable on itself, the researcher may rather be interested in finding the linear relationship that minimizes the *perpendicular* errors. The first principal component of PCA is precisely this linear relationship. Figure 3.14 shows the first and second principal components for a bivariate sample. As one can see, to the naked eye, the first principal component is very similar to a regression line.

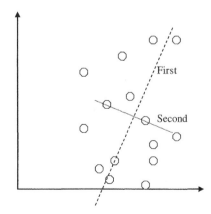

FIGURE 3.14   First and second principal components.

Formally, let us define the following linear combination of features:

$$z = a_0 + a_1 y_1 + \cdots + a_k y_k. \tag{3.66}$$

The sample variance of this linear combination is

$$s_z^2 = a' Sa, \tag{3.67}$$

where $\mathbf{a}$ is a $k \times \mathbf{1}$ vector of weights and $\mathbf{S}$ is the variance–covariance matrix of the $\mathbf{k}$ features. The PCA problem consists of finding a characterization of what would be the set of weights of vector $\mathbf{a}$ such that the linear combination y would have the maximum variance.

A moment of reflection shows that this problem is ill-defined: If we do not constrain the elements of $\mathbf{a}$ in some sort, then we can always find infinity as the maximum variance, because all the variances of the features are positive by definition. To circumvent this problem, we limit ourselves to vectors $\mathbf{a}$ having unit length.

The first principal component is, therefore, the linear combination of features y whose weight of vector $\mathbf{a}$ solves the following problem:

$$\max_a \frac{a' Sa}{a' a}. \tag{3.68}$$

Let us call this ratio λ. The first-order condition is

$$\frac{\partial \lambda}{\partial \alpha} = \frac{a' a(2Sa) - a' Sa(2a)}{(a' a)^2} = 0, \tag{3.69}$$

or

$$Sa - \frac{a' Sa}{a' a} a = 0, \tag{3.70}$$

which is, by definition of λ,

$$Sa - \lambda a = 0. \tag{3.71}$$

This expression is familiar; it is the problem of finding eigenvalues. Therefore, the λ that we are looking for, namely the maximum variance of the linear combination, is none other than the maximum eigenvalue of the variance–covariance matrix of the features. Accordingly, the weight vector $\mathbf{a}$ corresponding to the first component is the eigenvector associated with this eigenvalue. The first principal component is therefore

$$z = a' y. \tag{3.72}$$

How do we define the second principal component? Once we have found the combined feature $z_1 = a'_1 y$ explaining the maximum amount of variance that is present in the original set of features, the second principal component is the combined feature $z_2 = a'_2 y$, explaining the maximum amount of variance achievable by any linear combination such that $a'_1 a_2 = 0$, that is, in an orthogonal direction to the first principal component.

Showing this property of the second principal component is straightforward: using a Lagrange multiplier γ, we differentiate with respect to a_2

$$\frac{\partial}{\partial a_2} \left(\frac{a'_2 S a_2}{a'_2 a_2} + \gamma a'_1 a_2 \right) = 0, \tag{3.73}$$

which gives the solutions $\gamma = 0$ and $S a_2 = \lambda 2 a_2$.

As we can place the eigenvalues in decreasing order $\lambda_1 > \lambda_2 > \cdots > \lambda_k$, we can readily see how to find the lesser order principal components; they are simply constructed using the same formula as for the first principal component $\mathbf{z}$ above, but instead using the corresponding eigenvector. Any given principal component of order n represents the linear combination of features having the largest variance among all linear combinations orthogonal to $z_1, z_2, \ldots, z_{n-1}$. There is, of course, a maximum of $\mathbf{k}$ eigenvectors for any variance–covariance matrix S. A nonsingular matrix S will have $\mathbf{k}$ different principal components along $\mathbf{k}$ dimensions.

There is an important relationship between the size of the eigenvalues and the percent of variance explained by the respective principal components. We have already shown that each eigenvalue is equal to the variance of each corresponding component. As we know from elementary linear algebra that

$$\sum_{j=1}^{k} s_{z_j}^2 = \sum_{j=1}^{k} \lambda_j = tr(S), \tag{3.74}$$

we can report the percent of variance explained by the p first components as

$$\frac{\sum_{j=1}^{p} \lambda_j}{\sum_{j=1}^{k} \lambda_j}. \tag{3.75}$$

Another way to show the same information is by displaying a so-called "scree" plot (Fig. 3.15).

A nice feature of principal components is immediately visible from such a plot. The first few principal components generally explain most of the variance that is present in the data, whereas the cumulative impact of the principal components that are on the portion of the screen plot with a small slope add little to the explanation. This is the reason why principal components are usually an effective way of reducing the dimensionality of a feature set. We can use the first principal components (up to where contribution becomes negligible) in lieu of the whole data set.

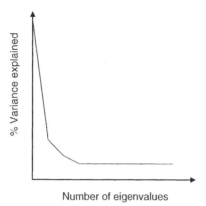

FIGURE 3.15   A typical scree plot.

3.1.10.2   *Linear Discriminant Analysis (LDA)*   The principal component analysis results in a linear subspace of maximum variance for a set of data points. When the data points (also called vectors, examples, or samples) are attached with class labels, it is often hoped that the projections of points onto the subspace are maximally separated into different classes. PCA cannot fulfill this purpose because it does not consider the class labels in estimating the subspace. The class of techniques for learning subspace by maximizing separability is generally called discriminant analysis. The simplest and most popular one, linear discriminant analysis (LDA), also called as Fisher discriminant analysis (FDA), learns a linear subspace that maximizes the Fisher criterion [33].

LDA is a parametric feature extraction method, assuming Gaussian density functions with equal covariance for all classes. The extracted subspace performs fairly well even in situations where the classes undergo non-Gaussian distributions, or the covariance matrices are not equal. Particularly, LDA is a good feature extraction or dimensionality reduction method in character recognition. The subspace features (projected points in subspace) can be classified by either linear or nonlinear classifiers.

Assume that each point belongs to one of M classes $\{\omega_i | i = 1, \ldots, M\}$. The data points are in an original D-dimensional space, and a d-dimensional subspace ($d < D$) is to be extracted. Each class has a Gaussian density function

$$p(\mathbf{x}|\omega_i) = \frac{1}{(2\pi)^{D/2}|\Sigma_i|^{1/2}} \exp\left[-\frac{1}{2}(\mathbf{x} - \mu_i)^T \Sigma_i^{-1}(\mathbf{x} - \mu_i) \right], \qquad (3.76)$$

where the mean vector μ_i and the covariance matrix Σ_i can be estimated by maximum likelihood (ML) from the points of class ω_i:

$$\mu_i = \frac{1}{N_i} \sum_{\mathbf{x}_n \in \omega_i} \mathbf{x}_n, \qquad (3.77)$$

$$\Sigma_i = \frac{1}{N_i} \sum_{\mathbf{x}_n \in \omega_i} (\mathbf{x}_n - \mu_i)(\mathbf{x}_n - \mu_i)^T, \tag{3.78}$$

where N_i denotes the number of points in class ω_i. A within-class scatter matrix S_w and a between-class scatter matrix S_b are then defined by

$$S_w = \sum_{i=1}^{M} P_i \Sigma_i, \tag{3.79}$$

$$S_b = \sum_{i=1}^{M} P_i (\mu_i - \mu_0)(\mu_i - \mu_0)^T, \tag{3.80}$$

where $P_i = \frac{N_i}{N}$ (N is the total number of points), and $\mu_0 = \sum_{i=1}^{M} P_i \mu_i$.

It can be proved that S_w and S_b sum up to a total scatter matrix S_t:

$$S_t = \frac{1}{N} \sum_{n=1}^{N} (\mathbf{x}_n - \mu_0)(\mathbf{x}_n - \mu_0)^T = S_w + S_b, \tag{3.81}$$

and S_b is equivalently computed by

$$S_b = \sum_{i=1}^{M} \sum_{j=i+1}^{M} P_i P_j (\mu_i - \mu_j)(\mu_i - \mu_j)^T. \tag{3.82}$$

Denoting the axis vectors of the d-dimensional subspace by $\mathbf{w}_1, \ldots, \mathbf{w}_d$, as the columns of a $D \times d$ transformation matrix W, the task of subspace learning is to optimize W under a criterion. In the subspace, the matrices of within-class scatter, between-class scatter, and total scatter are transformed to $W^T S_w W$, $W^T S_b W$, and $W^T S_t W$, respectively. Whereas PCA maximizes the total variance in subspace, which equals $\text{tr}(W^T S_t W)$, LDA maximizes a separability measure

$$J_1 = \text{tr}\big((W^T S_w W)^{-1} W^T S_b W\big). \tag{3.83}$$

It is shown that the selected subspace vectors that maximize J_1 are the eigenvectors of matrix $S_w^{-1} S_b$ corresponding to the d largest eigenvalues [33]. Also, when replacing S_b with S_t in J_1, the eigenvectors remain the same with those of $S_w^{-1} S_b$.

Though both S_w and S_b are symmetric, the matrix $S_w^{-1} S_b$ is generally not. Hence, to compute the eigenvectors of $S_w^{-1} S_b$ is a little more complicated than a PCA that diagonalizes a symmetric positively definite matrix. Equivalently, the subspace vectors of LDA can be computed in two steps by diagonalizing symmetric matrices: whitening S_w followed by PCA on the whitened space. Denote the eigenvectors of S_w by the columns of an orthonormal matrix P and the eigenvalues by the diagonal elements of a diagonal matrix Λ_1. By diagonalization,

$$S_w = P\Lambda_1 P^T. \tag{3.84}$$

Using a transformation matrix $W_1 = P\Lambda_1^{-1/2}$, S_w is transformed to an indentity matrix:

$$W_1^T S_w W_1 = I. \tag{3.85}$$

Accordingly, S_b is transformed to

$$W_1^T S_b W_1 = S_2. \tag{3.86}$$

S_2 remains a symmetric matrix. Diagonalizing S_2 and retaining its d eigenvectors, corresponding to the largest eigenvalues, as the columns of a $D \times d$ matrix Q, the final transformation matrix of LDA is the combination of W_1 and Q:

$$W = W_1 Q = P\Lambda_1^{-1/2} Q. \tag{3.87}$$

A D-dimensional data vector $\mathbf{x}$ is projected onto the d-dimensional subspace by $\mathbf{y} = W^T \mathbf{x}$.

Figure 3.16 shows an example of LDA as compared to PCA. In 2D space, there are two classes of Gaussian distributions, the 1D subspace learned by LDA has better separability than that learned by PCA.

Though LDA performs fairly well in practical pattern recognition problems, it has some inherent drawbacks: (1) the dimensionality of subspace is limited by the number of classes because the rank of S_b is at most $\min(D, M - 1)$; (2) the matrix S_w may be singular when the number of samples is small; (3) by assuming equal covariance, it ignores the difference of covariance between different classes; (4) when the data distribution of each class deviates largely from Gaussian, LDA performs inferiorly.

In recent years, many works have been done to overcome the above drawbacks. Kimura proposed a hybrid of LDA and PCA for extracting more than $M - 1$ features for small number of classes [53]. Heteroscedastic discriminant analysis (e.g., [78])

FIGURE 3.16   Subspace axes of PCA and LDA for data points of two classes.

considers the class means as well as the difference of covariance in estimating the subspace. Nonparametric discriminant analysis assumes that the distribution of each class is arbitrary and computes a between-class scatter matrix from the vectors connecting near points of different classes [33]. The decision boundary feature extraction method [70] and the Gaussian mixture discrminant analysis [42] can be viewed as the special cases of nonparametric discriminant analysis.

3.1.11   Kernels

Kernel methods have enjoyed considerable popularity recently. They can be described as a way to transpose a hard, nonlinear problem in lower dimensions to an easier, linearly separable problem in higher dimensions. This passage to higher dimensions is not innocuous; It should raise a red flag in the mind of readers familiar with the statistical learning literature, as transposing a problem to a higher dimension is usually a bad idea. This is the proverbial "curse of dimensionality." The reason for this problem lies in the fact that data become scarce in higher dimensions, and meaningful relationships become harder to identify. However, kernel methods offer in many cases, not all cases, unfortunately, a reasonable compromise because of their use of clever data representations and computational shortcuts.

3.1.11.1 Basic Definitions   Suppose first that our n-dimensional data is **x**. Let us first consider an embedding map Φ such that

$$\Phi : x \in \mathfrak{R}^n \mapsto \Phi(x) \in F \subseteq \mathfrak{R}^N, \tag{3.88}$$

where the embedding dimension **N** can be bigger than the data dimension **n**, and Φ is general (nonlinear). The choice of the map depends on the problem at hand. Let us immediately give an example of this. For instance, if we know that our data consist of bidimensional points of coordinates (x_1, x_2) belonging to two classes that are quadratically separable, we can use the following Φ:

$$\Phi : \mathfrak{R}^2 \mapsto \mathfrak{R}^3, \tag{3.89}$$

$$(x_1, x_2) \mapsto (z_1, z_2, z_3) = (x_1^2, \sqrt{2}x_1x_2, x_2^2). \tag{3.90}$$

Consider the problem of finding the separating hyperplane in this three-dimensional feature space: an expression of this hyperplane is $w'z + b = 0$; substituting, we get

$$w_1x_1^2 + w_2\sqrt{2}x_1x_2 + w_3x_2^2 + b = 0, \tag{3.91}$$

which is the equation of an ellipse with respect to the original coordinates. It is, therefore, possible to use algorithms such as perceptron learning to find this boundary in three-dimensional space and separate the classes. This is known as the "kernel trick."

Let us define now what we mean by a kernel function:
Definition 1: A kernel is a function κ that for all x, y, $\in X$ satisfies

$$\kappa(x, y) = \langle \Phi(x), \Phi(y) \rangle, \tag{3.92}$$

where the angular brackets denote the inner product and Φ is a mapping from X to an inner product feature space F:

$$\Phi : x \in \Re^n \mapsto \Phi(x) \in F \subseteq \Re^N. \tag{3.93}$$

Let us verify that our proposed mapping corresponds to a kernel

$$\begin{aligned} \kappa(x, y) &= \langle \Phi(x), \Phi(y) \rangle = \left\langle (x_1^2, \sqrt{2}x_1x_2, x_2^2), (y_1^2, \sqrt{2}y_1y_2, y_2^2) \right\rangle \\ &= x_1^2 y_1^2 + 2x_1x_2y_1y_2 + x_2^1 y_2^2 = (x_1y_1 + x_2y_2)^2 = \langle x, y \rangle^2. \end{aligned}$$

Note the important fact that once we know that this kernel function corresponds to the square of the inner product in data space, we do not need to evaluate the function Φ in feature space in order to evaluate the kernel. This is an important computational shortcut.

Definition 2: If we have **p** observations in the original space, the kernel matrix K (which is the Gram matrix in the feature space) is defined as the $p \times p$ matrix whose entries are the inner products of all pairs of elements in feature space:

$$K_{ij} = \kappa(x_i, x_j) = \left\langle \Phi(x_i), \Phi(x_j) \right\rangle. \tag{3.94}$$

An important property of kernel matrices is semipositive definitiveness. As a matter of fact, an important theorem states that any positive semidefinite matrix can be considered as a kernel matrix. This is given by Mercer's theorem, a fundamental result in the field. This is interesting, as we can use a matrix K without caring about how to compute the embedding function Φ, which is a good thing in the cases where the embedding is infinite dimensional (as for the Gaussian kernel). What is important in a particular kernel is the fact that it defines a *measure of similarity* to see that it may be useful to remember the relationship of the ordinary dot product and the cosine of the angle between two finite-dimensional vectors in real space.

An illustration of this is to consider the two extreme cases of kernel matrices. Suppose that we consider the identity matrix $I_{p \times p}$ as a kernel. It is obviously positive semidefinite, so it satisfies the conditions of Mercer's theorem. It defines the following trivial measure of similarity: any vector is similar to itself and is different from all others. This is not very helpful. The other extreme case is with a quasi-uniform matrix; in this case, all vectors have the same degree of similarity to all others, and this is also not very helpful.

Kernel matrices are convenient, but the flipside is that they are also information bottlenecks. Kernel matrices are the interfaces between the data and the learning algorithms that operate in feature space. The original data is, therefore, not accessible to the learning algorithm. If a structure that is present in the data is not reflected in the kernel matrix, then it cannot be recovered. The *choice* of the kernel matrix is therefore crucial to the success of the analysis.

This evacuation of the original data may be an advantage in certain cases, for instance, when the original data objects are nonvectorial in nature (strings, texts, trees, graphs, etc.), and are therefore not amenable to direct analysis using vector methods.

3.1.11.2 *Kernel Construction*   How do researchers select kernels? Ideally, kernels should be tailored to the needs of the application and be built according to the characteristics of the particular domain under examination. Remember how we came up with our quadratic embedding above, and how we found out it corresponded to a squared dot product in data space; it is because we knew that the underlying relationship was quadratic. The ideal kernel for learning a function $\mathbf{f}$ is therefore $\kappa(\mathbf{x}, \mathbf{y}) = \mathbf{f}(\mathbf{x})\mathbf{f}(\mathbf{y})$. This suggests a way of ascertaining the value of a particular kernel matrix by comparing it to this ideal kernel, but in most cases we do not know the form of the function we want to learn.

Another route is to start from simple kernels and combine them to get more complex kernels. There are often families of kernels that are recognized to be useful for some applications, and prototypical members of a family can be used as a starting point. To illustrate and close this section, let us state this important theorem about closure properties of kernels:

Theorem: Let κ_1 and κ_2 be kernels over $X \times X$, $\mathbf{f}$ be a real-valued function on X, A be a positive semi-definite matrix, and $\mathbf{p}$ be a polynomial. Then the following functions are kernels:

$$\kappa(x, y) = \kappa_1(x, y) + \kappa_2(x, y), \tag{3.95}$$

$$\kappa(x, y) = a\kappa_1(x, y), \tag{3.96}$$

$$\kappa(x, y) = \kappa_1(x, y) \odot \kappa_2(x, y), \tag{3.97}$$

$$\kappa(x, y) = f(x)f(y), \tag{3.98}$$

$$\kappa(x, y) = x'Ay, \tag{3.99}$$

$$\kappa(x, y) = p(\kappa_1(x, y)), \tag{3.100}$$

$$\kappa(x, y) = \exp(\kappa_1(x, y)), \tag{3.101}$$

$$\kappa(x, y) = \exp\left(\frac{-\|x - y\|^2}{2\sigma^2}\right). \tag{3.102}$$

Note: In (3.97), the operator represents the Hadamard product, which is the term-by-term product of the elements of each matrix. The kernel in (3.102) is known as the Gaussian kernel, used in radial basis function networks.

3.1.11.3 An Application: Kernel PCA   We now outline the nonlinear principal components methods defined using kernels. The reader will immediately see the similarity with the derivation that was explained above in the linear case. However, we suppose that in data space the direction explaining the most variance is curvilinear. Therefore, applying ordinary PCA will result in a model error. Again, we use the "kernel trick": We define a feature space (the embedding Φ) that is a nonlinear transformation of the data space, and it is in this feature space that the ordinary version of PCA is then applied.

As in the linear case, we need the variance–covariance matrix of the data this time in feature space. It is

$$C = \frac{1}{n} \sum_{i=1}^{n} \Phi(x_i)\Phi(x_i)', \tag{3.103}$$

We next compute the principal components by setting the eigenvalue problem:

$$\lambda v = Cv = \frac{1}{n} \sum_{i=1}^{n} \Phi(x_i)\Phi(x_i)'v. \tag{3.104}$$

Barring the case $\lambda = 0$, this says that all eigenvectors v must lie in the span of $\Phi(x_1), \ldots, \Phi(x_n)$. This allows us to define the equivalent system by multiplying both sides by $\Phi(x_k)$, and we get

$$\lambda \Phi(x_k)v = \Phi(x_k)Cv \text{ with } v = \sum_{i=1}^{n} a_i \Phi(x_i), \tag{3.105}$$

where $\Phi(a_i)$ are the expansion coefficients of $\Phi(v)$. We can therefore rewrite the problem as

$$\lambda a = Ka. \tag{3.106}$$

One thing to note is that the nonlinearity is entirely taken into account by the kernel matrix K. It is therefore essential to keep in mind that again, the choice of the kernel must be made wisely. This is after all the perennial dilemma of nonlinear methods: they explode the limitations of linear methods, but then one runs into the question of *which* nonlinear relationship to specify.

Another important remark is that although the problem of linear PCA has the dimensionality of the data space, kernel nonlinear PCA has a much larger dimensionality, that is, of the number of elements in the data space. This can clearly become a problem and has prompted lines of enquiry along the idea of using sparse kernel matrices set within a Bayesian framework.

3.2 FEATURE SELECTION FOR PATTERN CLASSIFICATION

Given a feature set F, which contains all potential N features and a data set D, which contains data points to be described using available features from F, a feature selection process selects a subset of features G from F, such that the description of D based on G is optimized under certain criterion J, for example, best predictive accuracy of unseen data points within D or minimum cost of extracting features within G.

In many real-world applications, such as pattern recognition, image processing, and data mining, there are high demands for the capability of processing data in high-dimensional space, where each dimension corresponds to a feature, and data points are described with hundreds or even thousands of features. However, the curse of dimensionality, first termed by Bellman [10], is commonly observed: The computational complexity and number of training instances to maintain a given level of accuracy grow exponentially with the number of dimensions.

Feature selection, also called dimensionality reduction, plays a central role in mitigating this phenomenon. It augments the performance of the induction system with respect to three aspects [12]:

- Sample complexity: less training samples are required to maintain a desired level of accuracy;
- Computational cost: not only the cost of the inductive algorithm but also the effort of collecting and measuring features are reduced;
- Performance issues:
  - Overfitting is avoided since representation of data is more general with fewer features.
  - Accuracy is enhanced as noisy/redundant/irrelevant features are removed.

Section 3.2.1 describes four dimensions along which feature selection algorithms can be categorized. These are (a) objectives; (b) relationship with classifiers; (c) feature subset evaluation; and (d) search strategies. The latter three dimensions are of key importance and hence are studied in detail in Sections 3.2.1.1, 3.2.1.2, and 3.2.1.3.

3.2.1 Review of Feature Selection Methods

Feature selection techniques can be classified along four different dimensions.

- Objectives;
- Relationship with classifier;
- Evaluation criteria (of feature subsets);
- Search strategies.

Feature selection algorithms have three different objectives [62]: (a) minimum error (ME): given a subset with fixed size or a range of different sizes, seek a minimum error rate; (b) minimum subset (MS): given acceptable performance degradation, seek a minimum subset; (c) multicriteria compromise (CP) of the above two objectives. Each objective corresponds to a type of optimization problem. The first objective is an unconstrained combinatorial optimization problem; the second objective is a constrained combinatorial optimization problem; where the third objective, as indicated by its name, is a multicriteria optimization problem.

Besides objectives, feature selection approaches adopt different strategies based on (a) how feature selection interacts with classifiers; (b) how feature subsets are evaluated; and (c) how the search space is explored. These strategies play crucial roles in any feature selection algorithm and hence will be studied in detail in the sequel.

3.2.1.1 *Relationship with Classifiers* Based on the relationship with classifiers, feature selection approaches can generally be divided into "filters" and "wrappers." A filter algorithm operates independently of any classification algorithm. It evaluates feature subsets *via* their intrinsic properties. Irrelevant/redundant features are "filtered" out before the features are fed into the classifier. On the contrary, a wrapper algorithm is normally tied to a classifier, with the feature selection process wrapped around the classifier, whose performance directly reflects the goodness of the feature subsets. Typical filter algorithms include [3, 16, 54]. Typical wrapper algorithms include [47, 56, 57, 75, 113].

Blum and Langley [12] grouped feature selection approaches into three categories: filter, wrapper, and embedded. In the third category, feature selection is embedded into the classifier. Both of them are implemented simultaneously and cannot be separated [66]. Typical embedded algorithms are decision trees [92, 69, 13] as well as algorithms used for learning logic concepts [82, 114].

Some researchers adopt a hybrid (not an embedded) form of feature selection that utilizes aspects of both the filter and wrapper approaches in the hope of enjoying the benefits of both [15, 20, 99].

We merge the taxonomies proposed by [12, 77], resulting in a division of the feature selection into four categories: embedded, filter, wrapper, and hybrid, as shown in Figure 3.17.

Note that in embedded approaches, feature selection is tied tightly to, and is performed simultaneously with, a specific classifier. When the classifier is constructed, the feature selection, hence dimensionality reduction, is achieved as a side effect. Decision trees are typical examples. However, the decision tree induction algorithm ID3 cannot be relied upon to remove all irrelevant features present in Boolean concepts [4]. Hence, appropriate feature selection should be carried out before developing a decision tree. Also, due to the inseparability of feature selection from classification in embedded approaches, we will only deal with the other three approaches in detail.

(1) *Filter approaches*
A filter feature selection algorithm, as indicated by its name, filters out features before their use by a classifier. It is independent of any classifier and evaluates feature subsets

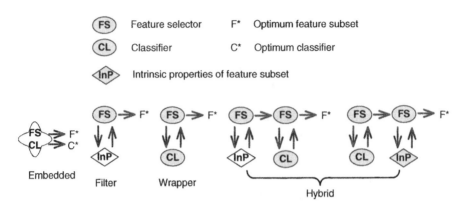

FIGURE 3.17 The four approaches to feature selection.

via their intrinsic properties. These properties act as an approximation of the value of the features from the point of view of a "generic classifier."

Relief [55] is a typical filter approach as it selects features that are statistically relevant to the target concept and feeds them into the classifier. Both FOCUS [4, 3] and its variant—C-FOCUS [7] are typical filter approaches. In FOCUS, the filtered features are given to ID3 [93] to construct decision trees. In C-FOCUS, a neural network was used as a classifier. Cardie [16] used decision trees to specify input features for a nearest neighbor classifier. Branch and bound and most of its improved versions [19, 85, 102, 121] also employ a filter framework.

Any function used for feature subset evaluation must satisfy a monotonicity condition; that is, given two candidate subsets f_1 and f_2, if f_1 is a subset of f_2, $J(f_1)$ must be worse than $J(f_2)$. Bhattacharya distance [19, 102] and divergence [102] are popular evaluation criteria. Other metrics used to evaluate feature subsets include consistency [68, 76] and correlation [39, 40, 123], all of which will be studied in the next section.

(2) Wrapper approaches
In a wrapper, the feature selection process is wrapped around a learning algorithm. The feature subsets are evaluated *via* the performance of the learning algorithm. A wrapper method naturally entails training and implementation of learning algorithms during the procedure of feature selection.

Kohavi and John [56] tested two classifiers: decision trees and naive Bayes. They reported that their wrapper is able to remove correlated features that impair the performance and hence achieve significant accuracy improvement. They also observed two problems with wrappers: high computational cost and overfitting. They stated that the latter problem occurs for small training sets and suggested enlarging the training set to alleviate the effect. Other wrappers use various forms of classifiers such as neural networks [43, 87, 88, 98] and nearest neighbor classifiers [1, 45, 65].

The main problem of a wrapper is that its cost may be prohibitively high because it involves repeated training and implementation of the classifier. Various solutions have been suggested. Kohavi and John [56] suggested replacing k-fold cross-validation with hold-out validation. k-fold cross-validation partitions available data into k sets. In each run $k - 1$ sets are used for training and the remaining kth set is used for testing. Finally, an average accuracy over k runs is obtained. The hold-out test, on the contrary, only involves one run in which 50–80% of total data is used for training and the remainder for testing. Although more efficient, the hold-out test is not statistically optimal.

The training of neural networks is known to be computationally expensive. Hence, wrappers with neural networks as classifiers employ various tricks to avoid repeated training. In [43], the authors incorporated a weight analysis-based heuristic, called artificial neural net input gain measurement approximation (ANNIGMA), in their wrapper model. Oliveira et al. [88] used sensitivity analysis to estimate the relationship between the input features and the performance of the network.

(3) *Hybrid approaches*
As both filters and wrapper have their own limitations, some researchers propose to hybridize them to adopt the benefits of both.

Zhang et al. [127] proposed a ReliefF-GA-wrapper method. Features were first evaluated by ReliefF [58], the extension of Relief [55]. The resulting scores of the evaluated features were fed into a genetic algorithm (GA) so that features with higher rankings receive higher probabilities of selection. The fitness is a compromise between the size of the feature subset and the accuracy of a nearest mean classifier [33]. The authors demonstrated, through their empirical study, that this hybrid filter-wrapper model outperformed some pure filters and wrappers, for example, ReliefF or GA-wrapper alone.

In contrast to Zhang et al.'s work [127] in which the wrapper amends the filter, Yu et al. [122] developed a two-pass feature selection with the wrapper followed by the filter. In the first pass, wrapper models with forward search were used, with multilayer perceptron neural network and support vector machines as the classifiers. The size of the feature set after this pass was shrunk dramatically, whereas the classification accuracy was increased slightly. In the second pass, the Markov blanket filter method was used to select highly relevant features from the feature subset obtained from pass 1. The resulting feature set was further minimized while maintaining the classification accuracy. Hence, the algorithm achieved the smallest possible feature subset that provided the highest classification accuracy. Other papers worthy of mention include [15, 20, 99].

(4) *Summary*
Some researchers [40, 76, 123] believe that the filter approaches are preferable to wrapper approaches because (a) they are not biased toward any specific classifier; and (b) they are computationally efficient. Although the filters are more efficient and generic than the wrappers, especially in high-dimensional spaces, they are criticized by another school of researchers [1, 47, 71] in that they may yield a bias that is completely different from that required by a specific classifier. The resulting feature subset is, therefore, not optimal in the sense that the classifier does not fulfill its best potential with the given inputs. A reasonable compromise (discussed in previous

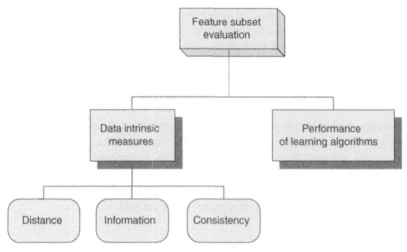

FIGURE 3.18   Feature subset evaluation criteria.

section) is to hybridize a filter with a wrapper. For example, a filter can be used first to shrink the search space somewhat. Then a wrapper follows by selecting features that best suit a specific classifier.

3.2.1.2 Evaluation Criteria   The goodness of a feature subset may or may not be tied to the performance of a classifier. Filter feature selectors, being independent of any classifier, utilize the intrinsic properties of the data. In [23], the data intrinsic measures include interclass distance, probabilistic distance, probabilistic dependence, and entropy. Ben-Bassat [11] proposes three categories: information or uncertainty measures, distance measures, and dependence measures. Besides the classifier's error rate, Dash and Liu [21] divide the evaluation functions into four categories: distance, information, dependence, and consistency. Dependence denotes the correlation between two random variables. Although dependence is considered as an independent category, as suggested by Ben-Bassat [11] and Dash and Liu [21], it is always in the form of distance or information theoretic measures. Hence, we divide data intrinsic measures into three categories: (a) distance measures; (b) information measures; and (c) consistency measures (see Fig. 3.18).

(1) *Data intrinsic distance measures*
Distance measures the degree of similarity/dissimilarity between sets of data. In feature selection, the greater the distance, the better the class divergence, and hence the feature subset. Although Kira and Rendell used distance between training instances [55], most distance metrics come in two broad categories:

- Interclass distance;
- Probabilistic distance.

Interclass distance is the average distance of distances between different classes, for example, the Minkowski distance. A Minkowski distance of order m between two classes is

$$d_m = \left(\sum_{i=1}^{N} |x_i - y_i|^m \right)^{\frac{1}{m}},$$ (3.107)

where $(x_1, x_2, \ldots, x_N)$ and $(y_1, y_2, \ldots, y_N)$ are feature subsets used to describe classes and N is the number of features in the set.

Specifically, when m equals 1, the Minkowski distance reduces to the city block distance

$$d_1 = \sum_{i=1}^{N} |x_i - y_i|.$$ (3.108)

When m equals 2, the Minkowski distance reduces to the Euclidean distance

$$d_2 = \sqrt{\sum_{i=1}^{N} |x_i - y_i|^2}.$$ (3.109)

The Euclidean distance d can be computed from the ratio of scatter matrices [90]. Scatter matrices denote how feature vectors are scattered in the feature space. The between-class scatter matrix S_b is

$$S_b = \sum_{i=1}^{m} P(\omega_i) v_i v_i^T.$$ (3.110)

The within-class scatter matrix S_ω is

$$S_\omega = \sum_{i=1}^{m} P(\omega_i) W_i.$$ (3.111)

Here $P(\omega_i)$ is the a priori probability of class ω_i, v_i is the mean vector of class ω_i and W_i is the covariance matrix for class ω_i.

The Euclidean distance can then be computed as

$$d = \frac{|S_b + S_\omega|}{|S_\omega|}.$$ (3.112)

Note that d is to be maximized as larger d means more class separation (S_b) and more compact classes (S_ω). An example of the application of this concept for the evaluation of signal-based features is [38].

Probabilistic distance is the distance between probability density functions of the classes. Typical probabilistic distance metrics are divergence and Bhattacharya distance. Divergence between classes, ω_1 and ω_2 in a form of a Kullback–Leibler distance is given as [11, 86]

$$J_d(\omega_1, \omega_2) = \frac{1}{2}(v_2 - v_1)^T(W_1^{-1} + W_2^{-1})(v_2 - v_1) \\ + \frac{1}{2}\text{trace}(W_1^{-1}W_2 + W_2^{-1}W_1 - 2I), \tag{3.113}$$

where trace(x) is the trace of the matrix $\mathbf{x}$ and is equal to the sum of its eigenvalues. The divergence generally assumes a Gaussian distribution of samples. Bhattacharya distance [19, 79, 102, 103] is

$$J_b(\omega_1, \omega_2) = \frac{1}{8}(v_2 - v_1)^T \left(\frac{W_1^{-1} - W_2^{-1}}{2}\right)^{-1} \times (v_2 - v_1) + \frac{1}{2}\ln\frac{|W_1 + W_2|}{4\sqrt{|W_1 W_2|}}. \tag{3.114}$$

In [90], Piramuthu performed feature selection using SFS on decision trees. Several interclass distances (e.g., Minkowski distance, city block distance, Euclidean distance, and Chebychev distance) and probabilistic distances (e.g., Bhattacharyya distance, Matusita distance, divergence distance, Mahalanobis distance, and Patrick–Fisher measure) were tested. Results on some real data sets showed that interclass distances generally yielded better performance than probabilistic distances.

(2) *Information theoretic measures*
A basic concept in information theoretic measures is entropy. The entropy of a random variable X denotes the amount of uncertainty in X. It is computed as

$$H(X) = -\sum_{\omega_i} P(\omega_i) \log_2 P(\omega_i), \tag{3.115}$$

where $P(\omega_i)$ is the proportion of X belonging to class ω_i.

As shown in Table 3.1, suppose there are eight samples with Boolean feature F_1 and discrete feature F_2 of values a, b, c, d. There are altogether two classes: C_1 and C_2. As there are three samples belonging to C_1 and three belonging to C_2, the entropy for the random class variable C is

$$H(C) = -\frac{3}{6}\log_2\frac{3}{6} - \frac{3}{6}\log_2\frac{3}{6} = 1. \tag{3.116}$$

Note that the range of $H(C)$ is $(0, \infty)$. A larger value means more uncertainty and 0 means no uncertainty at all.

If one of the features is known (i.e., selected), the amount of remaining uncertainty in C is called the conditional entropy of C, denoted as $H(C|F)$

$$H(C|F) = \sum_f P(F = f)H(C|F = f). \tag{3.117}$$

TABLE 3.1   Classes and features.

Samples	F_1	F_2	Classes
1	True	a	C_1
2	False	b	C_2
3	False	a	C_2
4	True	d	C_2
5	False	c	C_1
6	True	b	C_1

It is expected that the uncertainty in C is reduced, given either one of the features. Hence, the goodness of a feature F can be quantified by the reduction in the entropy of class C, given that F is known. This quantity is called mutual information or information gain:

$$I(F;C) = H(C) - H(C|F). \tag{3.118}$$

Larger mutual information or information gain indicates better features.

Mutual information is a popular metric in feature selection [27, 35, 71]. However, it favors features with many values. As seen from the example, F_2 has four values, whereas F_1 only has two values. Hence, mutual information is alternatively normalized as follows:

$$G(F;C) = \frac{I(F;C)}{-\sum_F P(F)\log_2 P(F)}. \tag{3.119}$$

$G(F;C)$ is called the gain ratio. Note that the term in the denominator is actually the entropy of feature variable F.

Both information gain and gain ratio reflect the correlation between a feature and the class. However, they disregard the interaction between features. Hall and Smith proposed a correlation-based feature selection algorithm [41] that seeks maximum correlation between features and classes and minimum correlation between the features themselves. Similarly, Wu and Zhang [116] extended the information gain by taking into account not only the contribution of each feature to classification but also correlation between features.

(3) *Consistency*

Consistency is the opposite of inconsistency. A feature subset is inconsistent if there are at least two training samples that have the same feature values but do not belong to the same class [22]. The inconsistency count is computed as follows.

Suppose there are n training samples with the same set of feature values. Among them, m_1 samples belong to class 1, m_2 samples belong to class 2, and m_3 samples belong to class 3. m_1 is the largest among the three. Hence the inconsistency count is $n - m_1$. The inconsistency rate is the sum of all inconsistency counts divided by the size of the training set N.

The authors in [22] showed that the time complexity of computing the inconsistency rate is approximately $O(N)$. The rate is also monotonic and has some tolerance to noise. However, also as the authors stated, it only works for features with discrete values. Therefore, for continuous features, there is a need for a discretization process.

Finally, consistency bears a min-features bias [4]; that is, it favors the smallest feature subset as long as the consistency of the hypothesis is satisfied. The inconsistency rate is used in [4, 76, 98].

(4) *Measures tied to classifiers*
When measures of feature subsets are tied to the classifiers, they are dependent on the performance of the classifier that uses the features.

When seeking minimum error (objective ME), the evaluation criterion relies on the intrinsic properties of the data (filter approach) or classification accuracy of the classifier (wrapper approach) [85, 102, 121].

If the goal is to seek a minimal feature subset with acceptable classification performance (objective MS), this is a constrained optimization problem. A user-specified constant is required as a threshold for the classification performance. For a given feature subset f, the objective function usually takes the following form:

$$J(f) = L(f) + P(f), \tag{3.120}$$

where $L(f)$ is the dimensionality of f and $P(f)$ is a performance measure (e.g., accuracy or error rate) of the classification algorithm using f.

Siedlecki and Sklansky [100] proposed to minimize a criterion function that combines both dimensionality and error rate of f. In their work, $L(f)$ is the dimensionality of f and $P(f)$ is a penalty function given by

$$P(f) = \frac{\exp((\text{err}_f - t)/m) - 1}{\exp(1) - 1}, \tag{3.121}$$

where t is a feasibility threshold, m is a scaling factor, and err_f is the error rate of the classifier using f. The authors set t to $\geq 12.5\%$ and m to 1% in different experiments in order to avoid an excessively large region of acceptable performance. These two parameters are application dependent and demand empirical tuning. If err_f is less than t, the candidate receives a small reward (a negative value). However, as err_f exceeds t, the value of $P(f)$ (and hence $J(f)$) rises rapidly. Similar functions are also employed in [29, 124].

If a compromise between classification performance and the size of the feature subset is to be pursued (objective CP), the fitness of feature subset f is usually a weighted linear combination of $L(f)$ and $P(f)$, as in [9, 28, 30, 91, 107],

$$J(f) = \lambda L(f) + \beta P(f), \tag{3.122}$$

where $L(f)$ is generally a linear function of the dimensionality of f and is given by

$$L(f) = \frac{|f|}{N},\qquad\qquad(3.123)$$

where N is the total number of available features. $P(f)$ is the error rate. Constants λ and β satisfy different conditions, depending on the relative importance of their associated terms, in the specific application. In [28], $\lambda + \beta = 1$. In [30], λ and β are assigned to 2 and 0.2, respectively.

If $J(f)$ is to be maximized, $L(f)$ is the number of features that are not selected, and $P(f)$ is the accuracy, as in [107].

Rather than weighting $L(f)$ and $P(f)$ to form a composite objective function, Oliveira et al. [88] suggest the employment of a multiobjective optimization procedure, where more than one objectives are present and a set of Pareto optimal solutions are sought.

3.2.1.3 Search Strategies
Feature selection can be seen as a search process, where the search space is made up of all available solutions, that is, feature subsets. Figure 3.19 illustrates a search space. Suppose there are three available features. Each point in the search space is a feature subset in which dark circles denote features selected and white circles denote features removed. Starting from an initial point within the space, the space is explored either exhaustively or partially with some heuristics until a final winner is found.

Similar to a search process, the core of a feature selection process contains four components:

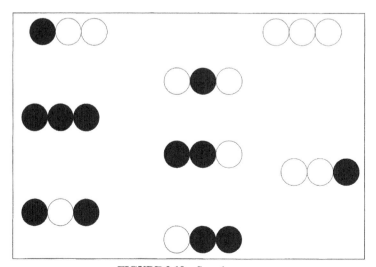

FIGURE 3.19   Search space.

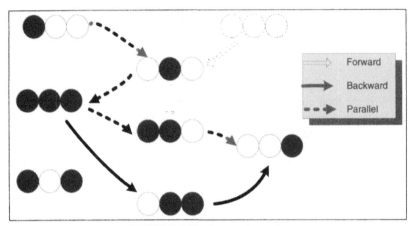

FIGURE 3.20   Forward search.

- Initial points where the search starts from;
- Search strategies used to control the fashion of the exploration;
- Evaluation criteria used to provide feedback with respect to solutions;
- Termination criteria to halt the search.

The selection of initial points generally depends on the specific search strategies utilized. Sequential search usually starts from an empty or a full feature set, whereas parallel search starts from a random subset.

The search strategies can be generally divided into two broad categories: sequential search and parallel search. The sequential search constructs the next states to be visited either incrementally (i.e., forward search) or decrementally (i.e., backward search) from the previous state, whereas a parallel search explores the space in a more randomized fashion. Figure 3.20 demonstrates a potential path of forward, backward, and parallel search, respectively.

Typical examples of sequential search include sequential forward selection, sequential backward elimination [2, 56, 62], and branch and bound [83, 57, 81, 96]; whereas random mutation hill climbing [28, 87, 101], genetic algorithms [62, 96, 123, 115, 118, 120], simulated annealing [80], and tabu search [126] are typical parallel searches.

Features subsets are evaluated using the intrinsic properties of the data or the performance of a certain classifier. The former generally entails a "filter" approach, and the latter is tied to a "wrapper" approach.

The search stops if the termination criterion is satisfied, which can be as follows:

- All available subsets have been evaluated;

- A sufficiently good candidate is found, for example, the smallest subset with classification error below a threshold or a fixed sized subset with the best predictive accuracy;
- A maximum number of iterations have been reached.

In the literature, the resulting feature subset may be validated on a separate set of data from training samples [77]. This validation process, however, is not within the domain of feature selection.

Search strategies are generally classified into three types: complete, heuristic, and randomized.

A complete search explores the whole search space in a direct or indirect fashion. The former typically involves exhaustive enumeration of feature subsets. The latter, on the contrary, prune the search space, resulting in more efficient searches, while implicitly guaranteeing complete coverage of the search space, and hence the optimality of the results. Branch and bound is a typical example of implicitly complete search algorithms [19, 85, 102, 121].

In contrast to complete search, heuristic search is guided toward promising areas of the search space by heuristics. Only part of the search space is visited, and optimality is generally not guaranteed. Typical examples of heuristic search approaches include sequential forward selection, backward elimination [56], and beam search [1, 2].

In randomized search, the algorithm executes a stochastic process in which different training samples are used and different feature subsets are obtained for each independent run. Given enough running time, randomized search may yield better results than heuristic search, while requiring less computational time than complete search. Typical randomized search techniques include random mutation hill climbing [28, 87, 101], genetic algorithms [62, 96, 111, 112, 118, 120], simulated annealing [80], and tabu search [126].

Taking into account the fact that any search is made up of a sequence of basic moves within the search space, and that each move denotes a certain relationship between successive states, we divide feature selection methods into two broad categories: sequential search and parallel search.

Starting from an empty or full feature subset, or both, the sequential search typically moves from the previous state(s) to the next state(s) incrementally or decrementally, by adding or by removing one or more features, respectively, from the previous feature subset(s). On the path traversed from the starting point to the end point, each state is either the superset or the subset of its successors. In contrast, parallel search starts from one or more randomly generated feature sets. The next states to be visited are not generated sequentially from previously visited states, as in sequential search, but rather in a randomized fashion. Movement in search space is implicitly guided by meta-heuristics so that future states are better (i.e., closer to optimal states) than past states. Therefore, parallel search exhibits an overall implicitly directional behavior. Its search is guided, by meta-heuristics, toward promising areas in the search space. Sequential search, however, bears an explicit search direction.

Full feature set $F = \{X_0, X_1, X_2, \ldots, X_n\}$
Empty set: Φ
Evaluation function: $J(X)$
Optimum value of $J(X) : J^*$

Sequential search	Parallel search
Initial feature subset:	Initial feature subset:
$\quad X = \Phi \ (forward)$	$\quad X = \{X' \mid X' \subseteq F\}$
$\quad$ or $X = F(backward)$	While true
While true	$\quad X = G(X)$
$\quad X = X + \{X' \mid X' \subseteq F\}$ (forward)	$\quad$ compute $J(X)$
$\quad$ or $X = X - \{X' \mid X' \subseteq F\}$ (backward)	$\quad$ if $J(X) = J^*$ or convergent
$\quad$ compute $J(X)$	$\quad\quad$ exit
$\quad$ if $J(X) = J^*$ or convergent	end while
$\quad\quad$ exit	
end while	

FIGURE 3.21   The basic schemes of sequential search and parallel search.

Figure 3.21 shows the basic schemes of sequential and parallel search approaches. In both schemes, x' is a subset of F. $G(x)$ in parallel search is a function used to generate new feature subsets. Unlike sequential search, which yields new samples from their predecessors, $G(x)$ may or may not be applied to previous samples. The program exits when it reaches the optimum of the evaluation function or when other termination conditions are satisfied, for example, there is no noticeable improvement in $J(x)$, or the maximum number of iterations has been reached.

It can be seen that the major difference between sequential searches and parallel searches is the way in which new feature subsets are generated. Table 3.2 summarizes the differences between these two methodologies with respect to: starting points, movement in the search space, relationship between successive states, and overall search direction.

TABLE 3.2   Major differences between sequential search and parallel search.

	Sequential search	Parallel search
Starting points	Empty, or full feature subsects, or both subsets	Randomly generated feature subsets
Move between successive states	Incremental or decremental operations	Implicitly guided
Relationship between succesive states	Subsets or supersets of successor	Depends on specific techniques
Overall search direction	Forward or backward	Toward promising areas of the search space

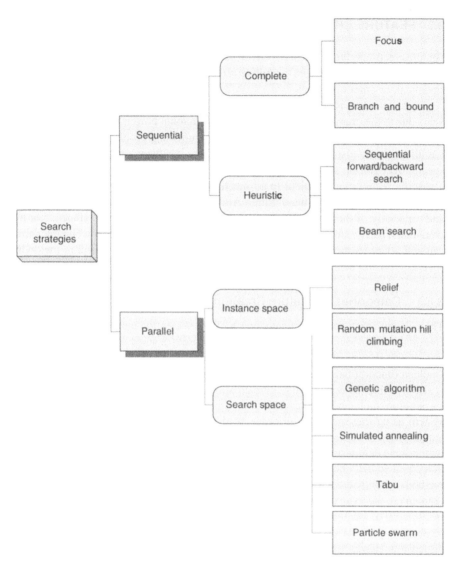

FIGURE 3.22   A taxonomy of feature selection search strategies.

In Figure 3.22, we provide a tree-based taxonomy of major search algorithms used for feature subset discovery and optimization.

Note that sequential heuristic approaches are generally greedy, whereas parallel search methodologies are generally random. Parallel search can be further divided into two categories, according to its sampling methods. Although most parallel approaches extract samples in the search space, Relief [55] samples the training space that is made up of training instances.

3.3 FEATURE CREATION FOR PATTERN CLASSIFICATION

Feature creation is the synthesis of novel, usually more complex features, from a set of older, usually simpler features (primitives), through some combinational, algorithmic, statistical, or other operation.

Feature creation is not a new science, but it has gained increased interest in recent years due to the dramatic increase in raw computational power, and the large number of pattern processing (as opposed to the number crunching) problems that this increase has allowed scientists to tackle. In addition, there has been a proliferation of class-specific recognition solutions that resulted in systems that work extremely well for a specific recognition problem (e.g., machine-printed English), but fail to transfer to any other domain (e.g., handwritten Arabic) without considerable reengineering, and especially of the feature extraction part of the original solution. Hence, feature creation is, partly, an attempt to invent more general feature creation techniques that can self-adapt to specific pattern recognition tasks via machine learning methods, such as GAs and neural networks (NN).

3.3.1 Categories of Feature Creation

There is more than one way of categorizing feature creation methods. The method of creation is an obvious criterion: GA based, NN based, or some other method. Another criterion is the specific form of representation of candidate features: mask-based, Boolean function-based or polynomial function-based, string-based or tree-based. It is also possible to break down feature creation methods by application domain: word recognition, fault detection, and the like. Finally, one could also look at the complete system using these methods in order to identify the level of realism or real-world applicability of the overall application: Some applications introduce seriously novel methods but demonstrate them using toy problems; others involve serious applications that could actually be used in, for example, a machine shop or a factory. Hence, (a) method of feature creation, (b) representation of candidate feature, and (c) application domain are the main dimensions of the space of research and applications developed in the area of feature creation. The real-world applicability of such endeavors is likely to improve with time and expertise, leading to the elimination of this whole dimension. Hence, we choose not to include real-world applicability as a dimension in Figure 3.23.

We choose to apply a hybrid method for division of the surveyed works in order to ensure a minimum level of overlap between the (historical) themes of research that we were able to identify. In the next section, we divide the surveyed publications into four categories:

- Approaches that do not employ genetic algorithms or genetic programing: we call these nonevolutionary methods;
- Approaches that employ genetic algorithms or genetic programing and use either a string-like or tree-like notation to represent a polynomial feature function: we call these polynomial-based methods;

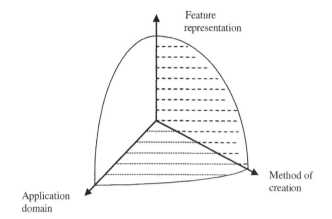

FIGURE 3.23   The space of feature creation methods.

- Approaches that utilize the full power of genetic programing and evolve both the feature functions and the program that uses them: we call these full genetic programing methods;
- Approaches that employ genetic algorithms or/and genetic programing to synthesize all/most of a complete real-world pattern recognition system. We call these evolvable pattern recognizers.

The last category is perhaps of special interest to people working in pattern recognition, for it points the way toward the inevitable development of evolvable generic pattern recognition software systems that require minimal human intervention (perhaps only at setup time) and use theoretically sound techniques.

3.3.2   Review of Feature Creation Methods

We review four categories of feature creation methods: nonevolutionary, polynomial based, genetic programing, evolvable pattern recognizers, and finally give a historical summary.

3.3.2.1   *Nonevolutionary Methods*   In this category of feature creation methods, we outline some concrete examples as below.

(1) *Uhr* 1963
Uhr's main motivation is to show that feature-extraction operators can arise from the recognition problem at hand and be adapted, in form as well as weighting, in response to feedback from the program's operation on the pattern samples provided to it [110].

Uhr describes a complete pattern recognition system. However, our focus is its feature extraction and generation capabilities. There are two sets of features here: built-in features and created features. In the following, we shall explain the process of feature extraction and creation.

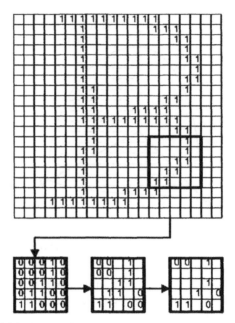

FIGURE 3.24 Automatic operator (feature) creation procedure [110].

An unclassified input pattern is digitized into a 20 × 20 binary pixel matrix. The pattern is drawn in black pixels on a white background. A bounding box is drawn around the black input pattern: It is called a mask. A number of feature-extracting operators are applied to the input pattern.

An operator is a 4 × 4 matrix of cells. Each cell can contain 0, 1, or blank. An operator is generated (a) randomly (with some restrictions on the number of 1s in the operator), (b) manually by the experimenter, or more interestingly, (c) through an automated process described below (see Fig. 3.24):

- A 5 × 5 matrix of cells is randomly positioned within the input pattern: the black pixels in the pattern are copied into 1s and the white pixels into 0s;
- All 0s connected to 1s are replaced with blanks;
- All remaining cells are replaced with blanks, with probability 0.5;
- If it turns out that the set of 1s in the operator is identical to the set of 1s in an exiting or previously rejected operator, then this operator is rejected and the process is restarted.

The program was tested using seven different sets of five hand-printed characters, namely A, B, C, D, and E. The best accuracy of recognition returned by the program was about 80%, which is unacceptable by today's standards. However, this was not intended as an engineering prototype, but rather as a proof of concept, and in that sense it succeeded.

Binary video

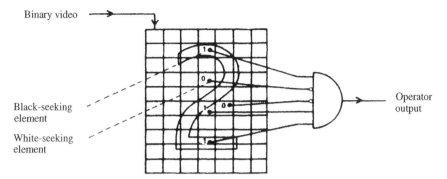

Black-seeking element

White-seeking element

Operator output

FIGURE 3.25 A feature detection operator [105].

(2) *Stentiford* 1985

The purpose of this work is to show that an automated evolutionary approach can succeed in the creation of the right set of features for an optical character recognition system [105].

Here, a feature is defined as a conjunction of a number of black-seeking or white-seeking elements (see Figure 3.25). Each feature had a random arrangement of those seeking elements that serve as an initial state. N such features were used. The patterns to be recognized were a set of 34 different uppercase letters and digits. These patterns were sampled from addresses on printed British mail. The alphanumeric characters were digitized to lie within a 16×24 binary grid. A set of K reference patterns were defined through the selection of one character per class. The vector of reference patterns is called C_i.

The reference pattern vector C_i gives an N-dimensional vector $\mathbf{F}_i$ in feature space, where

$$\mathbf{F}_i = \{f_{i1}, f_{i2}, \ldots, f_{ij}\}, \quad i = 1, 2, \ldots, K, \tag{3.124}$$

and where f_{ij} is the response of the jth feature to the ith reference pattern. Also, the jth feature relates to the vector response f_j for the K reference patterns, where

$$f_j = \{f_{1j}, f_{2j}, \ldots, f_{Kj}\}, \quad j = 1, 2, \ldots, K. \tag{3.125}$$

When trying to recognize the C_i, the best discrimination is achieved when the $\mathbf{F}_i$ have the maximum angle from each other. An orthogonal arrangement of the $\mathbf{F}_i$ comes very close to optimum. It can further be shown that when $K = N$, a necessary and sufficient condition for such an arrangement is that f_j are mutually orthogonal.

Results were obtained using a training set of about 30,000 characters and a test set of 10,200 characters. A 1% testing accuracy was obtained with 3166 features and 319 reference characters.

(3) *Gillies* 1990

To our knowledge, Gillies was the first to propose a mechanism that automatically generates morphological template-based feature detectors for use in printed character recognition [36]. And, if characters can be recognized, then so can be the printed words made of segmentable characters.

To create a feature detector, the following steps are followed:

1. From a large database of 67,000 isolated character images, a random sample of 20 images per distinct character (class) is chosen. The number of distinct characters is 57;

2. Each of the images is normalized (without trying to maintain the aspect ratio) to 16 × 24 pixel window;

3. The sum images for each character are formed. A sum image of a character is formed by adding, pixel-by-pixel, the 20 sample images of this character;

4. The pixel-sum images are used to generate morphological template detectors:

    (a) One of the pixel-sum images is chosen at random;

    (b) A randomly sized and positioned window is placed on top of the sum-image: this window frames the feature detector (under construction);

    (c) Each pixel value in the sum-image lying within the window is tested: if it is below a *l*ow threshold, then the corresponding pixel in the feature detector is made a part of the "background" template; however, if the pixel value is greater than a *h*igh threshold, then it is made a part of the "Foreground" template. The rest of the pixels can be considered as don't cares, as they do not play a role in future processing.

Every feature detector generated above has a detection zone associated with it. It is a 3 × 5 rectangular window placed at the origin of the feature detector window. When a feature detector is applied to a character image, it reports either a match or a no match. In doing so, it only takes into consideration the 3 × 5 detection zone at the center of the feature detector window. In morphological terminology, a match occurs if

$$(Z \cap (I \ominus F) \cap (\bar{I} \ominus B)) \neq 0, \tag{3.126}$$

where Z is the detection zone, I is the input character image, F is the foreground template, and B is the background template.

Features were generated through this process and compared against other features that were manually created. Both sets were fed into one identically configured classifier, which was used for word classification. The features that were automatically generated performed better in every test.

The main problem with this method is that it only works for binary images. Another problem is the way in which windows and feature detectors are generated, which is essentially random. Finally, the generalizability of this approach to other recognition problems, where the patterns are not as easily isolated and summed is doubtful. Nevertheless, this is a pioneering effort and should be recognized as such.

(4) *Gader* 1996

These authors assert that their work is based on the work of Stentiford and of Gillies (described above). Both Stentiford and Gillies dealt with *printed* characters, although in his work Gader proposes a somewhat novel technique for automatic generation of features for *handwritten* digit recognition [34]. Furthermore, he evaluates the quality of the features using both orthogonality *and* information.

The method can be divided into the following stages: pixel-sum image construction, feature generation, and feature selection. Our focus here is on the first two processes.

Pixel-sum Image Construction: From a database of binary images of digits, retrieve 1000 images per digit (from 0 to 9) and moment-normalize them so that they fit in a 24×18 pixel standard window. Sum the normalized binary images for each class of digits (e.g., all the 1000 images of a "1") and normalize that sum, so that the value of each summed pixel is in $[-1, 1]$. Call the resulting summed and normalized image for each class the sum image for that class.

Feature Generation: For each class $K(K = 0-9)$, generate P_K random features:

1. To generate a single feature for a given class, a rectangular window is randomly generated, with the height between 6 and 12 pixels and the width between 4 and 8 pixels. The position of the window is also randomly generated and is expressed in terms of horizontal and vertical shifts from the upper left corner of the sum image. The part of the sum image for a class that lies within this window is considered a feature detector for this class.

2. A detection zone is assigned to each feature detector. The detection zone is a rectangular window with the same center as that of the feature detector, but it is 2 pixels taller and 2 pixels wider.

3. The probability $P((F_j \mid X_{Ki})$ that feature F_j present in digit image X_{Ki} is computed by taking the maximum linear correlation between the feature detector and the detection zone, and then scaling it between $[0, 1]$.

4. A response vector is associated with each digit image X_{Ki}, with dimension P representing the probability of the presence of the various features in that image. P is given by the sum of all the P_K features generated for each class K. It is worth noting here that the probability $P(F_j \mid K)$ of the presence of a feature F_j in a class K is equal to the average of the response of the features over $N(= 50)$ digit image samples from the class K.

One hundred features were generated independently of any classifier. These features were fed into a 100-input, 10-class output, 2-hidden layer neural net, with 25 nodes in the first layer and 15 in the second. Sets of experiments were carried out using two standard sets of data. The features selected via the orthogonality measure performed better that the ones selected using the information measure. Nevertheless, in both cases the accuracy rate was high and compared favorably to other results published at the time.

There are a number of weaknesses with this method. The features are extracted from summed images, which require (a) that we have a sufficiently large and diverse sample of images to sum, per pattern; (b) that we have a correct and feasible method for summing the images; (c) that the pattern in the image be isolated and free from any distracting attributes not belonging to the pattern itself (e.g., background).

3.3.2.2 Polynomial-Based Methods   We give three concrete examples of this category of methods.

(1) *Chang* 1991
Chang et al. [17] pick up where Siedlecki et al. stopped. Their work explores the application of genetic algorithms for both feature selection and feature creation.

New features are generated as polynomials of original features taken two at a time. For example, if *feature_1* has identification bit string 01 and *feature_2* has identification bit string 10, then string 01110101 represents a new feature $feature\_1^1 + feature\_2^1$. The first two bits identify the first feature, the third bit identifies the power of the first feature, and the second set of three bits has the same purpose with respect to the second function. The last two bits identify the operation (+), which belongs to the set $\{+, -, *, /\}$. Each member of the population is made of the original features (which are implicit) and a fixed number of new features, which are presented in the way just explained.

The authors claim that the GA found a better set of features than those they were originally looking for. In an experiment with 300 training patterns, the GA not only found a useful set of features but also enabled a nearest-neighbor classifier to achieve higher accuracy of recognition than two other classifiers.

The main problems we see with this work are (a) the individual can only contain a set of number of new features (in addition to all the original features); (b) the way in which a new feature is created (from old features) only allows for a small section of the large space of possible polynomial forms to be generated; (c) the linear string-based fashion in which features are combined is quite limiting: The more recent tree-based representations are much more flexible and allow for a much wider range and complexity of expression.

(2) *Chen* 2001
This paper appears to be the first to introduce a technique for automatic creation of functions of symptom parameters (SPs), to be used in the diagnosis of faults in machinery [18]. Prior to this paper, the following process was done by an expert hand:

1. Measure signals of normal and afflicted machinery;
2. Using intuition, define several candidate SPs;
3. Evaluate the values of the SPs using the measured signals;
4. Check the sensitivity of each SP;
5. Adopt the most sensitive ones for fault diagnosis or else return to step 1.

The paper lists eight SPs (or primitive features) that are among the most widely used in the field of signal-based pattern recognition. These are variation rate, standard deviation, mean, absolute mean, skewness, kurtosis and four other functions of peak and valley values. Each one of these values is normalized by subtracting the mean (of the normal signal) from it, and then dividing that by the standard deviation (again, of the normal signal).

A new SP is synthesized from these primitive SPs by evolving a population of functions coded as trees of operators and terminals. The operators are the four basic arithmetic operators plus an implied parameter: power, which is expressed as a number. The terminals are the primitive SPs.

In experimentation on faulty bearing, the best individuals generated automatically by the GA showed high discriminative abilities. Its performance was measured against that of the linear discriminative method, and it showed better discrimination abilities.

(3) *Guo* 2005
Machine condition monitoring (MCM) is an area of increasing importance in the manufacturing industry; disastrous faults can be preempted, and condition-based rather than periodic maintenance can reduce costs. The objective of this research is the automatic extraction of a set of features (using genetic programing) in order to predict, with high accuracy, the condition (or class) of rolling bearings [38]. By running the machine and recording the sound made by suspect bearings, one (expert) could tell the condition of that bearing. There are five different faulty conditions and one normal condition. The features are all extracted from an acoustic signal detected by a microphone placed at a prespecified distance from the suspected bearings. Many features can be extracted from the audio signal. Hence, the extraction and selection of features can greatly affect classifier performance (Table 3.3).

A population of agents (initially generated at random) goes through an evolutionary process that results in individuals with competent discriminative ability. Bearing data recorded from machines is used as input to the initial population. Each individual represents a transformation network, which accepts input data and outputs a specific class. A fitness value, which represents the discriminative ability of an individual (feature), is computed and assigned to that individual. Only those that have the best fitness (the elite) survive the selection process. Crossover, mutation, and reproduction are used to create new and potentially different individuals from older ones. Then, the process of fitness evaluation and selection repeats until the termination criterion is met.

What is new in this work is that (a) all the features are automatically created without any human intervention; (b) a novel, fast method is used to evaluate the fitness of candidate features, without resorting to the classifier; (c) genetic programing automatically decides, during evolution, whether to carry out feature creation or selection.

The main limitations of this work are that the terminals are fixed in number. Though the whole feature expression (tree) can change, the terminals that are used in constructing the trees cannot. This would have been tolerable if the terminals were simple and generic, but they are not. They are all high-level statistical features (moments) that have worked very well for this application but may not work as well for other signal-based classification tasks. This highlights a need for creating new terminals when applying this approach to another problem. Last, but not least, the operators themselves, though judiciously chosen, are fixed in nature and in number.

TABLE 3.3 Operator set for the GP [38].

Symbol	No. of Inputs	Description
+,-	2	Addition, substraction
*,/	2	Multiplication, division
square, sqrt	1	Square, square root
sin, cos	1	Trigonometric functions
asin, acos	1	Trigonometric functions
tan, tanh	1	Trigonometric functions
reciprocal	1	Reciprocal
log	1	Natural logarithm
abs, negator	1	Absolute, change sign

3.3.2.3 *Full Genetic Programing Methods*   In this category of methods, we give two concrete examples.

(1) *Koza* 1993

This paper describes an approach for both generating (feature) detectors and combining these detectors in a binary classifier. Genetic programing is used for both tasks: feature generation and classifier synthesis. Each detector is represented as an automatically defined function (ADF), and the main program embodies the classification procedure (or classifier) and calls upon any/all of the ADFs as often as it requires [61].

The task here is to distinguish between an L and an I pattern by generating appropriate detectors as well as a classifier. Given that a pattern is represented as 6×4 map of black and white pixels, the objective is to automatically synthesize a computer program that accepts any one of the 2^{24} possible input patterns, and then returns the correct output classification: L, I, or NIL (for neither).

The general approach makes use of two interesting concepts: (a) the use of small (e.g., 3×3) masks for detection of local features; (b) the use of movement as an integrative means of relating features to each other.

A mask is implemented as a short function that inspects the Moore neighborhood of its central pixel (including the central pixel) to confirm the presence of a local feature, such as a three-pixel vertical line.

Such feature detecting functions are not hard coded by the user but are ADFs. Each ADF takes the form of a tree, representing a composition of Boolean disjunctions, conjunctions, and negations of nine pixel-value sensors. A sensor, say S (for south) returns a 1 if it matches a black pixel in the underlying input pattern and a 0 otherwise. The sensors are X (for the central pixel), N for north, S for south, E for east, W for west, NE for north east, NW for north west, SE for south east, and SW for south west.

The main program takes the form of a tree representing the composition of Boolean disjunctions, conjunctions, and negations of feature detecting functions.

Hence, an individual is a classifier made of five feature detecting functions and one program that utilizes them. We have described the structure of an individual, but how is genetic programing used to evolve, evaluate, and finally select individuals that execute the required classification task, and with a high degree of accuracy?

In order to write a GP, one needs: a set of terminals and a set of (primitive) functions that use them; an initialization technique; a way for assessing fitness and for selecting those individuals that are fittest; means for diversification (e.g., crossover); values for the various parameters of a GP (e.g., population size); and a termination criterion.

The ADFs have the terminal set T_f ={X, N, S, W, E, NE, NW, SE, SW} and the function set F_f = {AND, OR, NOT}. The main program has the terminal set T_c = {I, L, NIL} and a set of functions (F_c) containing movement functions, logical operators, four ADFs and the HOMING operator.

The formulation of a feature is so general that it is doubtful whether the current design is directly scalable to real-world pattern recognition problems, with millions of pixels per image. In addition, the approach appears to be suited to applications where there is a single example (or prototype) per pattern. However, real-world applications require an agent to learn a concept for a pattern (class) from many examples of noisy instances. It is not clear to us how the capability to learn from multiple examples can be incorporated into this technique. Finally, this approach assumes that enough discriminating features of a class can be (efficiently) extracted from the original input image. However, it is sometimes useful for a mapping (or more) to be applied before the distinguishing features of a class of patterns can be confidently and efficiently extracted. Nevertheless, this is a pioneering effort that has shown the way to other pattern recognition applications of genetic programing.

(2) *Andre* 1994

This work builds on Koza's results (above), but in a way that is more applicable to the problem of multicategory character recognition. In Koza's work, the feature detectors are Boolean ADFs of pixel-value sensors. In contrast, Andre uses hit–miss matrices that can be moved over a pattern until a match is found [6]. These matrices are evolved using a custom-made two-dimensional genetic algorithm. In Koza's work, the ADF-encoded feature detectors are used by a main program that is capable of moving over the whole pattern, while attempting to detect local features, until a final classification is reached. This is very similar to the way the Anrdre's *control code* moves and matches, using the hit–miss matrices, until a final conclusion is reached. However, a GA is used to evolve the matrices, whereas a GP is used to evolve the control code.

The purpose of this work appears to be the extension of Koza's work to more realistic character recognition problems, though the problems attempted remain simplistic. In addition, the use of a two-dimensional GA and associated 2D crossover operator is a significant innovation that may prove quite useful for image/pattern processing applications.

This approach has the distinction of using two representations: one for the hit–miss matrices representing the feature detectors and another for the control code carrying out the classification task using the hit–miss matrices.

The overall representation of an individual consists of five hit–miss matrices and a tree-based expression (main program) implemented in Lisp.

A hit–miss matrix of, for example, 3 × 3 elements is shown below: a 1-element seeks a black pixel, a 0-element seeks a white pixel, and a do not care element matches either pixel. A hit–miss matrix, such as the one in Figure 3.26, convoluted with an image, returns another image.

Functions and terminals are used in the evolution of the Lisp-style control code. The terminal set (for the hardest problem attempted by Andre) is T = {0, 1, moveE, moveW, moveN, and moveS}. The first two terminals represent the boolean values, false and true. The last four terminals in the set are actually functions that change only the X and Y variables representing the current horizontal and vertical coordinates, respectively, of the pattern recognition agent on the input pattern.

The function set of the control code is F = {progn2, Ifdf0, Ifdf1, Ifdf2, Ifdf3, Ifdf4}. progn2 is a macro that accepts two arguments, evaluates in the order of the first and the second arguments, and then returns the value of the second argument. An Ifdf$x$ (where x is an integer between 0 and 4) is a two-argument function that checks to see if its associated df$x$ hit–miss matrix matches the input pattern, and if it does, then the first argument is evaluated, else the second argument is evaluated.

The whole approach is far from delivering an even restricted pragmatic pattern recognition system. Real-world patterns come in many forms and at a much higher resolution than the test sets used in this work.

3.3.2.4 Evolvable Pattern Recognizers
Again, we give two concrete examples for this category of methods.

(1) *Rizki* 2002

According to Rizki, two approaches to the definition of features are commonplace. One uses a growth process, synthesizing a small set of complex features, incremen-

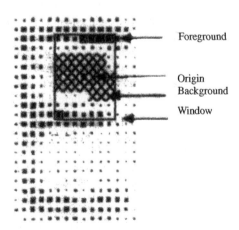

FIGURE 3.26   A hit–miss matrix and its application to a T pattern [6].

tally. The other uses a selection process in which a large set of small features are generated and then filtered [94].

Some researchers have evolved a single binary classifier (or detector), as well as a set of binary classifiers used by a higher level process to carry out multiple classifications. This work, however, proposes a method to evolve and train a complete multiclassifier for use in the classification of high-resolution radar signals. The system described here is called HELPR. It is the latest incarnation of a series of evolvable classifiers (by the same authors), including MORPH [94], which evolves a binary classifier and E-MORPH [94], which evolves a multiclassifier as well.

The system has two major parts, a set of feature detectors and a perceptron classifier. A single feature detector is made of a transformation network and a capping mechanism.

The transformation network is a network of morphological arithmetic and conditional operators, which is used to enhance the most discriminating regions of the one-dimentionsal input signal. A digitized input signal of n elements will produce an n-element output, which is then capped to produce a set of k scalars. These go into the classifier.

The capping mechanism is a single-layer perceptron with inputs equal to $k \times M$, where M is the number of feature detectors, and the number of outputs equal to the number of classes (to be distinguished). The perceptron cap functions as a template filter that acts to increase the separation between the various classes.

HELPR succeeds in creating new complex features and produces a recognition system that is highly accurate on real-world recognition tasks. However, evolution in HELPR is actually limited to the set of (feature) detectors. The only evolutionary activity that takes place at feature generation level, the classifier, is trained separately once the features are evolved. HELPR is a partially evolvable autonomous pattern recognizer: It only evolves the sets of feature detectors; however, their perceptron caps are trained and not evolved. In addition, there are still many parameters that are set manually (e.g., M). Perhaps, a completely hands-off parameterless evolutionary platform is in sight, but it is certainly not in hand.

(2) *Kharma* 2005

The long-term goal of project CellNet is the inclusion of methodologies for the simultaneous evolution of both feature and classifiers (Cells and Nets); at present, however, the set of primitive features is hard-coded [50].

In its latest full version, CellNet-CoEv is capable of evolving classifiers (or hunters) for a given set of images of patterns (or prey) that are also coevolving (getting harder to recognize). A Hunter is a binary classifier—a structure that accepts an image as input and outputs a classification. A hunter consists of cells, organized in a net. A Cell is a logical statement—it consists of the index of a feature function along with bounds. Prey are primarily images, drawn from the CEDAR database of handwritten digits. In addition to the image, prey disguise themselves via a set of camouflage functions controlled genetically [60].

All hunters were developed using identical system parameters, although training and validation images for each run were chosen randomly from a pool of 1000 images.

The hunters were placed in a genetic algorithm, using fitness-proportional selection and elitism.

Each run was executed for a maximum of 250 generations, outputting data regarding validation accuracy at 10 generations each. Training and validation accuracies are all reasonably high (close to or exceeding 90%) and very close to each other. Indeed, the average difference between training and validation accuracies was -0.6%, which indicates a touch of underfitting!

These results represent a significant step forward for the goal of an autonomous pattern recognizer: competitive coevolution and camouflage are expected to aid in the problem of overfitting and reliability without expert tuning, and also in the generation of a larger and more diverse data set.

3.3.2.5 Historical Summary

3.3.2.5 Historical Summary Here, we present a historical summary of feature creation for pattern recognition. The main publications in feature creation literature are divided into four categories.

Category I is preevolutionary methods that include [110, 105, 36, 34]. All of these techniques use mask- or pixel-based features; evaluate them using some kind of reference pattern; have no or limited higher level feature generation; and are not serious real-world applications. In 1963, Uhr and Vossler [110] became the first to propose the autogeneration, evaluation, and weighting of features for character recognition. Unusual for a work of such age, it actually used handwritten patterns and allowed for the creation of higher level features from simpler ones. In 1985, Stentiford [105] revived the idea of automatic feature design (or creation), with a paper that described the use of global hit–miss matrices as features and proposed orthogonality between feature vectors as a means for assessing the quality of these features. In a 1990 paper, based on his PhD thesis, Gillies [36] presented the idea of pixel-sum images as the bases for the automatic creation of morphological templates; templates that are used as feature detectors. Gillies also used orthogonality as a measure of quality of features, though not exactly in the same way as Stentiford did. In 1996, Gader and Khabou [34] proposed the use of a pixel-sum image, which is the result of summing 1000 (as opposed to 20 for Gillies) images per class—in effect producing a probabilistic distribution model for each class of character. He clearly separates the notion of feature from that of feature detector and assesses his features, which are mask based, using both orthogonality and information.

Category II is polynomial-based methods that include [17, 18, 38]. All of these techniques employ primitive features that are statistical functions of discrete (or discretizable) data signals; create complex features in the form of polynomial functions of the primitive features; and most have applications in the area of machine fault diagnosis. In 1991, Chang et al. [17] were, to the best of our knowledge, the first to use a genetic algorithm to create new features as polynomials of existing features, taken two at a time. As they used a GA and not a GP, they represented new features in terms of linear strings. They evaluated the feature sets using a k-nearest neighbor classifier. In 1999, Kotani et al. [59] devised a tree-based representation for the feature functions. Their representation was more appropriate for encoding polynomials than strings, but it only allowed for $+$ and $-$ operators in the nonterminal nodes of the expression tree. This must have contributed to the usual number of six different genetic operators that their technique used. In 2001, Chen et al. [18] took the next logical step and used expression trees that allowed for the basic four arithmetic operators and power in their nonterminal nodes; the genetic operators used were typical. Finally in 2005, Guo et al. [38] used moment-based statistical features as terminals and an extensive—perhaps too extensive—set of arithmetic trigonometric and other functions as operators in the making of trees representing new features. They also used the Fisher criterion (as opposed to final classification accuracy) to speed up the process of fitness evaluation of the new features. The resulting features were fed into various classifiers, including artificial neural nets and support vector machines.

Category III is full genetic programing methods that include [61, 6]. Both methods evolve a Lisp-style program that controls a roving agent that uses any/all of five evolvable Boolean functions/masks to correctly identify a character. What distinguishes these works is that they evolve both the features and the classifier that uses them simultaneously. It was 1993 when Koza [61] (also, the acknowledged father of genetic programing) proposed the use of an evolvable program-controlled turtle, with five equally evolvable Boolean functions (representing feature detectors), to autogenerate a complete recognition system for a toy problem: distinguishing between an I and an L. In 1997, Andre [6] built on Koza's work, by amending it to use two-dimensional hit–miss matrices (instead of Boolean functions) and introduce appropriate two-dimensional genetic operators. Andre tested his modified approach to three sets of "toy" problems, with the last being the synthesis of binary classifiers for single low-resolution digits between 0 and 9. The potential of this approach is considerable, for it is quite general, but is still to be realized.

Category IV is evolvable pattern recognizers. These are perhaps the most ambitious ongoing projects. Two systems described below do this. CellNet [50] blurs the line between feature selection and the construction of binary classifiers out of these features. HELPR [94] also evolves feature detectors, but the classification module is completely separate from the feature extraction module. Other differences exist, but, both attempts are the only systems, that we know of, that aim at using artificial evolution to synthesize complete recognition systems (though currently for different application domains), with minimum human intervention .

HELPR is composed of two modules: a features extraction module and a classification module. The classification system is not evolvable; only the feature extractor is [94]. The feature extraction module is made of a set of feature detectors. Its input is the raw input pattern and its output is a feature vector. The system is designed to handle signals (not visual patterns or patterns in general). CellNet is an ongoing research project aiming at the production of an autonomous pattern recognition software system for a large selection of pattern types [50]. Ideally, a CellNet operator would need little-to-no specialized knowledge to operate the system. To achieve this, CellNet divides the problem (and hence the solution) into two parts: feature creation and classifier synthesis.

3.3.3   Future Trends

In the above section, we have reviewed various methods for automatic feature creation. Next, we discuss the future trends in this field from three perspectives: generic applications, automatic complexification, and real-world technologies.

3.3.3.1   *Generic Applications*   The first instinct of any design engineer when faced with a specific pattern recognition problem is to think locally; that is, to identify the right representation for the problem and solution and to look for those most distinctive features of the various patterns to be recognized. This approach has, to some degree, carried over to the various feature creation endeavors reviewed in previous sections. For example, those applications in the area of machine fault diagnosis have all used polynomial-based representations that are well suited to signal processing. On the contrary, visual pattern recognition applications have all moved toward mask-based feature representations and matching genetic operators. It may well be a good idea to do this when one is driven by ultimate performance in terms of effectiveness and efficiency. However, this does not necessarily encourage the designer to devise representations and operators that have generic or at least generalizable capabilities.

Pattern recognition research should take a serious look at recognition problems where more than one category of patterns is to be recognized. An example would be devising feature extraction methods that succeed in extracting distinctive features for both a human emotional state detection task and a biological cell type classification task. The two problems appear distinct but are indeed quite similar. A lot of research on emotional state detection has been focused on the discovery of those statistical features of various physiological signals (e.g., blood pressure) that correlate most reliably with (pronounced) emotional states (e.g., anger). On the contrary, the type of a certain cell can be determined, in certain applications, from statistical features (e.g., high-frequency components) of the gray-level intensity map of the cell image. Both cases use statistical features that may be coded as polynomials, with the proviso that the input pattern in one case is an one-dimensional signal, whereas in the other case, it is a two-dimensional gray-scale image.

If researchers intentionally take on recognition tasks that involve two or more categories of input patterns, such as both Arabic words and Chinese characters, or both tree images and military target profiles, then we are bound to move in the direction of more generic pattern classification and feature creation methodologies.

3.3.3.2   *Automatic Complexification*   One weakness of current techniques used for feature creation is the limited ability of all surveyed techniques to create high-level features autonomously. It is true that some polynomial-based techniques allow for the created functions to be functions of primitive features as well as previously created ones. It is also a fact that all genetic programing techniques employ automatic function definitions from a set of feature functions. However, creating new functions by using previously created ones does not necessarily change the type of function that will result; a polynomial function of any number of polynomial functions will always be a polynomial function (and not, for example, a sinusoidal one). In

genetic programing approaches, on the contrary, automatic function definition has been limited to the automatic creation of a preset number of Lisp-type subroutines, each defining a feature extraction function. The significant advances in automatic function definition, achieved by John Koza and others, have not yet been applied to the problem of automatic creation of feature functions for pattern recognition purposes.

There is a long way, not yet traveled, toward applying and advancing many computer science techniques, such as hierarchical genetic programing [8], layered learning [106], Adaptive representations of learning [95] and, of course, automatically defined functions [61] to the problem of feature creation for pattern recognition.

We envisage a number of future developments that will allow feature creation programs based on GP methodologies to generate highly complex features with little human intervention—explicit or implied. First, starte with the simplest possible (primitive) feature extracting functions. Second, the various moves that a pattern recognition agent can make over an image are representative of the range of relationships between features (e.g., above or below). As such, it is also important to allow for a variable number of moves (or relationships). Third, a mechanism such as ADFs should be included in the approach to allow for the automatic definition (in the form of a program) of new features from primitive or created features and primitive or created relationships (between features). This mechanism should allow for recursion and could use discovered repeating patterns in the input images as bases for creating new features. Fourth, a new mechanism that allows for the automatic generation of programs that embody topological relationships between features. Such a mechanism can use repeating relationships between features in discovered patterns as bases for coding new relationships. Fifth, there should also be a mechanism that examines existing programs, defining new features or new relationships, to see if there are any commonalities between these functions (i.e., function invariants) that would allow for the creation of a single parameterized abstract function in place of two or more similar features or relationship functions. Finally, it is important that the right environment be created for the agent such that it has the right "motivation" not only to find features but also to discover the smallest number of simplest possible features suited for a given recognition task.

3.3.3.3 Real-World Technologies

3.3.3.3 Real-World Technologies   One problem with any new field of research is the lack of strong techniques with direct applicability to real-world problems. Feature creation is no exception, and all of the surveyed papers do not offer a readily applicable technology or even general design methodology. We mean that if a researcher is faced with a feature creation challenge tomorrow, then that researcher will not be able to directly plug-and-play any software component into his/her system or even cut-and-paste a method into his overall approach. In other words, we do not yet have a standard feature creation technology (e.g., image processing toolbox) or standard feature creation methods (e.g. the Fourier transform) that would allow practitioners and researchers alike to reuse existing solutions to feature creation challenges, with little or no configuration.

It is important that some of the research done in the future be directed toward (a) creation of stand-alone tool boxes with proper interfaces that would allow a worker in pattern recognition to choose the right feature creation component for his particular problem type, and then deploy it, as part of a larger system, with confidence in both the theoretical and pragmatic capabilities of the component; (b) invention of methods that are of general applicability to a large set of pattern classification tasks. Such methods need to be theoretically sound as well as practical.

It is not yet clear whether feature creation research is going to create its own attractor, around which various research efforts revolve, or if the various disciplines that require or could benefit from feature creation will maintain their hold on their own sphere of feature creation ideas and lead to highly specialized sets of tools and methodologies. Whatever happens, it is clear to us that, given the considerable increase in computational power, the use of computerized generate-and-test techniques will likely to increase, and with it the use of evolutionary algorithms for feature creation and selection.

3.4 CHAPTER SUMMARY

This chapter talks about three important aspects of the process of pattern recognition. Section 3.1 covered feature extraction. It presented a large number of features commonly used for character recognition. Some of the features are structural, such as the various discrete features, and others are statistical, such as moments and fourier descriptors.

In Section 3.2, feature selection was examined through a three-sided prism. One side dealt with the relationship between selection and classification; the other side dealt with the various criteria used for evaluating the different feature subsets; the third side presented a summary of the various search strategies used for exploring the space of possible feature subsets.

The final section of Chapter 3 talked about the slowly emerging field of feature creation. Ultimately, we would like to build machines that are similar to human beings in their ability to discover or invent features, on the fly, for any recognition task and do so without external guidance. The most recent efforts dedicated to the goal of feature creation have used some form of genetic algorithms or programing to evolve two-dimensional mask-based features or arithmetic feature functions. However, there is still much room that we hope will be filled with research in the near future.

REFERENCES

1. D. W. Aha and R. L. Bankert. Feature selection for case-based classification of cloud types: an empirical comparison. In *Proceedings of the AAAI Workshop on Case-Based Reasoning*. Seattle, WA, 1994, pp. 106–112.

2. D. W. Aha and R. L. Bankert. A comparative evaluation of sequential feature selection algorithms. In D. Fisher and H.-J. Lenz, editors, *Learning from Data*. Springer, New York, 1996, pp. 199–206.

3. H. Almuallim and T. G. Dietterich. Learning with many irrelevant features. In *Proceedings of the 9th National Conference on Artificial Intelligence*. Anaheim, 1991, Vol. 2, pp. 547–552.

4. H. Almuallim and T. G. Dietterich. Learning boolean concepts in the presence of many irrelevant features. *Artificial Intelligence*. **69**(1–2), 279–305, 1994.

5. A. Amin. Prototyping structural description using an inductive learning program. *IEEE Transactions on System, Man Cybernetics, Part C*, **30**(1), 150–157, 2000.

6. D. Andre. Automatically defined features: the simultaneous evolution of 2-dimensional feature detectors and an algorithm for using them. In K. E. Kinnear, editor, *Advances in Genetic Programming*. The MIT Press, Cambridge, MA, 1994, Chapter 23.

7. A. Arauzo, J. M. Benitez, and J. L. Castro. C-FOCUS: a continuous extension of FOCUS. In J. M. Benitez, et al., editors, *Advances in Soft Computing—Engineering, Design and Manufacturing*. Springer, London, 2003, pp. 225–232.

8. W. Banzhaf, D. Banscherus, and P. Dittrich. Hierarchical genetic programing using local modules. In *Proceedings of the International Conference on Complex Systems*. Nashua, NH, 1998.

9. B. Bhanu and Y. Lin. Genetic algorithm based feature selection for target detection in SAR images. *Image and Vision Computing*. **21**, 591–608, 2003.

10. R. E. Bellman. *Adaptive Control Processes*. Princeton University Press, 1961.

11. M. Ben-Bassat. Use of distance measures, information measures and error bounds in feature evaluation. In P. R. Krishnaiah and L. N. Kanal, editors, *Handbook of Statistics*, North Holland, 1982, pp. 773–791.

12. A. L. Blum and P. Langley. Selection of relevant features and examples in machine learning. *Artificial Intelligence*, **97**(1–2), 245–271, 1997.

13. L. Brieman, J. H. Friedman, R. A. Olshen, and C. J. Stone. *Classification and Regression Trees*, Wadsworth, 1984.

14. L. G. Brown. A survey of image registration techniques. *ACM Computing Surveys*, **24**(4), 325–363, 1992.

15. E. Cantu-Paz, S. Newsam, and C. Kamath. Feature selection in scientific applications. In *Proceedings of the 10th ACM International Conference on Knowledge Discovery and Data Mining*, Seattle, WA, 2004, pp. 788–793.

16. C. Cardie. Using decision trees to improve case-based learning. In *Proceedings of the 10th International Conference on Machine Learning*, Amherst, 1993, Morgan Kaufmann, pp. 25–32.

17. E. I. Chang, R. P. Lippmann, and D. W. Tong. Using genetic algorithms to select and create features pattern classification. *Proceedings of the International Joint Conference on Neural Networks*, San Diego, CA, 1990, Vol. 3, pp. 747–753.

18. P. Chen, T. Toyota, and Z. He. Automated function generation of symptom parameters and application to fault diagnosis of machinery under variable operating conditions. *IEEE Transactions on System, Man Cybernetics Part A*, **31**(6), 775–781, 2001.

19. X.-W. Chen. An improved branch and bound algorithm for feature selection. *Pattern Recognition Letters*, **24**, 1925–1933, 2003.

20. S. Das. Filters, wrappers and a boosting-based hybrid for feature selection. In *Proceedings of the 18th International Conference on Machine Learning*, Williamstown, 2001, pp. 74–81.

21. M. Dash and H. Liu. Feature selection for classification. *Intelligent Data Analysis*, **1**(3), 131–156, 1997.

22. M. Dash and H. Liu. Consistency-based search in feature selection. *Artificial Intelligence*, **151**, 155–176, 2003.

23. P. A. Devijver and J. Kittler. *Pattern Recognition: A Statistical Approach*. Prentice Hall, 1982.

24. A. C. Downton and S. Impedovo, editors. *Progress in Handwriting Recognition*, World Scientific, Singapore, 1997.

25. R. O. Duda and P. E. Hart. Use of the Hough transformation to detect lines and curves in pictures. *Communications of ACM*, **5**(1), 11–15, 1972.

26. T. Elomaa and J. Rousu. General and efficient multisplitting of numerical attributes. *Machine Learning*, **36**(3), 201–244, 1999.

27. T. Eriksson, S. Kim, H.-G. Kang, and C. Lee. An information-theoretic perspective on feature selection in speaker recognition. *IEEE Signal Processing Letters*, **12**(7), 500–503, 2005.

28. M. E. Farmer, S. Bapna, and A. K. Jain. Large scale feature selection using modified random mutation hill climbing. In *Proceedings of the 17th International Conference on Pattern Recognition*, Cambridge, UK, 2004, Vol.2, pp. 287–290.

29. F. Ferri, P. Pudil, M. Hatef, and J. Kittler. Comparative study of techniques for large scale feature selection. In E. Gelsema and L. Kanal, editors, *Pattern Recognition in Practice IV*, Vlieland, The Netherlands, 1994, pp. 403–413.

30. H. A. Firpi and E. Goodman. Swarmed feature selection. In *Proceedings of the 33rd Applied Imagery Pattern Recognition Workshop*, Washington DC, 2004, pp. 112–118.

31. H. Freeman. Boundary encoding and processing. In B. Lipkin and A. Rosenfeld, editors, *Picture Processing and Pshcholopictorics*. Academic Press, New York, 1970, 241–266.

32. H. Fujisawa and C.-L. Liu. Directional pattern matching for character recognition revisited. *Proceedings of the 7th International Conference on Document Analysis and Recognition*, Edinburgh, Scotland, 2003, pp. 794–798.

33. K. Fukunaga. *Introduction to Statistical Pattern Recognition*, 2nd edition. Academic Press, New York, 1990.

34. P. D. Gader and M. A. Khabou. Automatic feature generation for handwritten digit recognition. *IEEE Transactions on Pattern Analysis and Machine Intelligence*, **18**(12), 1256–1261, 1996.

35. K. Giles, K.-M. Bryson, and Q. Weng. Comparison of two families of entropy-based classification measures with and without feature selection. In *Proceedings of the 34th Hawaii International Conference on System Sciences*, Maui, 2001, Vol. 3.

36. Andrew M. Gillies. Automatic generation of morphological template features. In *Proceedings of the SPIE—Image Algebra and Morphological Image Processing*, San Diego, CA, 1990, Vol. 1350, pp. 252–261.

37. R. C. Gonzalez and R. E. Woods. *Digital Image Processing*. Addison Wesley, Boston, MA, 1993.

38. H. Guo, L. B. Jack, and A. K. Nandi. Feature generation using genetic programing with application to fault classification. *IEEE Transactions on System, Man and Cybernetics Part B*, 35(1): 89–99, 2005.

39. M. A. Hall. Correlation-based feature selection for discrete and numeric class machine learning. In *Proceedings of the 17th International Conference on Machine Learning*, San Francisco, 2000, Morgan Kaufmann Publishers, pp. 359–366.

40. M. A. Hall and L. A. Smith. Feature subset selection: a correlation based filter approach. In *Proceedings of the 4th International Conference on Neural Information Processing and Intelligent Information Systems*, Dunedin, 1997, pp. 855–858.

41. M. A. Hall and L. A. Smith. Practical feature subset selection for machine learning. In *Proceedings of the 21st Australian Computer Science Conference*, Auckland, 1998, pp. 181–191.

42. T. Hastie and R. Tibshirani. Discriminant analysis by Gaussian mixtures. *Journal of the Royal Statistical Society, Series B*, **58**(1), 155–176, 1996.

43. C.-N. Hsu, H.-J. Huang, and D. Schuschel. The ANNIGMA-wrapper approach to fast feature selection for neural nets. *IEEE Transactions on System, Man, Cybernetics, Part B*, **32**(2), 207–212, 2002.

44. M.-K Hu. Visual pattern recognition by moment invariants. *IRE Transactions on Information Theory*, **IT-8**, 179–187, 1962.

45. F. Hussein, N. Kharma, and R. Ward. Genetic algorithms for feature selection and weighting, a review and study. In *Proceedings of the 6th International Conference on Document Analysis and Recognition*, Seattle, WA, 2001, pp. 1240–1244.

46. T. Iijima, H. Genchi, and K. Mori. A theoretical study of the pattern identification by matching method. In *Proceedings of the First USA–JAPAN Computer Conference*, Tokyo, Japan, 1972, pp. 42–48.

47. G. H. John, R. Kohavi, and K. Pfleger. Irrelevant features and the subset selection problem. In *Proceedings of the 11th International Conference on Machine Learning*, New Brunswick, 1994, pp. 121–129.

48. N. Kato, M. Suzuki, S. Omachi, H. Aso, and Y. Nemoto. A handwritten character recognition system using directional element feature and asymmetric Mahalanobis distance. *IEEE Transactions on Pattern Analysis and Machine Intelligence*, **21**(3), 258–262, 1999.

49. A. Kawamura, K. Yura, T. Hayama, Y. Hidai, T. Minamikawa, A. Tanaka, and S. Masuda. On-line recognition of freely handwritten Japanese characters using directional feature densities. In *Proceedings of the 11th International Conference on Pattern Recognition*, The Hague, 1992, Vol. 2, pp. 183–186.

50. N. Kharma, T. Kowaliw, E. Clement, C. Jensen, A. Youssef, and J. Yao. Project CellNet: evolving an autonomous pattern recogniser. *International Journal on Pattern Recognition and Artificial Intelligence*, **18**(6), 1–18, 2004.

51. A. Khotanzad and Y. H. Hong. Invariant image recognition by Zernike moments. *IEEE Transactions on Pattern Analysis and Machine Intelligence*, **12**(5), 489–490, 1990.

52. F. Kimura, K. Takashina, S. Tsuruoka, and Y. Miyake. Modified quadratic discriminant functions and the application to Chinese character recognition. *IEEE Transactions on Pattern Analysis and Machine Intelligence*, **9**(1), 149–153, 1987.

53. F. Kimura. On feature extraction for limited class problem. In *Proc. 13th Int. Conf. on Pattern Recognition*, Vienna, Austria, 1996, Vol. 2, pp. 25–29.

54. K. Kira and L. A. Rendell. The feature selection problem: Traditional methods and a new algorithm. In *Proceedings of the 10th National Conference on Artificial Intelligence*, San Jose, CA, 1992, pp. 129–134.

55. K. Kira and L. A. Rendell. A practical approach to feature selection. In D. Sleeman and P. Edwards, editors, *International Conference on Machine Learning*, Morgan Kaufmann Publishers, 1992, pp. 249–256.

56. R. Kohavi and G. H. John. Wrappers for feature subset selection. *Artificial Intelligence*, **97**(1–2), 273–324, 1997.

57. R. Kohavi and D. Sommerfield. Feature subset selection using the wrapper model: overfitting and dynamic search space topology. In *Proceedings of the 1st International Conference on Knowledge Discovery and Data Mining*, Montreal, Canada, 1995, pp. 192–197.

58. I. Kononenko. Estimating attributes: analysis and extension of relief. In *Proceedings of the European Conference on Machine Learning*, Catania, 1994, pp. 171–182.

59. M. Kotani, S. Ozawat, M. Nakai, and K. Akazawat. Emergence of feature extraction function using genetic programing. In *Proceedings of the 3rd International Conference on Knowledge-Based Intelligent Information Engineering Systems*, Adelaide, 1999, pp. 149–152.

60. T. Kowaliw, N. Kharma, C. Jensen, H. Moghnieh, and J. Yao. Using competitive co-evolution to evolve better pattern recognizers. *International Journal on Computational Intelligence and Applications*, **5**(3), 305–320, 2005.

61. J. R. Koza. Simultaneous discovery of detectors and a way of using the detectors via genetic programing. *Proceedings of the IEEE International Conference on Neural Networks*, 1993, Vol. 3, pp. 1794–1801.

62. M. Kudo and J. Sklansky. Comparison of algorithms that select features for pattern classifiers. *Pattern Recognition*, **33**, 25–41, 2000.

63. F. P. Kuhl and C. R. Giardina. Elliptic Fourier feature of a closed contour. *Computer Vision, Graphics and Image Processing*, **18**, 236–258, 1982.

64. P. Kultanen, L. Xu, and E. Oja. Randomized Hough transform. In *Proceedings of the 10th International Conference on Pattern Recognition*, Atlantic City, 1990, Vol. 1, pp. 631–635.

65. L. I. Kuncheva and L. C. Jain. Nearest neighbor classifier: Simultaneous editing and feature selection. *Pattern Recognition Letters*, **20**, 1149–1156, 1999.

66. T. N. Lal, O. Chapelle, J. Weston, and A. Elisseeff. Embedded methods. In I. Guyon, S. Gunn, M. Nikravesh, and L. Zadeh, editors, *Feature extraction, Foundations and Applications*, Springer, 2006, pp. 137–165.

67. L. Lam, S.-W. Lee and C. Y. Suen. Thinning methodologies—a comprehensive survey. *IEEE Transactions on Pattern Analysis and Machine Intelligence*, **14**(9), 869–885, 1992.

68. P. Langley. Selection of relevant features in machine learning. In *Proceedings of the AAAI Fall Symposium on Relevance*, New Orleans, 1994, pp. 140–144.

69. P. Langley and S. Sage. Oblivious decision trees and abstract cases. In *Proceedings of the AAAI-94 Workshop on Case-Based Reasoning*, Seattle, WA, 1994, pp. 113–117.

70. C. Lee and D. A. Landgrebe. Feature extraction based on decision boundary. *IEEE Transactions on Pattern Analysis and Machine Intelligence*, **15**(4), 388–400, 1993.

71. C. Lee and G. G. Lee. Information gain and divergence-based feature selection for machine learning-based text categorization. *Information Processing and Management*, **42**, 155–165, 2006.

72. S. Lee and J. C. Pan. Offline tracing and representation of signature. *IEEE Transactions on System, Man, and Cybernetics*, **22**(4), 755–771, 1992.

73. C.-L. Liu, Y.-J. Liu, and R.-W. Dai. Preprocessing and statistical/structural feature extraction for handwritten numeral recognition. In A. C. Downton and S. Impedovo, editors, *Progress of Handwriting Recognition*, World Scientific, Singapore, 1997, pp. 161–168.

74. C.-L. Liu, K. Nakashima, H. Sako, and H. Fujisawa. Handwritten digit recognition: benchmarking of state-of-the-art techniques, *Pattern Recognition*, **36**(10), 2271–2285, 2003.

75. H. Liu and R. Setiono. Feature selection and classification—a probabilistic wrapper approach. In *Proceedings of the 9th International Conference Industrial and Engineering Applications of AI and ES*, Fukuoka, Japan, 1996, pp. 419–424.

76. H. Liu and R. Setiono. A probabilistic approach to feature selection—a filter solution. In *Proceedings of the 13th International Conference on Machine Learning*, Bari, Italy, 1996, pp. 319–327.

77. H. Liu and L. Yu. Toward integrating feature selection algorithms for classification and clustering. *IEEE Transactions on Knowledge and Data Engineering*, **17**(4), 491–502, 2005.

78. M. Loog and R. P. W. Duin. Linear dimensionality reduction via a heteroscedastic extension of LDA: the Chernoff criterion. *IEEE Transactions on Pattern Analysis and Machine Intelligence*, **26**(6), 732–739, 2004.

79. H. A. Mayer and P. Somol. Conventional and evolutionary feature selection of SAR data using a filter approach. In *Proceedings of the 4th World Multi-Conference on Systemics, Cybernetics, and Informatics*, Orlando, 2000.

80. R. Meiri and J. Zahavi. Using simulated annealing to optimize the feature selection problem in marketing applications. *European Journal of Operational Research*, **171**(3), 842–858, 2006.

81. A. J. Miller. *Subset Selection in Regression*, 2nd edition. Chapman and Hall, 2002.

82. T. M. Mitchell. Generalization as search. *Artificial Intelligence*, **18**(2), 203–226, 1982.

83. S. Mori, C. Y. Suen, and K. Yamamoto. Historical review of OCR research and development. *Proceedings of IEEE*, **80**(7), 1029–1053, 1992.

84. R. Mukundan, S. H. Ong, and P. A. Lee. Image analysis by Tchebichef moments. *IEEE Transactions on Image Processing*, **10**(9), 1357–1364, 2001.

85. P. Narendra and K. Fukunaga. A branch and bound algorithm for feature subset selection. *IEEE Transactions on Computers*, **26**(9), 917–922, 1977.

86. K. R. Niazi, C. M. Arora, and S. L. Surana. Power system security evaluation using ANN: Feature selection using divergence. *Electric Power Systems Research*, **69**, 161–167, 2004.

87. C. M. Nunes, A. d. S. Britto Jr., C. A. A. Kaestner, and R. Sabourin. An optimized hill climbing algorithm for feature subset selection: evaluation on handwritten character recognition. In *Proceedings of the 9th International Workshop on Frontiers in Handwriting Recognition*, Tokyo, Japan, 2004, pp. 365–370.

88. L. S. Oliveira, R. Sabourin, F. Bortolozzi, and C. Y. Suen. Feature selection using multi-objective genetic algorithms for handwritten digit recognition. In *Proceedings of the 16th International Conference on Pattern Recognition*, Quebec City, 2002, pp. 568–571.

89. T. Petkovic and J. Krapac. Shape description with Fourier descriptors. Technical Report, University of Zagreb, Croatia, 2002.

90. S. Piramuthu. Evaluating feature selection methods for learning in data mining applications. *European Journal of Operational Research*, **156**, 483–494, 2004.

91. W. F. Punch, E. D. Goodman, M. Pei, L. Chia-Shun, P. D. Hovland, and R. J. Enbody. Further research on feature selection and classification using genetic algorithms. In S. Forrest, editor, *Proceedings of the 5th International Conference on Genetic Algorithms*, Morgan Kaufmann, San Mateo, 1993, pp. 557–564.

92. J. R. Quinlan. *C4.5 Programs for Machine Learning*, Morgan Kaufmann, San Mateo, 1993.

93. J. R. Quinlan. Induction of decision trees. *Machine Learning*, **1**(1), 81–106, 2003.

94. M. M. Rizki, M. A. Zmuda, and L. A. Tamburino. Evolving pattern recognition systems. *IEEE Trans. Evolutionary Computation*, **6**(6), 594–609, 2002.

95. J. P. Rosca. Towards automatic discovery of building blocks in genetic programing. In E. S. Siegel and J. R. Koza, editors, *Working Notes for the AAAI Symposium on Genetic Programing*, Menlo Park, 1995, pp. 78–85.

96. J. C. Schlimmer. Efficiently inducing determinations: a complete and systematic search algorithm that uses optimal pruning. In *Proceedings of the 10th International Conference on Machine Learning*, Amherst, 1993, pp. 284–290.

97. J. A. Schnabel. Shape description methods for medical images. Technical Report, University College London, UK, 1995.

98. D. Schuschel and C.-N Hsu. A weight analysis-based wrapper approach to neural nets feature subset selection. In *Proceedings of the 10th IEEE International Conference on Tools with AI.*, Taipei, 1998.

99. M. Sebbana and R. Nock. A hybrid filter/wrapper approach of feature selection using information theory. *Pattern Recognition*, **35**, 835–846, 2002.

100. W. Siedlecki and J. Sklansky. A note on genetic algorithms for large scale on feature selection. *Pattern Recognition Letters*, **10**, 335–347, 1989.

101. D. B. Skalak. Prototype and feature selection by sampling and random mutation hill climbing algorithms. In *Proceedings of the 11th International Conference on Machine Learning*, New Brunswick, 1994, pp. 293–301.

102. P. Somol, P. Pudil, and J. Kittler. Fast branch and bound algorithms for optimal feature selection. *IEEE Transactions on Pattern Analysis and Machine Intelligence*, **26**(7), 900–912, 2004.

103. P. Somol, P. Pudil, J. Novovicova, and P. Paclik. Adaptive floating search methods in feature selection. *Pattern Recognition Letters*, **20**, 1157–1163, 1999.

104. G. Srikantan, S. W. Lam, and S. N. Srihari. Gradient-based contour encoder for character recognition. *Pattern Recognition*, **29**(7), 1147–1160, 1996.

105. F. W. M. Stentiford. Automatic feature design for optical character recognition using an evolutionary search procedure. *IEEE Transactions on Pattern Analysis and Machine Intelligence*, **7**(3), 350–355, 1985.

106. P. Stone and M. M. Veloso. Layered learning. In *Proceedings of the 11th European Conference on Machine Learning*, Barcelona, 2000, pp. 369–381.

107. Z. Sun, G. Bebis, X. Yuan, and S. J. Louis. Genetic feature subset selection for gender classification: a comparison study. In *Proceedings of the IEEE Workshop on Applications of Computer Vision*, Orlando, 2002, pp. 165–170.

108. C.-H. Teh and R. T. Chin. On image analysis by the methods of moments. *IEEE Transactions on Pattern Analysis and Machine Intelligence*, **10**(4), 496–499, 1988.

109. O. D. Trier, A. K. Jain, and T. Taxt. Feature extraction methods for character recognition—a survey. *Pattern Recognition*, **29**(4), 641–662, 1996.

110. L. Uhr and C. Vossler. A pattern-recognition program that generates, evaluates and adjusts its own operators. In *Computers and Thought*, McGraw-Hill, New York, 1963, pp. 251–268.

111. H. Vafaie and I. F. Imam. Feature selection methods: genetic algorithms vs. greedy-like search. In *Proceedings of the International Conference on Fuzzy and Intelligent Control Systems*, Louisville, 1994.

112. H. Vafaie and K. D. Jong. Robust feature selection algorithms. In *Proceedings of the 5th Conference on Tools for Artificial Intelligence*, Boston, 1993, pp. 356–363.

113. H. Vafaie and K. A. D. Jong. Genetic algorithms as a tool for restructuring feature space representations. In *Proceedings of the International Conference on Tools with Artificial Intelligence*, Herndon, 1995, pp. 8–11.

114. S. A. Vere. Induction of concepts in the predicate calculus. In *Proceedings of the 4th International Joint Conference on Artificial Intelligence*, Tbilisi, 1975, pp. 351–356.

115. D. Wettschereck, D. W. Aha, and T. Mohri. A review and empirical evaluation of feature weighting methods for a class of lazy learning algorithms. *Artificial Intelligence Review*, 11(1–5), 273–314, 1997.

116. Y. Wu and A. Zhang. Feature selection for classifying high-dimensional numerical data. In *Proceedings of the IEEE International Conference on Computer Vision and Pattern Recognition*, Washington DC, 2004, Vol. 2, pp. 251–258.

117. Y. Yamashita, K. Higuchi, Y. Yamada, and Y. Haga. Classification of handprinted Kanji characters by the structured segment matching method. *Pattern Recognition Letters*, **1**, 475–479, 1983.

118. J. Yang and V. Honavar. Feature subset selection using a genetic algorithm. In J. R. Koza, et al., editors, *Genetic Programing*, Morgan Kaufmann, 1997.

119. M. Yasuda and H. Fujisawa. An improvement of correlation method for character recognition. *Transactions on IEICE Japan*, **J62-D(3)**, 217–224, 1979 (in Japanese).

120. H. Yoshida, R. Leardi, K. Funatsu, and K. Varmuza. Feature selection by genetic algorithms for mass spectral classifiers. *Analytica Chimica Acta*, **446**, 485–494, 2001.

121. B. Yu and B. Yuan. A more efficient branch and bound algorithm for feature selection. *Pattern Recognition*, **26**, 883–889, 1993.

122. J. Yu, S. S. R. Abidi, and P. H. Artes. A hybrid feature selection strategy for image defining. In *Proceedings of the 5th International Conference on Intelligent Systems Design and Applications*, Guangzhou, Chia, 2005, Vol. 8, pp. 5127–5132.

123. L. Yu and H. Liu. Feature selection for high-dimensional data: a fast correlation-based filter solution. In *Proceedings of the 20th International Conference on Machine Leaning*, Washington DC, 2003, pp. 856–863.

124. S. Yu, S. De Backer, and P. Scheunders. Genetic feature selection combined with composite fuzzy nearest neighbor classifiers for hyperspectral satellite imagery. *Pattern Recognition Letters*, **23**(1–3): 183–190, 2002.

125. D. Zhang and G. Lu, A comparative study on shape retrieval using Fourier descriptors with different shape signatures. In *Proceedings of the IEEE International Conference on Multimedia and Expo*, Tokyo, Japan, 2001, pp. 1139–1142.

126. H. Zhang and G. Sun. Feature selection using tabu search method. *Pattern Recognition*, **35**, 701–711, 2002.

127. L. X. Zhang, J. X. Wang, Y. N. Zhao, and Z. H. Yang. A novel hybrid feature selection algorithm: using Relief estimation for GA-wrapper search. In *Proceedings of the International Conference on Machine Learning and Cybernetics*, Xi'an, China, 2003, Vol. 1, pp. 380–384.

CHAPTER 4

PATTERN CLASSIFICATION METHODS

If the goal of feature extraction is to map input patterns onto points in a feature space, the purpose of classification is to assign each point in the space with a class label or membership scores to the defined classes. Hence, once a pattern is mapped (represented), the problem becomes one of the classical classification to which a variety of classification methods can be applied. This chapter describes the classification methods that have shown success or are potentially effective to character recognition, including statistical methods (parametric and nonparametric), structural methods (string and graph matching), artificial neural networks (supervised and unsupervised learning), and combination of multiple classifiers. Finally, some experimental results of classifier combination on a data set of handwritten digits are presented.

4.1 OVERVIEW OF CLASSIFICATION METHODS

The final goal of character recognition is to obtain the class codes (labels) of character patterns. On segmenting character patterns or words from document images, the task of recognition becomes assigning each character pattern or word to a class out of a predefined class set. As many word recognition methods also take a segmentation-based scheme with character modeling or character recognition embedded, as discussed in Chapter 5, the performance of character recognition is of primary importance for document analysis. A complete character recognition procedure involves the steps of preprocessing, feature extraction, and classification. On mapping the input pattern

Character Recognition Systems: A Guide for Students and Practitioner, by M. Cheriet, N. Kharma, C.-L. Liu and C. Y. Suen   Copyright © 2007 John Wiley & Sons, Inc.

to a point in feature space via feature extraction, the problem becomes one of the classical classification. For integrating the classification results with contextual information like linguistics and geometrics, the outcome of classification is desired to be the membership scores (probabilities, similarity or dissimilarity measurements) of input pattern to defined classes rather than a crisp class label. Most of the classification methods described in this chapter are able to output either class label or membership scores.

Pattern classification has been the main theme of pattern recognition field and is often taken as a synonym of "pattern recognition." A rigorous theoretical foundation of classification has been laid, especially to statistical pattern recognition (SPR), and many effective classification methods have been proposed and studied in depth. Many textbooks have been published and are being commonly referred by researchers and practitioners. Some famous textbooks are the ones of Duda et al. [17] (first edition in 1973, second edition in 2001), Fukunaga [23] (first edition in 1972, second edition in 1990), Devijver and Kittler [12], and so on. These textbooks mainly address SPR. Syntactic pattern recognition was founded by Fu and attracted much attention in 1970s and 1980s [22] but it has not found many practical applications. On the contrary, structural pattern recognition methods using string and graph matching have demonstrated effects in image analysis and character recognition.

From the late 1980s, artificial neural networks (ANNs) have been widely applied to pattern recognition due to the rediscovery and successful applications of the back-propagation algorithm [75] for training multilayer networks, which are able to separate class regions of arbitrarily complicated distributions. The excellent textbook of Bishop [4] gives an in-depth treatment of neural network approaches from SPR perspective. From the late 1990s, a new direction in pattern recognition has been with support vector machines (SVMs) [92, 7, 10], which are supposed to provide optimal generalization performance[1] via structural risk minimization (SRM), as opposed to the empirical risk minimization for neural networks.

This chapter gives an introductory description to the pattern classification methods that have been widely and successfully applied to character recognition. These methods are categorized into statistical methods, ANNs, SVMs, structural methods, and multiple classifier methods. Statistical methods, ANNs, and SVMs input a feature vector of fixed dimensionality mapped from the input pattern. Structural methods recognize patterns via elastic matching of strings, graphs, or other structural descriptions. By multiple classifier methods, the classification results of multiple classifiers are combined to reorder the classes. We will describe the concepts and mathematical formulations of the methods, discuss their properties, and provide design guidelines in application to character recognition, and finally show some experimental examples of character data classification.

---

[1]In general, the parameters of a classifier are estimated on a training sample set and its classification performance is evaluated on a separate test sample set. Generalization performance refers to the classification accuracy on test samples.

4.2 STATISTICAL METHODS

Statistical classification methods are based on the Bayes decision theory, which aims to minimize the loss of classification with given loss matrix and estimated probabilities. According to the class-conditional probability density estimation approach, statistical classification methods are divided into parametric and nonparametric ones.

4.2.1 Bayes Decision Theory

Assume that d feature measurements $\{x_1, \ldots, x_d\}$ have been extracted from the input pattern, the pattern is then represented by a d-dimensional feature vector $\mathbf{x} = [x_1, \ldots, x_d]^T$. $\mathbf{x}$ is considered to belong to one of M predefined classes $\{\omega_1, \ldots, \omega_M\}$. Given the a priori probabilities $P(\omega_i)$ and class-conditional probability distributions $p(\mathbf{x}|\omega_i)$, $i = 1, \ldots, M$, the a posteriori probabilities are computed by the Bayes formula

$$P(\omega_i|\mathbf{x}) = \frac{P(\omega_i)p(\mathbf{x}|\omega_i)}{p(\mathbf{x})} = \frac{P(\omega_i)p(\mathbf{x}|\omega_i)}{\sum_{j=1}^{M} P(\omega_j)p(\mathbf{x}|\omega_j)}. \tag{4.1}$$

Given a loss matrix $[c_{ij}]$ (c_{ij} is the loss of misclassifying a pattern from class ω_j to class ω_i), the expected loss (also called as conditional risk) of classifying a pattern $\mathbf{x}$ to class ω_i is

$$R_i(\mathbf{x}) = \sum_{j=1}^{M} c_{ij} P(\omega_j|\mathbf{x}). \tag{4.2}$$

The expected loss is then minimized by classifying $\mathbf{x}$ to the class of minimum conditional risk.

In practice, we often assume that the loss of misclassification is equal between any pair of classes and the loss of correct classification is zero:

$$c_{ij} = \begin{cases} 1, & i \neq j, \\ 0, & i = j. \end{cases} \tag{4.3}$$

The conditional risk then becomes the expected error rate of classification

$$R_i(\mathbf{x}) = \sum_{j \neq i}^{M} c_{ij} P(\omega_j|\mathbf{x}) = 1 - P(\omega_i|\mathbf{x}), \tag{4.4}$$

and the decision becomes selecting the class of maximum a posteriori (MAP) probability to minimize the error rate. The error rate $1 - \max_i P(\omega_i|\mathbf{x})$ is called Bayes error rate.

From Eq. (4.1) the a posteriori probability is proportional to the likelihood function $P(\omega_i)p(\mathbf{x}|\omega_i)$ because the denominator $p(\mathbf{x})$ is independent of class label. So, the MAP decision is equivalent to selecting the class of maximum likelihood:

$$\max_i g(\mathbf{x}, \omega_i) = \max_i P(\omega_i)p(\mathbf{x}|\omega_i), \quad i = 1, \dots, M, \tag{4.5}$$

or maximum log-likelihood:

$$\max_i g(\mathbf{x}, \omega_i) = \max_i \log P(\omega_i)p(\mathbf{x}|\omega_i), \quad i = 1, \dots, M. \tag{4.6}$$

The likelihood and log-likelihood functions are also called discriminant functions. For a pair of classes, ω_i and ω_j, the set of points in feature space with equal discriminant value $g(\mathbf{x}, \omega_i) = g(\mathbf{x}, \omega_j)$ is called decision surface or decision boundary.

To make Bayes decision (or simply minimum error rate decision) requires the a priori probabilities and the conditional probability density functions (PDFs) of defined classes. The a priori probability can be estimated as the percentage of samples of a class in the training sample set, or as often, assumed to be equal for all classes. The PDF can be estimated by variable approaches. Parametric classification methods assume functional forms for class-conditional density functions and estimate the parameters by maximum likelihood (ML), whereas nonparametric methods can estimate arbitrary distributions adaptable to training samples.

4.2.2   Parametric Methods

Parametric methods assume a functional form for the density of each class

$$p(\mathbf{x}|\omega_i) = f(\mathbf{x}|\theta_i), \quad i = 1, \dots, M, \tag{4.7}$$

where θ_i denotes the set of parameters for the density function of a class. In the following, we will describe an important class of classifiers that assume Gaussian density functions, parameter estimation by ML, and semiparametric classifier with Gaussian mixture density functions.

4.2.2.1   Gaussian Classifiers   Assuming multivariate Gaussian density (also called normal density) has some benefits: The theoretical analysis of properties and the analytical estimation of parameters are easy, and many practical problems have class-conditional densities that are approximately Gaussian. The parametric classifiers with Gaussian density functions are often called Gaussian classifiers or normal classifiers, which again have some variations depending on the degree of freedom of parameters.

The Gaussian density function of a class is given by

$$p(\mathbf{x}|\omega_i) = \frac{1}{(2\pi)^{d/2}|\Sigma_i|^{1/2}} \exp\left[-\frac{1}{2}(\mathbf{x} - \mu_i)^T \Sigma_i^{-1}(\mathbf{x} - \mu_i)\right], \tag{4.8}$$

where μ_i and Σ_i denote the mean vector and the covariance matrix of class ω_i, respectively. Inserting Eq. (4.8) into Eq. (4.6) the quadratic discriminant function (QDF) is obtained:

$$g(\mathbf{x}, \omega_i) = -\frac{1}{2}(\mathbf{x} - \mu_i)^T \Sigma_i^{-1}(\mathbf{x} - \mu_i) - \frac{1}{2} \log |\Sigma_i| + \log P(\omega_i). \qquad (4.9)$$

In the above equation, the term common to all classes has been omitted. The term $(\mathbf{x} - \mu_i)^T \Sigma_i^{-1}(\mathbf{x} - \mu_i)$ is also called Mahalanobis distance and can serve an effective discriminant function. In some publications, QDF is taken as the negative of Eq. (4.9), and sometimes the term $\log P(\omega_i)$ is omitted under the assumption of equal a priori probabilities. The negative of QDF can be viewed as a distance measure, and as for the Mahalanobis distance, the input pattern is classified to the class of minimum distance.

The decision surface of QDF, under $g(\mathbf{x}, \omega_i) = g(\mathbf{x}, \omega_j)$, can be represented by

$$\frac{1}{2}\mathbf{x}^T(\Sigma_j^{-1} - \Sigma_i^{-1})\mathbf{x} + (\mu_i^T \Sigma_i^{-1} - \mu_j^T \Sigma_j^{-1})\mathbf{x} + C = 0, \qquad (4.10)$$

where C is a term independent of $\mathbf{x}$. When $\Sigma_j^{-1} \neq \Sigma_i^{-1}$, the decision surface is a quadratic curvilinear plane; otherwise, it is a hyperplane.

The QDF assumes that the covariance matrix of Gaussian density of each class is arbitrary. When a constraint is imposed such that

$$\Sigma_i = \Sigma, \qquad (4.11)$$

the QDF becomes

$$\begin{aligned}
g(\mathbf{x}, \omega_i) &= -\tfrac{1}{2}(\mathbf{x} - \mu_i)^T \Sigma^{-1}(\mathbf{x} - \mu_i) - \frac{1}{2} \log |\Sigma| + \log P(\omega_i) \\
&= -\tfrac{1}{2}(\mathbf{x}^T \Sigma^{-1}\mathbf{x} - 2\mu_i^T \Sigma^{-1}\mathbf{x} + \mu_i^T \Sigma^{-1}\mu_i) - \tfrac{1}{2} \log |\Sigma| + \log P(\omega_i),
\end{aligned} \qquad (4.12)$$

where $\mathbf{x}^T \Sigma^{-1}\mathbf{x}$ and $\log |\Sigma|$ are independent of class label, so eliminating them does not affect the decision of classification. As a result, the discriminant function becomes a linear discriminant function (LDF):

$$\begin{aligned}
g(\mathbf{x}, \omega_i) &= \mu_i^T \Sigma^{-1}\mathbf{x} - \frac{1}{2}\mu_i^T \Sigma^{-1}\mu_i + \log P(\omega_i) \\
&= \mathbf{w_i}^T\mathbf{x} + w_{i0},
\end{aligned} \qquad (4.13)$$

where $\mathbf{w_i} = \Sigma^T \mu_i$ and $w_{i0} = \log P(\omega_i) - \frac{1}{2}\mu_i^T \Sigma^{-1}\mu_i)$ are the weight vector and the bias of LDF, respectively.

The QDF can be further simplified by restricting the covariance matrix of each class to be diagonal with an identical variance:

$$\Sigma_i = \sigma^2 I, \tag{4.14}$$

where I is the identity matrix. Inserting Eq. (4.14) into Eq. (4.9) and omitting a class-independent term, the QDF becomes

$$g(\mathbf{x}, \omega_i) = -\frac{\|\mathbf{x} - \mu_i\|^2}{2\sigma^2} + \log P(\omega_i). \tag{4.15}$$

Further assuming equal a priori probabilities, the decision is equivalent to selecting the class of minimum Euclidean distance $\|\mathbf{x} - \mu_i\|$.

The minimum Euclidean distance rule was commonly used in character recognition before 1980s, under the title of template matching or feature matching. The LDF generally outperforms the Euclidean distance because it copes with the correlation between features. The QDF has many more free parameters than the LDF, and therefore needs a large number of samples to estimate the parameters. If there are not more than d training samples for a class, the covariance matrix will be singular and the inverse cannot be computed directly. On a finite sample set, the QDF does not necessarily generalize better than the LDF because of its higher freedom of parameters. The following section presents some strategies that can improve the generalization performance of QDF.

4.2.2.2 Improvements to QDF

A straightforward strategy to improve the generalization performance of QDF is to restrict the freedom of parameters. The LDF reduces the freedom by sharing a common covariance matrix for all classes, but it will not perform sufficiently when the covariance matrices of different classes differ significantly. The regularized discriminant analysis (RDA) of Friedman [21] does not reduce the number of parameters of QDF, but it constrains the range of parameter values by interpolating the class covariance matrix with the common covariance matrix and the identity matrix:

$$\hat{\Sigma}_i = (1 - \gamma)[(1 - \beta)\Sigma_i + \beta\Sigma_0] + \gamma\sigma_i^2 I, \tag{4.16}$$

where $\Sigma_0 = \sum_{i=1}^{M} P(\omega_i)\Sigma_i$, $\sigma_i^2 = \frac{1}{d}\text{tr}(\Sigma_i)$, and $0 < \beta, \gamma < 1$. With appropriate values of β and γ empirically selected, the RDA can always improve the generalization performance of QDF.

Another strategy, proposed by Kimura et al. [41], called modified quadratic discriminant function (MQDF), can both improve the generalization performance and reduce the computational complexity of QDF. Conforming to the notations of Kimura et al., the negative QDF with the assumption of equal a priori probabilities is rewritten as

$$g_0(\mathbf{x}, \omega_i) = (\mathbf{x} - \mu_i)^T \Sigma_i^{-1} (\mathbf{x} - \mu_i) + \log |\Sigma_i|$$

$$= [\Phi_i^T (\mathbf{x} - \mu_i)]^T \Lambda_i^{-1} \Phi_i^T (\mathbf{x} - \mu_i) + \log |\Lambda_i|$$

$$= \sum_{j=1}^d \frac{1}{\lambda_{ij}} [\phi_{ij}^T (\mathbf{x} - \mu_i)]^2 + \sum_{j=1}^d \log \lambda_{ij}, \tag{4.17}$$

where $\Lambda = \text{diag}[\lambda_{i1}, \ldots, \lambda_{id}]$ with $\lambda_{ij}, j = 1, \ldots, d$, being the eigenvalues (ordered in nonincreasing order) of Σ_i, and $\Phi_i = [\phi_{i1}, \ldots, \phi_{id}]$ with $\phi_{ij}, j = 1, \ldots, d$, being the ordered eigenvectors.

In Eq. (4.17), replacing the minor eigenvalues ($j > k$) with a constant δ_i, the MQDF is obtained as

$$g_2(\mathbf{x}, \omega_i)$$

$$= \sum_{j=1}^k \frac{1}{\lambda_{ij}} [\phi_{ij}^T (\mathbf{x} - \mu_i)]^2 + \sum_{j=k+1}^d \frac{1}{\delta_i} [\phi_{ij}^T (\mathbf{x} - \mu_i)]^2$$

$$+ \sum_{j=1}^k \log \lambda_{ij} + (d - k) \log \delta_i \tag{4.18}$$

$$= \sum_{j=1}^k \frac{1}{\lambda_{ij}} [\phi_{ij}^T (\mathbf{x} - \mu_i)]^2 + \frac{1}{\delta_i} \epsilon_i(\mathbf{x}) + \sum_{j=1}^k \log \lambda_{ij} + (d - k) \log \delta_i,$$

where k denotes the number of principal axes and $\epsilon_i(\mathbf{x})$ is the residual of subspace projection:

$$\epsilon_i(\mathbf{x}) = \|\mathbf{x} - \mu_i\|^2 - \sum_{j=1}^k [(\mathbf{x} - \mu_i)^T \phi_{ij}]^2 = \sum_{j=k+1}^d [(\mathbf{x} - \mu_i)^T \phi_{ij}]^2. \tag{4.19}$$

The motivation of replacing minor eigenvalues is that they are prone to be underestimated on small sample size, so replacing with a constant stabilizes the generalization performance. Comparing with the QDF, the MQDF involves only the principal eigenvectors and eigenvalues, so both the storage of parameters and the computation of discriminant function are reduced.

4.2.2.3 *Parameter Estimation* The density parameters of parametric classifiers are generally estimated by the ML method. Denote the set of parameters of a class ω_i by θ_i. Given a set of training samples $\mathcal{X}_i = \{\mathbf{x}^1, \ldots, \mathbf{x}^{N_i}\}$, the likelihood function is

$$\mathcal{L}_i(\theta_i) = p(\mathcal{X}_i | \theta_i) = \prod_{n=1}^{N_i} p(\mathbf{x}^n | \theta_i). \tag{4.20}$$

The parameters are estimated such that $\mathcal{L}_i(\theta_i)$ is maximized. Equivalently, the log-likelihood is maximized:

$$\max \mathcal{LL}_i = \log \mathcal{L}_i(\theta_i) = \sum_{n=1}^{N_i} \log p(\mathbf{x}^n | \theta_i). \tag{4.21}$$

In the case of Gaussian density, the density function of Eq. (4.8) is inserted into Eq. (4.21). By setting $\nabla \mathcal{LL}_i = 0$, many textbooks have shown that the solution of ML estimation of Gaussian parameters is

$$\hat{\mu}_i = \frac{1}{N_i} \sum_{n=1}^{N_i} \mathbf{x}^n, \tag{4.22}$$

$$\hat{\Sigma}_i = \frac{1}{N_i} \sum_{n=1}^{N_i} (\mathbf{x}^n - \hat{\mu}_i)(\mathbf{x}^n - \hat{\mu}_i)^T. \tag{4.23}$$

For multiclass classification with Gaussian density parameters estimated on a labeled training sample set $\mathcal{X} = \{\mathbf{x}^1, \ldots, \mathbf{x}^N\}$, $N = \sum_{i=1}^{M} N_i$, if the covariance matrices are arbitrary, the parameters of each class can be estimated independently by Eqs. (4.22) and (4.23) on the samples of one class. If a common covariance matrix is shared as for the LDF, the common matrix is estimated by

$$\hat{\Sigma}_0 = \frac{1}{N} \sum_{i=1}^{M} \sum_{\mathbf{x}^n \in \omega_i} (\mathbf{x}^n - \hat{\mu}_i)(\mathbf{x}^n - \hat{\mu}_i)^T = \sum_{i=1}^{M} \hat{P}(\omega_i) \hat{\Sigma}_i, \tag{4.24}$$

where $\hat{P}(\omega_i) = \frac{1}{N} \sum_{n=1}^{N} I(\mathbf{x}^n \in \omega_i)$ is the estimate of a priori probability.

As for the QDF, the parameters of MQDF are estimated class by class. First, the mean vector and covariance matrix of each class are estimated by ML. Then the eigenvectors and eigenvalues of covariance matrix are computed by K-L (Karhunen-Loeve) transform. On artificially selecting a number k, the parameter δ_i can be set to be either class-dependent or class-independent provided that it is comparable to the eigenvalue λ_{ik}. In a different context, Moghaddam and Pentland [63] showed that the ML estimate of class-dependent δ_i is the average of minor eigenvalues:

$$\delta_i = \frac{\text{tr}(\Sigma_i) - \sum_{j=1}^{k} \lambda_{ij}}{d - k} = \frac{1}{d - k} \sum_{j=k+1}^{d} \lambda_{ij}. \tag{4.25}$$

However, as the estimate of minor eigenvalues is sensitive to small sample size, this ML estimate of δ_i does not give good generalization performance.

4.2.2.4 Semiparametric Classifier
The Gaussian density function is a reasonable approximation of pattern distributions in many practical problems, especially

for character recognition, where the distribution of each class in feature space is almost unimodal. Nevertheless, for modeling multimodal or unimodal distributions that considerably deviate from Gaussian, the Gaussian function is insufficient. A natural extension is to use multiple Gaussian functions for modeling a distribution. This way, called mixture of Gaussians or Gaussian mixture, can approximate arbitrarily complicated distributions. As the density function still has a functional form, this method is referred to as semiparametric.

For pattern classification, the Gaussian mixture is used to model the distribution of each class independently. In the following description, we ignore the class label in the density function. Suppose that the conditional density of a class is modeled by a mixture of m Gaussian functions (components), the density function is

$$p(\mathbf{x}) = \sum_{j=1}^{m} P_j p(\mathbf{x}|j) \quad \text{subject to} \quad \sum_{j=1}^{m} P_j = 1, \tag{4.26}$$

where $p(\mathbf{x}|j) = p(\mathbf{x}|\mu_j, \Sigma_j)$ is a Gaussian function:

$$p(\mathbf{x}|j) = \frac{1}{(2\pi)^{d/2}|\Sigma_j|^{1/2}} \exp\left[-\frac{1}{2}(\mathbf{x} - \mu_j)^T \Sigma_j^{-1}(\mathbf{x} - \mu_j)\right]. \tag{4.27}$$

The parameters of Gaussian mixture, $\{(P_j, \mu_j, \Sigma_j)|j = 1, \ldots, m\}$, are estimated by maximizing the likelihood on a training sample set $\mathcal{X} = \{\mathbf{x}^1, \ldots, \mathbf{x}^N\}$:

$$\max \mathcal{L} = \prod_{n=1}^{N} p(\mathbf{x}^n) = \prod_{n=1}^{N}\left[\sum_{j=1}^{m} P_j p(\mathbf{x}^n|j)\right]. \tag{4.28}$$

The maximum of this likelihood, or equivalently the log-likelihood, cannot be solved analytically because the hidden values of component labels are not available. An iterative algorithm, called expectation–maximization (EM) [11], was proposed to solve this kind of optimization problems with hidden values. Although the details of EM algorithm can be found in many references and textbooks, we give an outline in the following.

Given the number of components m, initial values of parameters $(P_j^{(0)}, \mu_j^{(0)}, \Sigma_j^{(0)})$, $j = 1, \ldots, m$, the EM algorithm updates the parameters iteratively until the parameter values do not change or a given number of iterations is finished. Each iteration (at time t) has an E-step and an M-step:

- E-step: The probabilities that each sample belongs to the components are computed based on the current parameter values:

$$P^{(t)}(j|\mathbf{x}^n) = \frac{P_j^{(t)} p(\mathbf{x}^n|\mu_j^{(t)}, \Sigma_j^{(t)})}{\sum_{i=1}^{m} P_i^{(t)} p(\mathbf{x}^n|\mu_i^{(t)}, \Sigma_i^{(t)})}, \quad n = 1, \ldots, m. \tag{4.29}$$

- *M*-step: The likelihood function is maximized based on the current component probabilities, such that the parameters are updated:

$$\mu_j^{(t+1)} = \frac{\sum_n P^{(t)}(j|\mathbf{x}^n)\mathbf{x}^n}{\sum_n P^{(t)}(j|\mathbf{x}^n)}, \tag{4.30}$$

$$\Sigma_j^{(t+1)} = \frac{\sum_n P^{(t)}(j|\mathbf{x}^n)(\mathbf{x}^n - \mu_j^{(t+1)})(\mathbf{x}^n - \mu_j^{(t+1)})^T}{\sum_n P^{(t)}(j|\mathbf{x}^n)}, \tag{4.31}$$

$$P_j^{(t+1)} = \frac{1}{N}\sum_n P^{(t)}(j|\mathbf{x}^n). \tag{4.32}$$

The Gaussian mixture model is prone to overfit the training data because it has more parameters than the multivariate Gaussian. A way to alleviate the overfitting and reduce the computational complexity is to use diagonal or even identity covariance matrices. Another way is to model mixture density in a low-dimensional subspace and combine the mixture density in the principal subspace and a unitmodal Gaussian in the residual subspace [63].

4.2.3 Nonparametric Methods

Nonparametric classifiers do not assume any functional form for the conditional distributions. Instead, the density can have arbitrary shape depending on the training data points. A simple way is to model the density using histograms, that is, partitioning the feature space into bins or grids and count the data points in each bin or grid. This is feasible for only very low-dimensional feature spaces. For high-dimensional space, the number of grids is exponential and the training data becomes sparse.

A general approach to density estimation is based on the definition of probability density: the frequency of data points in unit volume of feature space. Assume that the density $p(\mathbf{x})$ is continuous and does not vary considerably in a local region $\mathcal{R}$, the probability that a random vector $\mathbf{x}$, governed by the distribution of $p(\mathbf{x})$, fall in the region $\mathcal{R}$ can be approximated by

$$P = \int_{\mathcal{R}} p(\mathbf{x}')d\mathbf{x}' \approx p(\mathbf{x})V, \tag{4.33}$$

where V is the volume of $\mathcal{R}$. On the other hand, if there are N data points (N is sufficiently large) and k of them fall in the local region $\mathcal{R}$ around $\mathbf{x}$, the probability can be approximated by

$$P \approx \frac{k}{N}. \tag{4.34}$$

Combining Eqs. (4.33) and (4.34) results in

$$p(\mathbf{x}) \approx \frac{k}{NV}. \tag{4.35}$$

There are basically two approaches in applying formula (4.35) to practical density estimation. One is to fix the volume V and determine the value of k from the data, and another is to fix the value of k and determine the volume V from the data. The two approaches lead to Parzen window (or called kernel-based) method and k-nearest neighbor (k-NN) method, respectively.

4.2.3.1 Parzen Window Method
The Parzen window method assumes that the local region around data point $\mathbf{x}$ is a hypercube with edge length h, then the volume of region is $V = h^d$. The window function of unit hyercube is defined by

$$H(\mathbf{u}) = \begin{cases} 1, & |u_i| < 1/2, \ i = 1, \ldots, d \\ 0, & \text{otherwise.} \end{cases} \tag{4.36}$$

Given a set of training data points $\mathcal{X} = \{\mathbf{x}^1, \ldots, \mathbf{x}^N\}$, the number of points falling in the window around a new vector $\mathbf{x}$ is

$$k = \sum_{n=1}^{N} H\left(\frac{\mathbf{x} - \mathbf{x}^n}{h}\right). \tag{4.37}$$

Substituting this number into Eq. (4.35) leads to the density estimate around $\mathbf{x}$:

$$\hat{p}(\mathbf{x}) = \frac{1}{N} \sum_{n=1}^{N} \frac{1}{h^d} H\left(\frac{\mathbf{x} - \mathbf{x}^n}{h}\right). \tag{4.38}$$

It is shown in textbooks [17, 4] that the expectation of the estimated density in Eq. (4.38) is a smoothed version of the true density:

$$\begin{aligned} \mathcal{E}[\hat{p}(\mathbf{x})] &= \frac{1}{N} \sum_{n=1}^{N} \mathcal{E}\left[\frac{1}{h^d} H\left(\frac{\mathbf{x} - \mathbf{x}^n}{h}\right)\right] \\ &= \mathcal{E}\left[\frac{1}{h^d} H\left(\frac{\mathbf{x} - \mathbf{x}'}{h}\right)\right] \\ &= \int \frac{1}{h^d} H\left(\frac{\mathbf{x} - \mathbf{x}'}{h}\right) p(\mathbf{x}') d\mathbf{x}'. \end{aligned} \tag{4.39}$$

The kernel width h plays the role of a smoothing parameter and should be carefully selected to compromise the bias and variance. On finite training sample size, a small value of h leads to an overfitting of training data, whereas a large value leads to a biased estimation of density.

The window function can be extended to other forms of function provided that $H(\mathbf{x}) \geq 0$ and $\int H(\mathbf{x})d\mathbf{x} = 1$ are satisfied. The standard Gaussian function is a natural choice of window function of infinite support:

$$H(\mathbf{u}) = \frac{1}{(2\pi)^{d/2}} \exp\left(-\frac{\|\mathbf{u}\|^2}{2}\right). \tag{4.40}$$

Accordingly, the density estimate of (4.38) is

$$\hat{p}(\mathbf{x}) = \frac{1}{N} \sum_{n=1}^{N} \frac{1}{(2\pi h^2)^{d/2}} \exp\left(-\frac{\|\mathbf{x} - \mathbf{x}^n\|^2}{2h^2}\right). \tag{4.41}$$

For classification, the formula (4.38) or (4.41) can be used to estimate the conditional density around $\mathbf{x}$ for each class, and then the Bayes decision rule is applied. The value of h can be selected by cross-validation such that the selected value leads to a high classification accuracy on a validation data set disjoint from the training set.

A serious problem with the Parzen window method is that all the training samples should be stored for density estimation and classification. A way to alleviate the storage and computation overhead is to model the density using a small number of kernel functions centered at selected points. This way is similar to the semiparametric density estimation method, and if the Gaussian window function is used, the center points can be estimated by the EM algorithm.

4.2.3.2 K-Nearest Neighbor Method

In the k-NN method, the volume of local region $\mathcal{R}$ for density estimation is variable whereas the number k of data points falling in $\mathcal{R}$ is fixed. Given N training data points from M classes, for estimating the local density around a new vector $\mathbf{x}$, a window centered at $\mathbf{x}$ is enlarged such that exactly k nearest data points fall in the window. Assume that the k nearest neighbors include k_i points from class $\omega_i, i = 1, \ldots, M, \sum_{i=1}^{M} k_i = k$, the class-conditional density around $\mathbf{x}$ is estimated by

$$\hat{p}(\mathbf{x}|\omega_i) = \frac{k_i}{N_i V}, \quad i = 1, \ldots, M, \tag{4.42}$$

where N_i is the number of training points of class ω_i. The a posteriori probabilities are computed by the Bayes formula:

$$p(\omega_i|\mathbf{x}) = \frac{P(\omega_i)\hat{p}(\mathbf{x}|\omega_i)}{\sum_{j=1}^{M} P(\omega_j)\hat{p}(\mathbf{x}|\omega_j)} = \frac{P(\omega_i)k_i/N_i}{\sum_{j=1}^{M} P(\omega_j)k_j/N_j}, \quad i = 1, \ldots, M. \tag{4.43}$$

Considering $\hat{P}(\omega_i) = \frac{N_i}{N}$, the computation of a posteriori probabilities is simplified to

$$p(\omega_i|\mathbf{x}) = \frac{k_i}{k}, \quad i = 1, \ldots, M. \tag{4.44}$$

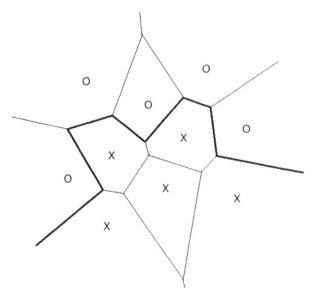

FIGURE 4.1 Voronoi diagram formed by training samples of two classes. The decision surface is denoted by thick lines.

This is to say, the a posteriori probability of a class is simply its fraction of k nearest neighbors in training samples, and the decision is to select the class with most nearest neighbors among k. This decision rule is called k-NN rule.

When k is set equal to one, the k-NN rule is reduced to the nearest neighbor (1-NN) rule that classifies the input pattern **x** to the class of the nearest training sample. When the distance measure is the Euclidean distance, the decision regions of the 1-NN rule in feature space form a Voronoi diagram (Voronoi tessellation), with each cell defined by a training sample and separated from each other by a hyperplane that is equidistant from two samples. An example of Voronoi diagram is shown in Figure 4.1.

It has been shown in textbook [17] that when the number of training samples approaches infinity, the classification error rate of 1-NN rule is bounded by two times the Bayes error rate. More exactly, denoting the Bayes error rate by P^*, the error rate of 1-NN rule is bounded by

$$P^* \leq P \leq P^* \left(2 - \frac{M}{M-1} P^*\right). \tag{4.45}$$

As for the Parzen window method, the drawback of k-NN and 1-NN methods is that all the training samples are required to be stored and matched in classification. Both the Parzen window and nearest neighbor methods are not practical for real-time applications, but because of their fairly high classification performance, they serve as good benchmarks for evaluating other classifiers. There exist many efforts aiming to reduce the storage and computation overhead of nearest neighbor methods.

Significant savings of storage and computation can be obtained by wisely selecting or synthesizing prototypes from training samples while maintaining or improving the classification performance. A supervised prototype learning method, called learning vector quantization (LVQ) [43], is often described in connection with ANNs, as addressed in the next section.

4.3 ARTIFICIAL NEURAL NETWORKS

ANNs were initially studied with the hope of making intelligent perception and cognition machines by simulating the physical structure of human brains. The principles and algorithms of ANNs have found numerous applications in diverse fields including pattern recognition and signal processing. A neural network is composed of a number of interconnected neurons, and the manner of interconnection differentiates the network models into feedforward networks, recurrent networks, self-organizing networks, and so on. This section focuses on the neural network models that have been widely applied to pattern recognition. We describe the network structures and learning algorithms but will not go into the details of the theoretical foundation. For a comprehensive description of neural networks, the textbook of Haykin [29] can be referred.

In neural networks, a neuron is also called a unit or a node. A neuronal model has a set of connecting weights (corresponding to synapses in biological neurons), a summing unit, and an activation function, as shown in Figure 4.2. The output of the summing unit is a weighted combination of input signals (features):

$$v = \sum_{i=1}^{d} w_i x_i + b. \tag{4.46}$$

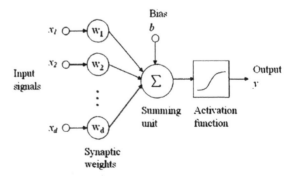

FIGURE 4.2 Model of a neuron.

The activation function can be either linear or nonlinear. A popular nonlinear activation function is the sigmoid or the logistic function:

$$g(a) = \frac{1}{1 + e^{-a}}. \tag{4.47}$$

The sigmoid function has two merits: (1) it is continuous and derivable such that gradient learning involving it is possible; (2) the value of sigmoid function acts like a posterior probability, which facilitates contextual integration of neuronal outputs. The derivative of sigmoid function is

$$g'(a) = g(a)[1 - g(a)]. \tag{4.48}$$

The probability-like feature of the sigmoid function can be seen from the a posteriori probability of a two-class Gaussian classifier, which is shown to be sigmoidal. Assume that two classes have Gaussian density functions as Eq. (4.8) with a common covariance matrix Σ, the a posteriori probability is

$$
\begin{aligned}
P(\omega_1|\mathbf{x}) &= \frac{P(\omega_1)p(\mathbf{x}|\omega_1)}{P(\omega_1)p(\mathbf{x}|\omega_1) + P(\omega_2)p(\mathbf{x}|\omega_2)} \\
&= \frac{P(\omega_1)\exp[-\frac{1}{2}(\mathbf{x}-\mu_1)^T\Sigma^{-1}(\mathbf{x}-\mu_1)]}{\begin{array}{l}P(\omega_1)\exp[-\frac{1}{2}(\mathbf{x}-\mu_1)^T\Sigma^{-1}(\mathbf{x}-\mu_1)]\\ \quad + P(\omega_2)\exp[-\frac{1}{2}(\mathbf{x}-\mu_2)^T\Sigma^{-1}(\mathbf{x}-\mu_2)]\end{array}} \\
&= \frac{1}{1 + \dfrac{P(\omega_2)}{P(\omega_1)}e^{-[(\mu_1-\mu_2)^T\Sigma^{-1}\mathbf{x}+C]}} \\
&= \frac{1}{1 + e^{-[\mathbf{w}^T\mathbf{x}+b]}},
\end{aligned} \tag{4.49}
$$

where C is a constant independent of $\mathbf{x}$, $\mathbf{w} = \Sigma^{-1}(\mu_1 - \mu_2)$, and $b = C + \log\frac{P(\omega_1)}{P(\omega_2)}$.

It can be seen from Eq. (4.46) that a single neuron performs as a linear discriminator. A network with multiple neurons interconnected in a sophisticated manner can approximate complicated nonlinear discriminant functions. Feedforward networks, including single-layer networks, multilayer networks, radial basis function (RBF) networks, and higher order networks, are straightforward for approximating discriminant functions via supervised learning. We will introduce the basic principle of supervised learning in the context of single-layer networks and then extend to other network structures. Unsupervised learning algorithms, including competitive learning and self-organizing map (SOM), can find the cluster description of a data set without class labels. They can be applied to classification by finding a description for each class. We will also describe a hybrid learning algorithm, learning vector quantization (LVQ), that adjust cluster centers in supervised learning.

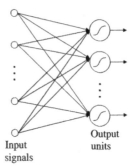

FIGURE 4.3   Single-layer neural network. The bias terms are not shown.

4.3.1   Single-Layer Neural Network

The structure of a single-layer neural network (SLNN) for multiclass classification is shown in Figure 4.3. The input signals (features) are connected to all the output neurons, each corresponding to a pattern class. In the case of linear activation function, the output of each neuron is a linear discriminant function:

$$y_k(\mathbf{x}) = v_k = \sum_{i=1}^{d} w_{ki} x_i + w_{k0} = \sum_{i=0}^{d} w_{ki} x_i = \mathbf{w}_k^T \mathbf{x}', \quad k = 1, \ldots, M, \quad (4.50)$$

where w_{k0} is a bias term, which can be viewed as the weight to a fixed signal $x_0 = 1$. The weights of each neuron form a $(d + 1)$-dimensional weight vector $\mathbf{w}_k$, and $\mathbf{x}'$ is an enhanced $(d + 1)$-dimensional vector. The input pattern is classified to the class of maximum output. Using sigmoid activation function (which is monotonic) does not affect the decision of classification because the order of output values remains unchanged. However, the sigmoid function can be explained as a posteriori probability and can improve the performance of weight learning.

The connecting weights of SLNN are estimated in supervised learning from a labeled training sample set, with the aim of minimizing the classification error or an alternative objective. The learning algorithms include the perceptron algorithm, pseudo inverse, gradient descent, and so on. The perceptron algorithm feeds the training samples iteratively and adjusts the weights in a fixed step whenever a sample is misclassified. It aims to minimize the classification error on training samples but cannot guarantee convergence when the samples are not linearly separable. The pseudo inverse and gradient descent methods are described below.

4.3.1.1 Pseudo Inverse   The pseudo inverse is a noniterative method that estimates the weights of single-layer network with linear output units (linear activation functions). Given a training sample set $\mathcal{X} = \{(\mathbf{x}^n, c^n) | n = 1, \ldots, N\}$ (c^n denotes the class label of sample $\mathbf{x}^n$), the output values of each sample, $y_k(\mathbf{x}^n)$, $k = 1, \ldots, M$, are computed by Eq. (4.50). The desired outputs (target values) of a labeled sample

$(\mathbf{x}^n, c^n)$ are set to

$$
t_k^n = \begin{cases} 1, & k = c^n, \\ -1, & \text{otherwise.} \end{cases} \tag{4.51}
$$

The sum of squared error on the training set is

$$
E = \frac{1}{2} \sum_{n=1}^{N} \sum_{k=1}^{M} [y_k(\mathbf{x}^n) - t_k^n]^2. \tag{4.52}
$$

The weights of network are optimized with the aim of minimizing the sum of squared error.

Representing all the weight vectors in a $M \times (d+1)$ matrix W (each row corresponding to a class) and all the data (enhanced feature vectors) in a $N \times (d+1)$ matrix X (each row corresponding to a sample), the outputs of all samples can be represented in a $N \times M$ matrix Y (each row contains the output values of a sample):

$$
Y = XW^T. \tag{4.53}
$$

Representing the target values of all samples in a $N \times M$ matrix T (each row composed of the target values of a sample), the sum of squared error of Eq. (4.52) can be rewritten as

$$
E = \frac{1}{2} \| XW^T - T \|^2. \tag{4.54}
$$

For minimizing E, take

$$
\frac{\partial E}{\partial W} = 0, \tag{4.55}
$$

which results in

$$
W^T = (X^T X)^{-1} X^T T = X^\dagger T, \tag{4.56}
$$

where $X^\dagger = (X^T X)^{-1} X^T$ is the pseudo inverse of X.

The advantage of pseudo inverse is that the weights can be solved analytically. The disadvantage is that the objective function is based on linear activation function of neurons, which does not explain a posteriori probability and cannot fit the 0–1 target values very well, though the objective function is minimized globally. Hence, the weights thus estimated do not necessarily give high classification performance.

4.3.1.2 *Gradient Descent* If the output neurons of SLNN take the sigmoid activation function:

$$y_k(\mathbf{x}) = g(v_k) = \frac{1}{1 + e^{-\mathbf{w}_k^T \mathbf{x}'}}, \quad k = 1, \ldots, M, \tag{4.57}$$

and accordingly, the desired outputs are set equal to

$$t_k^n = \begin{cases} 1, & k = c^n, \\ 0, & \text{otherwise,} \end{cases} \tag{4.58}$$

then the object function of Eq. (4.52) is a nonlinear function of weights and can no longer be minimized by analytical solution of linear equations. Nonlinear optimization problems can be solved generally by gradient search, provided that the objective function is derivable with respect to the free parameters. For the present problem of minimizing (4.52), the weights and biases are adjusted iteratively by searching along the negative direction of gradient in the space of parameters:

$$\mathbf{w}_k(t + 1) = \mathbf{w}_k(t) - \eta(t)\frac{\partial E}{\partial \mathbf{w}_k}, \quad k = 1, \ldots, M, \tag{4.59}$$

where $\eta(t)$ is the learning step, or called learning rate. The partial derivatives are computed by

$$\frac{\partial E}{\partial \mathbf{w}_k} = \sum_{n=1}^{N}\sum_{k=1}^{M}[y_k(\mathbf{x}^n) - t_k^n]y_k(\mathbf{x}^n)[1 - y_k(\mathbf{x}^n)]\mathbf{x}'^n, \quad k = 1, \ldots, M, \tag{4.60}$$

where $\mathbf{x}'^n$ is the enhanced vector of $\mathbf{x}^n$.

The complete gradient learning procedure is as follows. First, the parameters are initialized to be small random values, $w_{ki}(0)$ and $w_{k0}(0)$. Then the training samples are fed to the network iteratively. Once all the samples have been fed, compute the partial derivatives according to Eq. (4.60) and update the parameters according to Eq. (4.59). This cycle of feeding samples and updating parameters is repeated until the objective function does not change.

The updating rule of Eq. (4.59) updates the parameter values only once in a cycle of feeding all samples. Thus, the samples need to be fed for many cycles until the parameters converge. The convergence of learning can be accelerated by stochastic approximation [72], which updates the parameters each time a sample is fed. Stochastic approximation was originally proposed to find the root of univariate functional, and then extended to the optimization of univariate and multivariate objective functions by finding the roots of derivatives. By stochastic approximation, the squared error is calculated on each input pattern:

$$E^n = \frac{1}{2} \sum_{k=1}^{M} [y_k(\mathbf{x}^n) - t_k^n]^2. \tag{4.61}$$

Accordingly, the partial derivatives are computed by

$$\frac{\partial E^n}{\partial \mathbf{w}_k} = \sum_{k=1}^{M} [y_k(\mathbf{x}^n) - t_k^n] y_k(\mathbf{x}^n)[1 - y_k(\mathbf{x}^n)] \mathbf{x}'^n. \tag{4.62}$$

The parameters are then updated on each input pattern by

$$\mathbf{w}_k(t+1) = \mathbf{w}_k(t) - \eta(t)\frac{\partial E^n}{\partial \mathbf{w}_k} = \mathbf{w}_k(t) + \eta(t)\delta_k(\mathbf{x}^n)\mathbf{x}'^n, \tag{4.63}$$

where $\delta_k(\mathbf{x}^n) = [t_k^n - y_k(\mathbf{x}^n)]y_k(\mathbf{x}^n)[1 - y_k(\mathbf{x}^n)]$ is a generalized error signal.

It has been proved that the stochastic approximation algorithm converges to a local minimum of E (where $\nabla E = 0$) with probability 1 under the following conditions [72, 23]:

$$\begin{cases} \lim_{t\to\infty} \eta(t) = 0 \\ \sum_{t=1}^{\infty} \eta(t) = \infty \\ \sum_{t=1}^{\infty} \eta(t)^2 < \infty. \end{cases}$$

A choice of $\eta(t) = 1/t$ satisfies the above conditions. In practice, the parameters are updated in finite sweeps of training samples, and generally, setting the learning rate as a sequence starting with a small value and vanishing gradually leads to convergence if the objective function is continuously derivable.

Besides the squared error criterion, another criterion often used in neural network learning is the cross-entropy (CE):

$$CE = -\sum_{n=1}^{N} \sum_{k=1}^{M} \left\{ t_k^n \log y_k(\mathbf{x}^n) + (1 - t_k^n) \log[1 - y_k(\mathbf{x}^n)] \right\}, \tag{4.64}$$

which can be minimized as well by gradient descent for learning the network weights. Some researchers have proved that the minimization of either squared error or cross-entropy criterion leads to the estimation of Bayesian a posteriori probabilities by the output units of neural networks [71]. This is a desirable property for pattern recognition tasks.

The gradient descent optimization method with stochastic approximation, also referred to as stochastic gradient descent, can be used to estimate the parameters of any other classifier structures provided that the objective function is derivable with respect to the parameters. We will extend it to the training of multilayer and higher-order networks under the objective of squared error.

4.3.2 Multilayer Perceptron

The neuronal output of SLNN performs as a linear discriminant function (the activation function does not affect classification). No matter how the weights are learned, the SLNN provides a linear hyperplane decision boundary between any two classes. Hence, the SLNN is unable to separate classes with complicated distributions, for which the decision boundary is generally nonlinear. A way to enhance the separating ability is to use nonlinear functions of pattern features as the inputs of a linear combiner such that the linear combination, called generalized linear discriminant function (generalized LDF), is a nonlinear function of the original features. The intermediate nonlinear functions to combine can be either predefined or adaptively formed.

The neural network classifiers that we are going to describe hereof, namely, multilayer perceptron (MLP), RBF network, and higher order (polynomial) network, all fall in the range of generalized LDF. The intermediate functions of MLP are adaptive, those of higher order network are fixed, and those of RBF network can be either fixed or adaptive.

The MLP has an output layer of linear or sigmoidal neurons and one or more hidden layers of sigmoidal neurons. The sigmoid activation function at the output layer does not affect the decision of classification, but it helps in gradient descent learning of weights. Disregarding the sigmoid function at the output layer, the output values of MLP can be viewed as generalized LDFs because the outputs of hidden neurons are nonlinear. An example of MLP with one hidden layer (in total two layers) is shown in Figure 4.4.

Assume that a two-layer MLP has d input signals, m hidden units, and M output units, denoting the weights of the jth hidden unit by w_{ji}, its output as h_j, and the

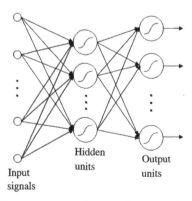

FIGURE 4.4 Multilayer perceptron. For simplicity, the bias terms are not shown.

weights of the kth output unit by w_{kj}, the output values are computed by

$$
\begin{aligned}
y_k(\mathbf{x}) = g[v_k(\mathbf{x})] &= g\left[\sum_{j=1}^{m} w_{kj}h_j + w_{k0}\right] \\
&= g\left[\sum_{j=1}^{m} w_{kj}g\left(\sum_{i=1}^{d} w_{ji}x_i + w_{j0}\right) + w_{k0}\right] \\
&= g\left[\sum_{j=1}^{m} w_{kj}g[u_j(\mathbf{x})] + w_{k0}\right], \quad k = 1, \ldots, M,
\end{aligned}
\tag{4.65}
$$

where $g(\cdot)$ is the sigmoid activation function, $v_k(\mathbf{x}) = \sum_{j=1}^{m} w_{kj}h_j + w_{k0}$, and $u_j(\mathbf{x}) = \sum_{i=1}^{d} w_{ji}x_i + w_{j0}$.

4.3.2.1 *Back-Propagation Learning*

The weights of MLP are generally trained on a labeled sample set to minimize the squared error, as in Eq. (4.52). This learning algorithm is commonly referred to as back propagation (BP). Using stochastic gradient descent, the samples are fed iteratively and the weights are adjusted on each input pattern:

$$
\begin{aligned}
w_i(t+1) &= w_i(t) + \Delta w_i(t), \\
\Delta w_i(t) &= -\eta(t)\frac{\partial E^n}{\partial w_i},
\end{aligned}
\tag{4.66}
$$

where w_i represents any of the weights or biases. The problem now amounts to computing the partial derivatives of E^n with respect to adjustable weights:

$$
\begin{aligned}
\frac{\partial E^n}{\partial w_i} &= \sum_{k=1}^{M}[y_k(\mathbf{x}^n) - t_k^n]\frac{\partial y_k(\mathbf{x}^n)}{\partial w_i} \\
&= \sum_{k=1}^{M}[y_k(\mathbf{x}^n) - t_k^n]y_k(\mathbf{x}^n)[1 - y_k(\mathbf{x}^n)]\frac{\partial v_k(\mathbf{x}^n)}{\partial w_i} \\
&= -\sum_{k=1}^{M}\delta_k(\mathbf{x}^n)\frac{\partial v_k(\mathbf{x}^n)}{\partial w_i}.
\end{aligned}
\tag{4.67}
$$

Specific to different weight terms, $\frac{\partial v_k(\mathbf{x}^n)}{\partial w_i}$ is computed by (denoting $h_0 = 1$ and $x_0 = 1$)

$$
\frac{\partial v_k(\mathbf{x})}{\partial w_{kj}} = h_j, \quad j = 0, \ldots, m,
\tag{4.68}
$$

and

$$
\frac{\partial v_k(\mathbf{x})}{\partial w_{ji}} = w_{kj}\frac{\partial h_j}{\partial w_{ji}} = w_{kj}h_j(\mathbf{x})[1 - h_j(\mathbf{x})]x_i,
\tag{4.69}
$$
$$
j = 1, \ldots, m, \ i = 0, \ldots, d.
$$

Combining Eqs. (4.66)–(4.69), the increments of weights on a training pattern **x** are

$$\Delta w_{kj}(t) = \eta(t)\delta_k(\mathbf{x})h_j, \quad k = 1, \ldots, M, \; j = 0, \ldots, m,$$
$$\Delta w_{ji}(t) = \eta(t)\sum_{k=1}^{M} w_{kj}\delta_k(\mathbf{x})h_j(\mathbf{x})[1 - h_j(\mathbf{x})]x_i, \qquad (4.70)$$
$$= \eta(t)\delta_j(\mathbf{x})x_i, \quad j = 1, \ldots, m, \; i = 0, \ldots, d,$$

where $\delta_j(\mathbf{x}) = \sum_{k=1}^{M} w_{kj}\delta_k(\mathbf{x})h_j(\mathbf{x})[1 - h_j(\mathbf{x})]$ is the error signal back-propagated from the output layer to the hidden layer. For a network with more than one hidden layer, the error signal can be similarly propagated to every hidden layer, and the increments of weights of hidden layers can be computed accordingly.

4.3.2.2 Acceleration to BP
The gradient descent learning algorithm guarantees that the weights converge to a local minimum of squared error criterion. In the parameter space, however, the surface of the squared error of MLP is complicated. It may have many local minima and plateaus. A way to overcome local minima is to train the network multiple times with different initial weight sets, which are expected to converge to different local minima, and then select the trained weights that have the lowest squared error. At a point on the plateau, the weights cannot move because the gradient of error surface is zero. An effective method to get rid of the plateau is to use a momentum in weight updating, which also results in skipping insignificant local minima.

The motivation of using momentum is to smooth the current increment of weights to the past increments such that in the parameter space, the weights move partially along the direction of past increments. By doing so, even if the gradient at current weight values is zero, the weights still move along past directions. With a momentum coefficient $0 < \gamma < 1$, the increment of a weight term is smoothed to

$$\Delta w_i(t) = -\eta(t)\frac{\partial E^n}{\partial w_i} + \gamma\Delta w_i(t-1). \qquad (4.71)$$

The stochastic gradient descent needs many iterations of feeding training samples and updating weights to guarantee converging to a local minimum. The number of iterations can be reduced significantly by using second-order derivatives, as in Newton's method:

$$\mathbf{w}(t+1) = \mathbf{w}(t) - \eta(t)H^{-1}\nabla E, \qquad (4.72)$$

where H is the Hessian matrix, composed of the second-order derivatives of error function with respect to the weight parameters:

$$H_{ij} = \frac{\partial^2 E}{\partial w_i \partial w_j}. \qquad (4.73)$$

The second-order derivative reflects the curvature of error surface at a point of weight values. The vector defined by $-H^{-1}\nabla E$ is called Newton direction. When the error function is quadratic, the Newton direction points exactly to the minimum in the parameter space. Assuming that the error function is quadratic in a local region, updating according to Eq. (4.72) can draw the weights closer to the local minimum of error surface than the negative gradient direction.

Though the Newton's method can reduce the number of iterations for converging, the computation of the Hessian matrix is expensive. Approximating the Hessian matrix to diagonal can significantly simplify computation and the convergence can still be accelerated to some extent. By doing this, only the diagonal elements of Hessian, $\frac{\partial^2 E}{\partial w_i^2}$, need to be computed, and the inversion of diagonal matrix is trivial. To compute the increments of weights, a constant λ is added for overcoming zero curvature:

$$\Delta w_i = -\eta \left(\frac{\partial^2 E}{\partial w_i^2} + \lambda \right)^{-1} \frac{\partial E}{\partial w_i}. \tag{4.74}$$

In this formula, $\eta \left(\frac{\partial^2 E}{\partial w_i^2} + \lambda \right)^{-1}$ can be viewed as a learning rate adaptive to weight terms. The adaptation of learning rate is important for accelerating the learning of weights or parameters that have appreciably different magnitude of second-order derivatives. This situation is not obvious for the weights of MLP, however.

Mostly, BP (stochastic gradient descent) learning with momentum works satisfactorily for MLP if the training samples are fed in a finite number of cycles and the learning rate gradually vanishes.

4.3.2.3 Discussions
Many papers have discussed the approximation ability of two-layer (one hidden layer) and three-layer (two hidden layer) MLPs. It has been shown that a three-layer network can map any continuous function exactly from d input variables to an output variable; and for classification, a two-layer network can approximate any decision boundary to arbitrary accuracy [4]. For a mapping or classification problem, the number of hidden nodes reflects the complexity and approximation ability of the network.

For practical pattern recognition problems, the complexity of neural network should be tuned to an appropriate size so as to generalize well to unseen new data. The appropriate size depends on the inherent separability of pattern classes and the number of training samples. A simple network will not be able to approximate the training data to sufficient accuracy and hence yields a large bias, whereas a complex network will fit the training data to an unnecessarily high accuracy and yields large variance to noisy data. The bias and variance should be suitably compromised for good generalization performance.

There are several ways to control the complexity of network or the freedom of weights for good generalization. A straightforward way is to try networks of different sizes and select the network that performs best on a validation data set disjoint from the training set. This strategy is called cross-validation. The validation set can be a holdout set extracted from the whole data set. Alternatively, the data set is partitioned into a number of subsets, and each subset is used for validation in rotation with the network trained on the union of the remaining subsets. This is called rotation method or leave-one-out (LOO).

The second strategy of tuning network complexity is to construct the network dynamically by adding or removing nodes until the network approximates the training data to an appropriate precision. Removing nodes, also referred to as pruning, is often adopted to reduce the network complexity to avoid overfitting.

An alternative way for avoiding overfitting is to constrain the freedom of weights instead of tuning the number of nodes. This can be achieved by two methods: weight sharing and weight decay. By binding the weights for some nodes, the number of free weights is reduced. Weight decay is to constrain the abstract values of weights. It is equivalent to minimizing a regularized error function:

$$E = \sum_{n=1}^{N} E^n + \frac{\beta}{2} \sum_i w_i^2, \qquad (4.75)$$

where $\beta > 0$ is a coefficient of regularization, and w_i is a connecting weight (not including the bias terms). In weight updating by stochastic gradient descent, the increment of weight is replaced by

$$\Delta w_i(t) = -\eta(t) \left(\frac{\partial E^n}{\partial w_i} + \frac{\beta}{N} w_i \right). \qquad (4.76)$$

The regularized squared error with weight decay in Eq. Eq. (4.75) can be applied to other networks (SLNN, RBF network, polynomial network, etc.) as well, and the weights are updated similarly.

The MLP has been frequently used in character recognition experiments and applications, and it is often used as a benchmark in pattern recognition research. In the character recognition field, some modifications to the structure of MLP have led to special success. In a general scenario, a number of predefined discriminating features are extracted from character patterns and input to a network for classification. In this way, a fully connected two-layer MLP performs fairly well. Using a modular network, with a subnet for each class, can further improve the accuracy [65]. A network working directly on character images, called convolutional neural network, has reported great success [51]. It has multiple hidden layers with local connection and shared weights. The hidden nodes, each connected to a local receptive field in the preceding layer, can be viewed as trainable feature extractors.

4.3.3 Radial Basis Function Network

The RBF network has one hidden layer, with each hidden node performing a localized nonlinear function, mostly a Gaussian function. The response values of basis functions are linearly combined by the output nodes. The activation function of output nodes can be either linear or sigmoidal. To facilitate minimum error training by gradient descent

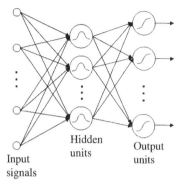

FIGURE 4.5   Radial basis function network. The bias terms of output nodes are not shown.

and make the output values approximate a posteriori probabilities, we use sigmoid functions in the output layer. A diagram of RBF network is shown in Figure 4.5.

For an RBF network with d input signals, m hidden units, and M output units, assume spherical Gaussian functions for the hidden nodes

$$\phi_j(\mathbf{x}) = \exp\left(-\frac{\|\mathbf{x} - \mu_j\|^2}{2\sigma_j^2}\right), \quad j = 1, \ldots, m, \tag{4.77}$$

where μ_j is the center of Gaussian, and σ_j^2 is the variance. The linear combination of Gaussian functions

$$y_k(\mathbf{x}) = v_k(\mathbf{x}) = \sum_{j=1}^{m} w_{kj}\phi_j(\mathbf{x}) + w_{k0} = \sum_{j=0}^{m} w_{kj}\phi_j(\mathbf{x}), \quad k = 1, \ldots, M \tag{4.78}$$

(denoting $\phi_0(\mathbf{x}) = 1$) can be viewed as generalized linear discriminant functions.

On a training sample set $\mathcal{X} = \{(\mathbf{x}^n, c^n)|n = 1, \ldots, N\}$, the parameters of RBF network (center parameters and weights) can be estimated by minimizing the sum of squared error, as in Eq. (4.52). By an early method, the parameters are estimated in two stages. First, the Gaussian centers are estimated as the cluster centers on clustering the training samples (either altogether or classwise), and the variances are averaged over the samples in each cluster. The centers and variances are then fixed and the weights of output layer are estimated by squared error minimization. In the case of linear outputs, the weights can be estimated by pseudo inverse, with the rows of data matrix storing the Gaussian values on all training samples.

The outputs of RBF network can better approximate a posteriori class probabilities by using the sigmoid activation function:

$$y_k(\mathbf{x}) = g(v_k(\mathbf{x})) = \frac{1}{1 + e^{-v_k(\mathbf{x})}}. \tag{4.79}$$

In this situation, the weights of output layer can no longer be estimated by pseudo inverse. On fixed Gaussian functions in hidden layer, the weights can be efficiently estimated by stochastic gradient descent:

$$w_{kj}(t + 1) = w_{kj}(t) - \eta(t)\frac{\partial E^n}{\partial w_{kj}}, \tag{4.80}$$

where

$$\begin{aligned}\frac{\partial E^n}{\partial w_{kj}} &= [y_k(\mathbf{x}^n) - t_k^n]y_k(\mathbf{x}^n)[1 - y_k(\mathbf{x}^n)]\phi_j(\mathbf{x}^n), \\ &= -\delta_k(\mathbf{x}^n)\phi_j(\mathbf{x}^n).\end{aligned} \tag{4.81}$$

The two-stage training method for RBF network is fast but will not necessarily lead to a high classification performance because the estimation of center parameters via clustering does not consider the separability of patterns of different classes. An alternative is to adjust all the parameters simultaneously in supervised learning by error minimization [4]. Under supervised training of all parameters, the RBF network can achieve a higher classification performance with much fewer hidden nodes. By doing this, the performance of RBF network can compete or even exceed the MLP. The training algorithm is described below.

As in BP training for MLP, the parameters of RBF network are updated simultaneously to minimize the squared error on training samples (Eq. (4.52)). The center parameters can be initialized by clustering as in two-stage training. By stochastic gradient descent, the weights of output layer are updated on each input pattern as in Eq. (4.80), and meanwhile, the center parameters (denoting $\tau_j = \sigma_j^2$) are updated by

$$\begin{aligned}\mu_j(t + 1) &= \mu_j(t) - \eta(t)\frac{\partial E^n}{\partial \mu_j}, \\ \tau_j(t + 1) &= \tau_j(t) - \eta(t)\frac{\partial E^n}{\partial \tau_j},\end{aligned} \tag{4.82}$$

where

$$\begin{aligned}\frac{\partial E^n}{\partial \mu_j} &= -\sum_{k=1}^{M} w_{kj}\delta_k(\mathbf{x}^n)\frac{\partial \phi_j(\mathbf{x}^n)}{\partial \mu_j}, \\ \frac{\partial E^n}{\partial \tau_j} &= -\sum_{k=1}^{M} w_{kj}\delta_k(\mathbf{x}^n)\frac{\partial \phi_j(\mathbf{x}^n)}{\partial \tau_j},\end{aligned} \tag{4.83}$$

and further,

$$\frac{\partial \phi_j(\mathbf{x}^n)}{\partial \mu_j} = \frac{\phi_j(\mathbf{x}^n)}{2\tau_j}(\mathbf{x}^n - \mu_j),$$
$$\frac{\partial \phi_j(\mathbf{x}^n)}{\partial \tau_j} = \frac{\phi_j(\mathbf{x}^n)}{2\tau_j^2}\|\mathbf{x}^n - \mu_j\|^2. \tag{4.84}$$

4.3.4 Polynomial Network

The higher order neural network (HONN) is also called as polynomial network or polynomial classifier [80]. Its output is a weighted combination of pattern features as well as their polynomial expansions. For example, the output of a third-order neuron (with linear activation function) is

$$y = \sum_i \sum_{j \geq i} \sum_{k \geq j} w_{ijk}^{(3)} x_i x_j x_k + \sum_i \sum_{j \geq i} w_{ij}^{(2)} x_i x_j + \sum_i w_i^{(1)} x_i + w_0. \tag{4.85}$$

The polynomial terms can be viewed as predefined nonlinear features that, together with the original pattern features, are the inputs to a single-layer neural network. Thus, the pattern features and the polynomial terms form an enhanced feature vector, and the outputs of polynomial network can be viewed as generalized linear discriminant functions.

A major concern with polynomial networks is the huge number of polynomial terms on high-dimensional features. For d features, the total number of polynomial terms up to rth order is [82]

$$D = \sum_{i=0}^{r} \binom{d+i-1}{i} = \binom{d+r}{r}. \tag{4.86}$$

With large d, the network will suffer from high computation complexity and will give degraded generalization performance. The complexity can be reduced by either reducing the number of input features or selecting expanded polynomial terms [80]. The former way (feature section or dimensionality reduction) is more computationally efficient and performs fairly well in practice. On the contrary, constrained polynomial structures with moderate complexity have been proposed, such as the pi-sigma network and the ridge polynomial network (RPN) [82]. They actually involve all the polynomial terms of input features up to a certain order, but the weights of polynomials are highly correlated. Therefore, they need polynomials of fairly high order (say, 5 or 6) to approximate complicated functions and cannot guarantee the precision of approximation in difficult cases.

For character recognition problems, which usually have hundreds of features, a second-order polynomial (binomial) classifier performs sufficiently. A binomial network with dimensionality reduction by principal component analysis (PCA) has shown superior classification performance in character recognition experiments

[46, 56]. In the following, we describe this type of network with sigmoidal output nodes and gradient descent learning.

We assume that the d original features are transformed to m ($m < d$) subspace features $z_j(\mathbf{x})$, $j = 1, \ldots, m$. For M-class classification, the outputs of binomial network are

$$y_k(\mathbf{x}) = g\left(\sum_i \sum_{j \geq i} w_{kij}^{(2)} z_i z_j + \sum_i w_{ki}^{(1)} z_i + w_{k0}\right), \quad k = 1, \ldots, M, \qquad (4.87)$$

where $g(\cdot)$ is the sigmoid activation function. By PCA, the subspace features are computed by

$$z_j = c \cdot \phi_j^T (\mathbf{x} - \mu), \quad j = 1, \ldots, M, \qquad (4.88)$$

where μ is the global mean of training vectors, ϕ_j, $j = 1, \ldots, M$, are the eigenvectors of the sample covariance matrix corresponding to the largest eigenvalues. A scaling factor c is used to normalize the size of feature values by, for example, scaling the largest variance of single subspace feature to one.

The connecting weights of binomial network (Eq. (4.87)) are optimized in supervised learning to minimize the squared error of training samples (Eq. (4.52)). By stochastic gradient descent, the weights are updated iteratively on each input pattern:

$$w(t + 1) = w(t) - \eta(t) \frac{\partial E^n}{\partial w}, \qquad (4.89)$$

where w represents $w_{kij}^{(2)}$, $w_{ki}^{(1)}$, or w_{k0}. The partial derivatives are specialized as

$$\begin{cases} \dfrac{\partial E^n}{\partial w_{kij}^{(2)}} = -\delta_k(\mathbf{x}^n) z_i(\mathbf{x}^n) z_j(\mathbf{x}^n), \\[2mm] \dfrac{\partial E^n}{\partial w_{ki}^{(1)}} = -\delta_k(\mathbf{x}^n) z_i(\mathbf{x}^n), \\[2mm] \dfrac{\partial E^n}{\partial w_{k0}} = -\delta_k(\mathbf{x}^n), \end{cases} \qquad (4.90)$$

where $\delta_k(\mathbf{x}^n) = [t_k^n - y_k(\mathbf{x}^n)] y_k(\mathbf{x}^n)[1 - y_k(\mathbf{x}^n)]$.

On fixing the polynomial terms, the polynomial network has one single layer of adjustable weights, so the gradient descent training is fast and mostly converges to a good solution though global optimum is not guaranteed in the case of sigmoid activation function.

4.3.5 Unsupervised Learning

Unsupervised learning networks, including autoassociation networks, competitive learning (CL) networks, and self-organizing map (SOM), have also been applied to pattern recognition. Whereas autoassociation learning aims to find a subspace or manifold of samples, competitive learning aims to find representative points from samples. Both result in a condensed representation of sample distribution, without considering the class labels of samples in learning. For classification, the training samples of each class are learned separately to give a class-specific representation, and a test pattern is then fitted to the representation of each class and classified to the class of minimum fitting error.

The autoassociation network is a multilayer feedforward network and can be viewed as a special structure of MLP. It has the same number of linear output nodes as input nodes. In training, the input values are used as target values such that the hidden nodes give a minimum error representation of samples and the output nodes give a reconstruction of input values. When the hidden nodes have linear activation functions, the network learns a linear subspace of samples just like PCA, no matter how many hidden layers and nodes are used. With nonlinear (sigmoidal or hyperbolic) activation functions in hidden layers, the network learns a nonlinear subspace or manifold of samples. Some results of character recognition using autoassociation networks can be seen in [40].

In the following, we go into some details of competitive learning and SOM.

4.3.5.1 *Competitive Learning*   The network structure underlying competitive learning has a single layer of output nodes. The weights of each output node connected with input features give a representative vector (called reference vector or codevector) of samples. The approximation of a set of samples with codevectors falls in the classical vector quantization (VQ) problem [26]. VQ is actually a clustering problem, where the codevectors are cluster centers. A sample vector is approximated by the closest codevector from a codebook $\{\mathbf{m}_i | i = 1, \ldots, L\}$. Assume that the sample vectors undergo distribution $p(\mathbf{x})$, then the aim of VQ is to find the codevectors such that the reconstruction error is minimized:

$$\min E = \int \|\mathbf{x} - \mathbf{m}_c\|^2 p(\mathbf{x}) d\mathbf{x}, \tag{4.91}$$

where $\mathbf{m}_c$, with $c = c(\mathbf{x})$, is the closest codevector to $\mathbf{x}$. On a finite set of training samples $\{\mathbf{x}^n | n = 1, \ldots, N\}$, the reconstruction error is computed by

$$E = \sum_{n=1}^{N} \|\mathbf{x}^n - \mathbf{m}_c\|^2, \tag{4.92}$$

where $c = c(\mathbf{x}^n)$.

For learning codevectors on samples to minimize the error (4.92), the codevectors are initially set to random values or randomly selected from sample vectors. Then in competitive learning, the codevectors are updated by stochastic gradient descent. At time t, the closest codevector $\mathbf{m}_c$ to sample $\mathbf{x}(t)$ is found, and the codevectors are updated by

$$\begin{aligned} \mathbf{m}_c(t+1) &= \mathbf{m}_c(t) + \alpha(t)[\mathbf{x}(t) - \mathbf{m}_c(t)], \\ \mathbf{m}_i(t+1) &= \mathbf{m}_i(t), \quad i \neq c \end{aligned} \tag{4.93}$$

where $\alpha(t)$ is a monotonically decreasing learning step. We can see that only the closest codevector (the winner of competition) to the input vector is adjusted. Thus, this competitive learning scheme is also called winner-take-all (WTA) learning or hard competitive learning.

The WTA competitive learning can be viewed as the online version of a classical partitional clustering algorithm, the k-means algorithm [35]. Both competitive learning and k-means clustering are susceptible to local minimum of error function. Competitive learning may also suffer from the node underutilization problem (some codevectors are never adjusted in training). The so-called soft competitive learning (SCL) algorithms can alleviate these two problems. Generally, SCL adjusts the codevectors in different degrees according to the distance of each codevector to the input vector. In the SCL scheme of Yair et al. [99], the probability of each codevector on an input vector $\mathbf{x}(t)$ is computed based on soft-max:

$$P_i(t) = \frac{e^{-\gamma(t)\|\mathbf{x}(t)-\mathbf{m}_i\|^2}}{\sum_{j=1}^{L} e^{-\gamma(t)\|\mathbf{x}(t)-\mathbf{m}_j\|^2}}, \tag{4.94}$$

where $\gamma(t)$ is a monotonically increasing parameter to control the hardness. The codevectors are then adjusted by

$$\mathbf{m}_i(t+1) = \mathbf{m}_i(t) + \alpha(t)P_i(t)[\mathbf{x}(t) - \mathbf{m}_i(t)]. \tag{4.95}$$

Another soft competition scheme, called "neural-gas" [60], adjusts the codevectors according to the ranks of distances. On an input vector $\mathbf{x}(t)$, the codevectors are assigned ranks $k_i(\mathbf{x}(t)) \in \{0, \ldots, L-1\}, i = 1, \ldots, L$, with 0 for the closest one and $L-1$ for the farthest one. The codevectors are then adjusted by

$$\mathbf{m}_i(t+1) = \mathbf{m}_i(t) + \alpha(t)e^{-k_i(\mathbf{x}(t))/\lambda}[\mathbf{x}(t) - \mathbf{m}_i(t)], \tag{4.96}$$

where λ is a constant. The "neural-gas" is more computationally efficient than (4.95) as the probability-like coefficient $e^{-k_i(\mathbf{x}(t))/\lambda}$ has L fixed values. The experimental results of ref. [60] also show that "neural-gas" has better convergence toward the minimum of quantization error.

4.3.5.2 Self-Organizing Map   The SOM of Kohonen is a topology-preserving VQ algorithm. The nodes corresponding to codevectors are spatially ordered in a topology structure (a lattice, a graph, or a net). Unlike in soft competitive learning, the codevectors are adjusted according to degrees depending on the distance of codevector

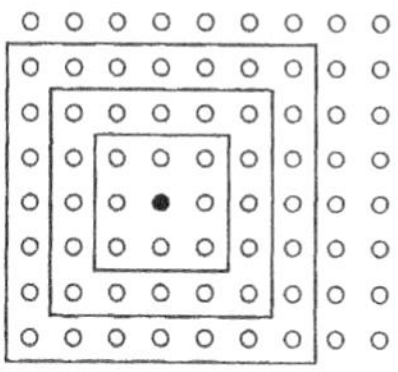

FIGURE 4.6   Two-dimensional array of nodes for SOM.

to input vector, in SOM, the degree of adjustment depends on the adjacency of the node with the winning node (corresponding to the winning codevector) in the topology.

Figure 4.6 shows a two-dimensional array of nodes for approximating high-dimensional sample vectors. The filled circle denotes the winning node, and the rectangles denote the neighborhood areas of three sizes. The learning rule of SOM is outlined as follows. On an input vector $\mathbf{x}(t)$, denoting the winning codevector as $\mathbf{m}_c$, the codevectors are adjusted by

$$\begin{aligned}
\mathbf{m}_i(t+1) &= \mathbf{m}_i(t) + \alpha(t)[\mathbf{x}(t) - \mathbf{m}_i(t)], \quad i \in N_c(t), \\
\mathbf{m}_i(t+1) &= \mathbf{m}_i(t), \qquad\qquad\qquad\qquad\quad \text{otherwise,}
\end{aligned} \tag{4.97}$$

where $N_c(t)$ denotes the neighborhood of the winning node in the topology. Initially, the size of neighborhood is large such that the neighboring nodes are drawn to represent proximate points in the feature space. The neighborhood shrinks gradually in learning such that the codevectors are adjusted to minimize the quantization error.

An alternative to the learning rule (4.97) uses "soft" neighborhood. In this scheme, the nodes are weighted according to the topological distance to the winning node. $h_{ci}(t) = e^{-\|\mathbf{r}_i - \mathbf{r}_c\|/\lambda}$, where $\mathbf{r}_i$ and $\mathbf{r}_c$ denote the positions of the node and the winning node in the topology, and λ decreases gradually. Accordingly, the codevectors are adjusted by

$$\mathbf{m}_i(t+1) = \mathbf{m}_i(t) + \alpha(t)h_{ci}(t)[\mathbf{x}(t) - \mathbf{m}_i(t)]. \tag{4.98}$$

The SOM yields the effect that the codevectors of adjacent nodes in the topology are also proximate in the feature space. It can be used for VQ as hard and soft competitive learning but is more suitable for exploring the adjacency structure of data. When used for classification, the SOM or competitive learning algorithm generates a number of codevectors (to serve prototypes) for each class. The input pattern is compared with the prototypes of all classes and is classified according to the k-NN or 1-NN rule. As the prototypes are generated class by class without considering the boundary between classes, they do not necessarily lead to high classification accuracy. The LVQ algorithm, to be described in the following, adjusts the prototypes with the aim of improving classification accuracy. The initial class prototypes of LVQ can be generated by SOM, competitive learning, or k-means clustering.

4.3.6   Learning Vector Quantization

The LVQ algorithm of Kohonen [43, 44] adjusts class prototypes with the aim of separating the samples of different classes. The 1-NN rule with Euclidean distance metric is generally taken for classification. Kohonen proposed several versions of LVQ that adjusts selected prototypes heuristically. Some improvements of LVQ learn prototypes by minimizing classification or regression error [55]. In the following, we will review some representative algorithms of LVQ, including LVQ2.1, LVQ3, the minimum classification error (MCE) method of Juang and Katagiri [37], and the generalized LVQ (GLVQ) of Sato and Yamada [77].

As for neural training, the training patterns are fed to the classifier repeatedly to update the prototypes. For an input pattern $\mathbf{x}(t)$, find the two closest prototypes $\mathbf{m}_i$ and $\mathbf{m}_j$ to it, with distances $d_i = \|\mathbf{x}(t) - \mathbf{m}_i\|^2$ and $d_j = \|\mathbf{x}(t) - \mathbf{m}_j\|^2$. If $\mathbf{m}_i$ belongs to the genuine class of $\mathbf{x}(t)$ whereas $\mathbf{m}_j$ belongs to an incorrect class, and further the two distances are comparable:

$$\min\left(\frac{d_i}{d_j}, \frac{d_j}{d_i}\right) > \frac{1-w}{1+w}, \qquad (4.99)$$

where $0 < w < 1$, the prototypes are updated by

$$\begin{aligned}
\mathbf{m}_i(t) &= \mathbf{m}_i(t) + \alpha(t)[\mathbf{x}(t) - \mathbf{m}_i(t)], \\
\mathbf{m}_j(t) &= \mathbf{m}_j(t) - \alpha(t)[\mathbf{x}(t) - \mathbf{m}_j(t)],
\end{aligned} \qquad (4.100)$$

where $\alpha(t)$ is a monotonically decreasing learning step. By updating, $\mathbf{m}_i$ is drawn toward $\mathbf{x}(t)$ and $\mathbf{m}_j$ is pushed away from $\mathbf{x}(t)$.

The above algorithm is referred to as LVQ2.1. LVQ3 is enhanced based on LVQ2.1 with one more updating rule that when both the two closest prototypes $\mathbf{m}_i$ and $\mathbf{m}_j$ belong to the genuine class of $\mathbf{x}(t)$, they are updated by

$$\mathbf{m}_k = \mathbf{m}_k + \epsilon\alpha(t)(\mathbf{x} - \mathbf{m}_k), \quad k = i, j, \qquad (4.101)$$

where $0 < \epsilon < 1$. This rule enables the prototypes to be more representative of the training samples.

Juang and Katagiri proposed a MCE training method for the parameter learning of neural networks and statistical classifiers [37]. They defined a loss function based on discriminant functions, and the empirical loss on a training sample set is minimized by gradient descent to optimize the classifier parameters. For prototype classifiers, the discriminant function of a class is the negative of the minimum distance from the input pattern to this class (we now attach class label k to prototypes):

$$g_k(\mathbf{x}) = -\min_j \|\mathbf{x} - \mathbf{m}_{kj}\|^2. \qquad (4.102)$$

The misclassification measure of a pattern from class k is given by

$$\mu_k(\mathbf{x}) = -g_k(\mathbf{x}) + \left[\frac{1}{M-1} \sum_{j \neq k} g_j(\mathbf{x})^\eta \right]^{1/\eta}, \qquad (4.103)$$

which, as $\eta \to \infty$, becomes

$$\mu_k(\mathbf{x}) = -g_k(\mathbf{x}) + g_r(\mathbf{x}), \qquad (4.104)$$

where $g_r(\mathbf{x}) = \max_{i \neq k} g_i(\mathbf{x})$.

For prototype classifiers, the misclassification measure is specified as

$$\mu_k(\mathbf{x}) = \|\mathbf{x} - \mathbf{m}_{ki}\|^2 - \|\mathbf{x} - \mathbf{m}_{rj}\|^2, \qquad (4.105)$$

where $\mathbf{m}_{ki}$ is the closest prototype from the genuine class and $\mathbf{m}_{rj}$ is the closest prototype from incorrect classes. The misclassification measure is transformed to loss by

$$l_k(\mathbf{x}) = l_k(\mu_k) = \frac{1}{1 + e^{-\xi \mu_k}}, \qquad (4.106)$$

where ξ controls the hardness of 0–1 loss.[2] On a training sample set, the empirical loss is

$$L_0 = \frac{1}{N} \sum_{n=1}^{N} \sum_{k=1}^{M} l_k(\mathbf{x}^n) I(\mathbf{x}^n \in \omega_k), \qquad (4.107)$$

where $I(\cdot)$ is an indicator function that takes value 1 when the condition in the parentheses holds and 0 otherwise.

By stochastic gradient descent, the prototypes are updated on each input training pattern. It is noteworthy that only two selected prototypes are involved in the loss on a training pattern $\mathbf{x}(t)$. Their increments are computed by

$$\Delta \mathbf{m}_{ki}(t) = -\alpha(t) \frac{\partial l_k(\mathbf{x})}{\partial \mathbf{m}_{ki}} = 2\alpha(t)\xi l_k(1 - l_k)(\mathbf{x} - \mathbf{m}_{ki}),$$
$$\Delta \mathbf{m}_{rj}(t) = -\alpha(t) \frac{\partial l_k(\mathbf{x})}{\partial \mathbf{m}_{rj}} = -2\alpha(t)\xi l_k(1 - l_k)(\mathbf{x} - \mathbf{m}_{rj}). \qquad (4.108)$$

On an input pattern $\mathbf{x}$, the GLVQ also updates the closest genuine prototype $\mathbf{m}_{ki}$ and the closest incorrect prototype $\mathbf{m}_{rj}$, but the misclassification measure is slightly

[2]In implementation, ξ should be inversely proportional to the average magnitude of $|\mu_k|$, which can be estimated on training samples with the initial prototypes. Whether ξ is constant or decreasing appears to be not influential if the initial value is set appropriately.

different:

$$l_k(\mathbf{x}) = \frac{d_{ki}(\mathbf{x}) - d_{rj}(\mathbf{x})}{d_{ki}(\mathbf{x}) + d_{rj}(\mathbf{x})}, \tag{4.109}$$

where $d_{ki}(\mathbf{x})$ and $d_{rj}(\mathbf{x})$ are the distances from the input pattern to the two prototypes. The loss and the empirical loss are computed in the same way as in MCE and the empirical loss is minimized by stochastic gradient search, wherein the prototypes are updated on feeding a pattern as

$$\begin{aligned}\mathbf{m}_{ki} &= \mathbf{m}_{ki} + 4\alpha(t)l_k(1 - l_k)\frac{d_{rj}}{(d_{ki} + d_{rj})^2}(\mathbf{x} - \mathbf{m}_{ki}), \\ \mathbf{m}_{rj} &= \mathbf{m}_{rj} - 4\alpha(t)l_k(1 - l_k)\frac{d_{ki}}{(d_{ki} + d_{rj})^2}(\mathbf{x} - \mathbf{m}_{rj}).\end{aligned} \tag{4.110}$$

In practical implementation, the denominator $(d_{ki} + d_{rj})^2$ of Eq. (4.110) is replaced by $(d_{ki} + d_{rj})$ for simplification.

An experimental comparison of LVQ and other prototype learning algorithms in character recognition shows that the MCE and GLVQ algorithms are among the best performing ones [55]. These algorithms, updating only a few prototypes on a training sample, are feasible for large category set problems like Chinese character recognition.

The MCE learning method is applicable to any discriminant function-based classifiers for parameter optimization. The adoption of MCE learning to the modified quadratic discriminant function (MQDF) classifier, resulting the so-called discriminative learning quadratic discriminant function (DLQDF) [57], has shown superior performance in handwritten character recognition.

4.4 SUPPORT VECTOR MACHINES

The SVM [7, 10] is a new type of hyperplane classifier,[3] developed based on the statistical learning theory of Vapnik [92], with the aim of maximizing a geometric margin of hyperplane, which is related to the error bound of generalization. The research of SVMs has seen a boom from the mid-1990s, and the application of SVMs to pattern recognition has yielded state-of-the-art performance.

Generally, an SVM classifier is a binary (two-class) linear classifier in kernel-induced feature space and is formulated as a weighted combination of kernel functions on training examples.[4] The kernel function represents the inner product of two vectors in linear/nonlinear feature space. In linear feature space, the decision function, primarily as a weighted combination of kernel functions, can be converted to a linear combination of pattern features. Thus, it has the same form as LDF and single-layer

---

[3]SVM can be used for both classification and regression. We consider only SVM classification in this book.
[4]In machine learning literature, a training pattern is usually called an example, an instance, or a point.

neural network, but the weights are estimated in a totally different way. The nonlinear feature space (possibly of infinite dimensionality) is implicitly represented by the kernel function, and it is not necessary to access the nonlinear features explicitly in both learning and classification. Multiclass classification is accomplished by combining multiple binary SVM classifiers.

As there have been many papers and some textbooks (e.g., [10, 79]) for describing the details of SVMs, we will give only a brief introduction to the principle and learning method of maximal margin hyperplane classifier and its extension to nonlinear kernels, and summarize some issues related to practical implementation.

4.4.1 Maximal Margin Classifier

For binary classification in a d-dimensional feature space, a linear decision function is used:

$$f(\mathbf{x}) = \mathbf{w} \cdot \mathbf{x} + b, \qquad (4.111)$$

where $\mathbf{w}$ is the weight vector, and b is a bias ($-b$ is also called threshold). Classification is given by $\mathrm{sgn}[f(\mathbf{x})]$ ($+1$ and -1 correspond to two classes). In linearly separable case, the decision function specifies a hyperplane separating the points of two classes (Fig. 4.7). To obtain a canonical form of the hyperplane, $\mathbf{w}$ and b are rescaled such that the points of two classes closest to the hyperplane satisfy $|\mathbf{w} \cdot \mathbf{x} + b| = 1$. Thus, for all points $\mathbf{x}_i$, $i = 1, \ldots, \ell$, with labels $y_i \in \{+, -1\}$, $y_i(\mathbf{w} \cdot \mathbf{x} + b) \geq 1$ holds. The distance between the points of two classes closest to the hyperplane, $2/\|\mathbf{w}\|$, is called the margin of the hyperplane.

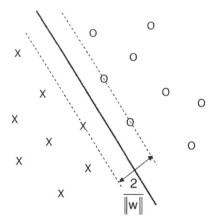

FIGURE 4.7   Hyperplane classifier for binary classification and the margin.

The decision function of SVM estimated from a training data set $\{(\mathbf{x}_i, y_i)|i = 1, \ldots, \ell\}$ is formulated by

$$f(\mathbf{x}) = \sum_{i=1}^{\ell} y_i \alpha_i \cdot k(\mathbf{x}, \mathbf{x}_i) + b, \qquad (4.112)$$

where $k(\mathbf{x}, \mathbf{x}_i)$ is a kernel function that implicitly defines an expanded feature space:

$$k(\mathbf{x}, \mathbf{x}_i) = \Phi(\mathbf{x}) \cdot \Phi(\mathbf{x}_i), \qquad (4.113)$$

where $\Phi(\mathbf{x})$ is the feature vector in the expanded feature space and may have infinite dimensionality. In the linear feature space, $k(\mathbf{x}, \mathbf{x}_i) = \mathbf{x} \cdot \mathbf{x}_i = \mathbf{x}^T \mathbf{x}_i$, and Eq. (4.112) is equivalent to

$$f(\mathbf{x}) = \sum_{i=1}^{\ell} y_i \alpha_i \cdot (\mathbf{x} \cdot \mathbf{x}_i) + b = \left(\sum_{i=1}^{\ell} y_i \alpha_i \mathbf{x}_i\right) \cdot \mathbf{x} + b = \mathbf{w} \cdot \mathbf{x} + b, \qquad (4.114)$$

where $\mathbf{w} = \sum_{i=1}^{\ell} y_i \alpha_i \mathbf{x}_i$.

The coefficients α_i, $i = 1, \ldots, \ell$, are estimated from the training points with the aim of maximizing the margin with constraints of linear separation:

$$\begin{aligned} \text{minimize } & \tau(\mathbf{w}) = \tfrac{1}{2} \|\mathbf{w}\|^2, \\ \text{subject to } & y_i(\mathbf{w} \cdot \mathbf{x}_i + b) \geq 1, \quad i = 1, \ldots, \ell. \end{aligned} \qquad (4.115)$$

This is a quadratic programing (QP) problem (primal problem) and can be converted to the following dual problem by introducing Lagrange multipliers:

$$\begin{aligned} \text{maximize } & W(\alpha) = \sum_{i=1}^{\ell} \alpha_i - \frac{1}{2} \sum_{i,j=1}^{\ell} \alpha_i \alpha_j y_i y_j \cdot (\mathbf{x}_i \cdot \mathbf{x}_j), \\ \text{subject to } & \alpha_i \geq 0, \ i = 1, \ldots, \ell, \quad \text{and } \sum_{i=1}^{\ell} \alpha_i y_i = 0. \end{aligned} \qquad (4.116)$$

The coefficients (multipliers) α_i, $i = 1, \ldots, \ell$, are obtained by solving the above QP problem. The bias b is determined by satisfying the Karush–Kuhn–Tucker (KKT) complementarity conditions:

$$\alpha_i[y_i(\mathbf{w} \cdot \mathbf{x}_i + b) - 1] = 0, \quad i = 1, \ldots, \ell. \qquad (4.117)$$

From the conditions, the coefficient α_i must be zero for those points with $y_i(\mathbf{w} \cdot \mathbf{x}_i + b) > 1$ (nonboundary points). The boundary points, with $y_i(\mathbf{w} \cdot \mathbf{x}_i + b) = 1$ and $\alpha_i > 0$, are called support vectors. The value of b is actually determined by the KKT conditions on the support vectors.

4.4.2 Soft Margin and Kernels

The assumption that all the training points are separable (in linear or nonlinear feature space) is often violated in practice. To deal with nonseparable points, the margin constraints of Eq. (4.115) is relaxed to

$$\text{subject to} \quad y_i(\mathbf{w} \cdot \mathbf{x}_i + b) \geq 1 - \xi_i, \quad \xi_i \geq 0, \ i = 1, \ldots, \ell. \tag{4.118}$$

The slack variable ξ_i is nonzero for nonseparable points. The objective of maximizing margin is accordingly modified to a soft margin:

$$\text{minimize} \quad \tau(\mathbf{w}, \xi) = \frac{1}{2}\|\mathbf{w}\|^2 + C\sum_{i=1}^{\ell} \xi_i. \tag{4.119}$$

On the contrary, for classification in nonlinear feature space, nonlinear kernel functions $k(\mathbf{x}, \mathbf{x}_i)$ are used to replace the inner product $\mathbf{x} \cdot \mathbf{x}_i$. The learning problem of SVM in a general kernel-induced feature space is then formulated as

$$\text{minimize} \ \tau(\mathbf{w}, \xi) = \frac{1}{2}\|\mathbf{w}\|^2 + C\sum_{i=1}^{\ell} \xi_i,$$
$$\text{subject to} \ y_i(\mathbf{w} \cdot \Phi(\mathbf{x_i}) + b) \geq 1 - \xi_i, \quad \xi_i \geq 0, \ i = 1, \ldots, \ell, \tag{4.120}$$

where

$$\mathbf{w} = \sum_{i=1}^{\ell} y_i \alpha_i \Phi(\mathbf{x}_i). \tag{4.121}$$

By introducing Lagrange multipliers, this QP problem is converted to a dual problem:

$$\text{maximize} \ W(\alpha) = \sum_{i=1}^{\ell} \alpha_i - \frac{1}{2}\sum_{i,j=1}^{\ell} \alpha_i \alpha_j y_i y_j \cdot k(\mathbf{x}_i, \mathbf{x}_j), \tag{4.122}$$
$$\text{subject to} \ 0 \leq \alpha_i \leq C, \ i = 1, \ldots, \ell, \quad \text{and} \ \sum_{i=1}^{\ell} \alpha_i y_i = 0.$$

In this formula, the constant C (also called noise parameter) serves the upper bound of multipliers and controls the tolerance of classification errors in learning.

For learning SVM from examples, the parameter C and the kernel function $k(\mathbf{x}, \mathbf{x}_i)$ are specified a priori, and the multipliers α_i, $i = 1, \ldots, \ell$, are estimated by solving the optimization problem of Eq. (4.122) using QP algorithms. The value of b is then determined from the KKT conditions on the support vectors. How to select parameter C and kernel functions and how to solve the QP problem are to be discussed in the next subsection.

The SVM has some attractive properties. First, the multipliers of training points are estimated in QP, which guarantees to find the global optimum because the convex quadratic objective function of Eq. (4.122), constrained in the region of the parameter space, has no local optimum. Second, the SVM realizes nonlinear classification via introducing kernel functions, which can be designed with some flexibility and can be generalized to nonvector representation of patterns. Third, the QP optimization of SVM results in a sparse representation of patterns because only a fraction of training points have nonzero multipliers and are used in decision. Further, the number of support vectors (the complexity of SVM classifier) is adaptable to the difficulty of separation of training points. This property leads to good generalization performance in classification.

Besides the inner product of vectors (linear kernel, which realizes linear classification), the frequently used nonlinear kernel functions include

- Sigmoid kernel:

$$k(\mathbf{x}, \mathbf{x}_i) = \tanh\left(\kappa \cdot (\mathbf{x} \cdot \mathbf{x}_i) + \theta\right). \tag{4.123}$$

- Polynomial kernel:

$$k(\mathbf{x}, \mathbf{x}_i) = \left(\kappa \cdot (\mathbf{x} \cdot \mathbf{x}_i) + 1\right)^p. \tag{4.124}$$

- RBF kernel:

$$k(\mathbf{x}, \mathbf{x}_i) = \exp\left(-\frac{\|\mathbf{x} - \mathbf{x}_i\|^2}{2\sigma^2}\right). \tag{4.125}$$

Incorporating the kernel functions into the decision function (Eq. (4.112)), the SVM with sigmoid kernel performs like a multilayer neural network, the one with polynomial kernel performs like a polynomial neural network, and the one with RBF kernel performs like an RBF neural network. The parameters of SVM are estimated in a different way from neural networks, however.

The selection of kernel parameters (κ, θ, p, and σ^2) will be discussed in the next subsection. The polynomial and RBF kernels have reported superior performance in pattern recognition applications.

4.4.3 Implementation Issues

For implementing SVMs for pattern classification of practical problems, we discuss four important issues: the implementation of QP, the selection of hyperparameters (model selection), complexity reduction, and multiclass classification. All these issues have attracted much attention in research. We will not describe the details but will summarize only some useful techniques.

4.4.3.1 Implementation of QP The objective of QP in Eq. (4.122) can be maximized using standard optimization techniques like gradient ascent. The complexity of storing the kernel matrix $k(\mathbf{x}_i, \mathbf{x}_j)$, $i, j = 1, \ldots, \ell$, and the complexity of computation increase quadratically with the size of the training data set. The storage of kernel matrix is infeasible for large-size problems with over tens of thousands of examples. To solve the particular QP problem of SVM learning, many special techniques have been developed. Mostly, the global optimization problem is decomposed into many local ones on subsets of data such that the storage of the kernel matrix on a subset is trivial and the optimization on a subset can be solved efficiently, sometimes analytically. The subset of data points (called "working set" or "active set") is iteratively selected and solved until some stopping conditions on the whole data set are satisfied.

Two useful criteria of stopping are the KKT conditions of the primal problem and the difference between the primal and dual objective functions (feasibility gap). Only at the optimum of multipliers, the KKT conditions are satisfied on all training points, and the feasibility gap vanishes. Certain level of error is tolerated in checking the stopping condition such that a nearly optimal solution is obtained in moderate computing time and the generalization performance is not influenced.

A seminal algorithm of SVM learning is the sequential minimal optimization (SMO) method of Platt [67]. The working set in SMO contains only two examples, on which the two multipliers that maximize the objective function can be computed analytically. The two examples are heuristically selected in two loops: the first example is selected from nonbound ($0 < \alpha_i < C$) points that violate KKT conditions, and the second one is selected such that the difference of decision error is maximized. Caches of decision error and kernel functions are used to reduce repeated computation.

The successive overrelaxation (SOR) algorithm of Mangasarian and Musicant [58] is even simpler than the SMO in that the working set contains only one example. This is done by reformulating the SVM problem as

$$
\begin{aligned}
&\text{minimize} \quad \frac{1}{2}\|\mathbf{w}\|^2 + C \sum_{i=1}^{\ell} \xi_i + b^2, \\
&\text{subject to} \quad y_i(\mathbf{w} \cdot \Phi(\mathbf{x}_i) + b) \geq 1 - \xi_i, \quad \xi_i \geq 0, \ i = 1, \ldots, \ell.
\end{aligned}
\tag{4.126}
$$

Its dual problem is

$$
\begin{aligned}
&\text{maximize} \quad \sum_{i=1}^{\ell} \alpha_i - \frac{1}{2}\sum_{i,j=1}^{\ell} \alpha_i\alpha_j y_i y_j \cdot k(\mathbf{x}_i, \mathbf{x}_j) - \frac{1}{2}(\sum_{i=1}^{\ell} y_i\alpha_i)^2, \\
&\text{subject to} \quad 0 \leq \alpha_i \leq C, \ i = 1, \ldots, \ell,
\end{aligned}
\tag{4.127}
$$

where $b = \sum_{i=1}^{\ell} y_i\alpha_i$. We can see that the major difference between Eqs. (4.127) and (4.122) is that the former removes the equality constraint $\sum_{i=1}^{\ell} \alpha_i y_i = 0$. Without the equality constraint, the objective can be maximized iteratively by solving one multiplier each time. The SOR algorithm is argued to have linear complexity with the number of data points and is able to learn with as many as millions of points.

Another method for learning SVM with huge data set, proposed by Dong et al. [13], follows the general SVM formulation with equality constraint. It quickly removes most of the nonsupport vectors and approximates the kernel matrix with block diagonal matrices, with each solved efficiently.

For practitioners who are interested in applying SVMs more than theoretical and algorithmic research, some publicly available softwares are very useful. The frequently used softwares include the SVM[light] [36] and the LIBSVM [8], and many softwares are available at the public Web site for kernel machines [39].

4.4.3.2 Model Selection

The selection of hyperparameters (constant C, parameters of kernel function) in SVM learning is very important to the generalization performance of classification. The linear kernel turns out to have no parameter, but actually the scaling of data vectors (equivalently, the scaling of inner product) is important. The scaling problem is also with the sigmoid and polynomial kernels (κ in Eqs. (4.123) and (4.124)). The width σ^2 of RBF kernel is also closely related to the scale of data. The value of C influences the convergence of learning and the generalization performance.

As for designing neural networks, the hyperparameters of SVM can be selected in the general strategy of cross-validation: LOO or holdout. On partitioning the available data into learning set and validation set, the SVM is learned multiple times with different choices of hyperparameters, and the choice that leads to the highest classification accuracy on the validation set is selected. As the number of combinations of hyperparameters (for a continuous parameter, multiple sampled values are taken) is large, this strategy is very time-consuming.

Unlike cross-validation that checks hyperparameters exhaustively, greedy search methods have been proposed to select hyperparameters by climbing up a classification performance measure. Chapelle et al. use an estimate of generalization error of SVM, which is a function of support vector multipliers, as the objective of optimization and search the hyperparameters by gradient descent [9]. Ayat et al. proposed an empirical error minimization method [1]. In SVM learning with a choice of hyperparameters, the empirical classification error is estimated on training points with the output of SVM transformed to a posteriori probability [68]. The hyperparameters are then updated toward the negative direction of gradient. It is noteworthy that in model selection by gradient search, the SVM is still learned multiple times.

In many cases, appropriate kernel models can be selected empirically via data rescaling. Consider that the hyperparameters κ and σ^2 are related to the scale of data vectors, normalizing the scale of data from the statistics of training points may help to find appropriate model easily, or largely reduce the range of hyperparameter search. The data vectors are rescaled uniformly: $\mathbf{x}' = \beta \mathbf{x}$. For the sigmoid or polynomial kernel, if the training vectors are rescaled such that the average self-inner product of all vectors is normalized to one, a value of $\kappa = 2^i$ with i selected from -4 to 4 will lead to reasonably good convergence and generalization. For the RBF kernel, consider the vector-to-center (square Euclidean) distance of each class. If the average distance over all classes is normalized to one by rescaling, a value of $\sigma^2 = 0.5 \times 2^i$ with i selected from -4 to 4 will perform well. If the data distribution of each class is multimodal, we can instead partition the data into clusters and estimate the scale from the within-cluster distances.

The order p of polynomial kernel has only a few choices: it is an integer and usually selected from $\{2, 3, 4, 5\}$. Empirically, the value of C can be sampled very sparsely without influencing the performance of SVM: It usually increases by a factor of ten: $C = 10^c$, and usually, $c \in \{-2, -1, 0, 1, 2\}$. From a few values of hyperparameters, the optimal one can be selected by cross-validation.

4.4.3.3 *Complexity Reduction*
Though SVM learning by constrained QP results in a sparse representation of patterns, the number of support vectors can still be very large in some cases. Depending on the distribution of data points, the fraction of support vectors in the training data set may be as small as 1% or as large as over 50%. In the latter case, the operation complexity of SVM classification is comparable to nonparametric (Parzen window or k-NN) classifiers. There have been some efforts toward the reduction of operation complexity, which can be categorized into three classes: sample reduction before SVM learning, complexity control in SVM learning, and support vector reduction after learning.

Reducing the number of training examples can both accelerate SVM learning and reduce the number of support vectors. This can be done in two ways: sample compression and sample selection. By sample compression, the data points are generally clustered (e.g., by k-means or vector quantization algorithms) and the cluster centers are taken as the examples of SVM learning. There is a trade-off between the sample reduction ratio and the generalization performance of SVM classifier. Examples can be selected using techniques like the editing of k-NN or 1-NN classifiers [12]. Consider that support vectors are mostly the points near classification boundary, it is reasonable to learn SVM with the points near the k-NN or 1-NN classification boundary. A method removes the examples of high confidence and of misclassification after fusing multiple SVMs learned on subsets of data [2].

The complexity can be controlled in SVM learning via directly modifying the objective of optimization. In the method of [15], the objective is formed by inserting Eq. (4.121) into Eq. (4.119) and adding a term controlling the magnitude of multipliers:

$$\text{minimize} \quad L(\alpha) = \frac{1}{2} \sum_{i,j=1}^{\ell} \alpha_i \alpha_j y_i y_j \cdot k(\mathbf{x}_i, \mathbf{x}_j) + C \sum_{i=1}^{\ell} \xi_i + D \sum_{i=1}^{\ell} \alpha_i. \quad (4.128)$$

Another method [59] directly uses the number of support vectors as a regularizing term of objective:

$$\text{minimize} \quad \|\alpha\|_0 + C \sum_{i=1}^{\ell} \phi(y_i f(\mathbf{x_i})), \quad (4.129)$$

where $\|\alpha\|_0$ denotes the number of nonzero multipliers, $\phi(z) = 1 - z$ if $z \leq 1$ and 0 otherwise.

After SVM learning, the support vectors can be compressed by synthesizing another smaller vector set or selecting a subset under the constraint that the classification performance of SVM is not sacrificed considerably. A method synthesizes a vector set with the aim that the squared error between the new weight vector and the weight vector before reduction is minimized [6]. The resulting vectors are not really support vectors but are used in the decision function for approximation. Another method exploits the linear dependency of support vectors in the feature space and removes the redundant ones without affecting the precision of decision [14]. By this method, the ratio of reduction is largely variable depending on the redundancy.

4.4.3.4 Multiclass Classification   SVMs are mostly considered for binary classification. For multiclass classification, multiple binary SVMs, each separating two classes or two subsets of classes, are combined to give the multiclass decision. Two general combination schemes are the one-versus-all and the one-versus-one. For an M-class problem, the one-versus-all scheme uses M SVMs, each separating one class from the union of the remaining ones. The SVM for class ω_i is learned with the examples of ω_i labeled as $+1$, and the examples of the other classes labeled as -1. After learning, it gives a decision function $f_i(\mathbf{x})$. The test pattern $\mathbf{x}$ is classified to the class of maximum decision $f_i(\mathbf{x})$.

In more sophisticated decision, the decision functions $f_i(\mathbf{x})$ are transformed to a posteriori probabilities by fitting sigmoid functions [67]:

$$P_i(y = +1|\mathbf{x}) = \frac{1}{1 + e^{-(Af_i(\mathbf{x})+B)}}, \tag{4.130}$$

where A and B are estimated in logistic regression[5] by minimizing a negative log-likelihood on a training data set of two classes $\{+1, -1\}$. The pattern $\mathbf{x}$ is then classified to the class of maximum a posteriori probability.

In the one-versus-one scheme, each SVM is used to discriminate between a pair of classes. Hence, for M classes, there should be $M(M-1)/2$ SVMs. Each SVM, learned with the examples of two classes (ω_i labeled as $+1$ and ω_j labeled as -1), gives a decision function $f_{ij}(\mathbf{x})$. The multiclass decision based on pairwise binary classifiers can be made by three rules. The first rule is the maximum votes of wins [45]. The decision function $f_{ij}(\mathbf{x})$ assigns a vote to ω_i when $f_{ij}(\mathbf{x}) > 0$ or to ω_j otherwise. The pattern $\mathbf{x}$ is classified to the class of maximum number of votes.

The directed acyclic graph (DAG) approach of Platt [69] does not activate the $M(M-1)/2$ binary classifiers simultaneously, but rather organize them in a DAG structure and calls them dynamically. The DAG has a triangular structure, with the number of nodes in each layer increasing by one, from one root node to $M-1$ leaf nodes. Each node (binary classifier) $f_{ij}(\mathbf{x})$ excludes the class ω_j if $f_{ij}(\mathbf{x}) > 0$ or

---

[5]Logistic regression will be described in the context of combining multiple classifiers at the measurement level.

excludes ω_i otherwise. On traversing from the root node to a leaf node, $M - 1$ classes are excluded and the pattern **x** is classified to the remaining class. The DAG approach activates only $M - 1$ binary classifiers for classifying a pattern, and hence is more efficient than the maximum votes decision. Its disadvantage is that it does not give a confidence score to the final decision.

The third rule of one-versus-one combination is based on MAP probability, via deriving M class probabilities from $M(M - 1)/2$ binary a posteriori probabilities. Denote the a posteriori probability of binary classifier $f_{ij}(\mathbf{x})$ as $r_{ij} = P(\omega_i|\mathbf{x}, \{\omega_i, \omega_j\})$, which can be transformed by, for example, Eq. (4.130). The class probabilities $p_i = P(\omega_i|\mathbf{x})$, $i = 1, \ldots, M$, can be derived from r_{ij} by pairwise coupling [28], which is outlined as follows. Assume that there are n_{ij} examples for a pair of classes $\{\omega_i, \omega_j\}$. Start with initial guess of class probabilities $\hat{p}_i$ (normalized to unity sum) and compute $\hat{\mu}_{ij} = \hat{p}_i/(\hat{p}_i + \hat{p}_j)$. Repeat updating probabilities until convergence:

$$\hat{p}_i \leftarrow \hat{p}_i \cdot \frac{\sum_{j \neq i} n_{ij} r_{ij}}{\sum_{j \neq i} n_{ij} \hat{\mu}_{ij}}, \qquad (4.131)$$

renormalize $\hat{p}_i$ and recompute $\hat{\mu}_{ij}$. The convergence condition is to maximize the average Kullback–Leibler divergence between r_{ij} and μ_{ij}:

$$\text{KL}(\mathbf{p}) = \sum_{i<j} n_{ij} \left[r_{ij} \log \frac{r_{ij}}{\mu_{ij}} + (1 - r_{ij}) \log \frac{1 - r_{ij}}{1 - \mu_{ij}} \right]. \qquad (4.132)$$

4.5 STRUCTURAL PATTERN RECOGNITION

Structural pattern recognition methods are used more often in online character recognition [85, 53] than in offline character recognition. Unlike statistical methods and neural networks that represent the character pattern as a feature vector of fixed dimensionality, structural methods represent a pattern as a structure (string, tree, or graph) of flexible size. The structural representation records the stroke sequence or topological shape of the character pattern, and hence resembles well to the mechanism of human perception. In recognition, each class is represented as one or more structural templates, the structure of the input pattern is matched with the templates and is classified to the class of the template of minimum distance or maximum similarity. The structural matching procedure not only provides an overall similarity but also interprets the structure of the input pattern and indicates the similarities of the components.

Despite the above merits of structural recognition, statistical methods and neural networks are more often adopted for the ease of feature extraction and learning from samples. Structural methods face two major difficulties: extracting structural primitives (strokes or line segments) from input patterns, and learning templates from samples. Primitive extraction from online character patterns (sequences of pen-down points) is much easier than from offline character images. Structural template learning from samples is undergoing study and has gained some progress. In practice, the templates are often selected from samples, constructed artificially or interactively.

Structural pattern recognition is often mentioned together with syntactic pattern recognition, which represents patterns and classes using formal linguistics and recognizes via grammatical parsing. Extracting linguistic representation from patterns is even more difficult than structural representation. This is why syntactic methods have not been widely used in practical recognition systems.

In the following, we describe two important types of structural recognition techniques that are useful for character recognition: attributed string matching and attributed graph matching. String matching techniques are often used in character string recognition as well, for matching string patterns with lexicon entries (see Chapter 5). In this section, we focus on the basic algorithms and take single character recognition as an example.

4.5.1 Attributed String Matching

When a string of symbols (e.g., letters from an alphabet) is to be matched (aligned) with a reference string, it is *deleted symbols* from, *inserted symbols* into, or *substituted symbols* such that the edited string is identical to the reference string. The total number of edit operations (deletions, insertions, and substitutions) is called edit distance or Levenstein distance. The edit distance can be efficiently computed by dynamic programing (DP) search, using an array (lattice) for representing the search space [94].

A pattern shape can be represented as a string of primitives, say, the sequence of stroke points, contour points, or line segments. When a symbol (pattern primitive) is represented by a number of attributes (features), the string is called an attributed string. The attributes are used to compute the cost of edit operations in string matching, and the matching distance is also called as weighted edit distance.

Consider two strings $A = a_1 \cdots a_m$ and $B = b_1 \cdots b_n$. The cost of edit operations $d(a_i, b_j) \geq 0$, $i = 0, 1, \ldots, m$, $j = 0, 1, \ldots, n$ $(i + j > 0)$, is defined as either fixed or attribute-dependent values. Let us take A as reference. When $i = 0$, $d(a_i, b_j)$ defines deletion cost; when $j = 0$, $d(a_i, b_j)$ is insertion cost; and otherwise, $d(a_i, b_j)$ is substitution cost. All the possible edit operations of B matching with A can be represented in the array $M(i, j)$, $i = 0, 1, \ldots, m$, $j = 0, 1, \ldots, n$. Each point (i, j) in the array can be viewed as a node in a network. An example of array is shown in Figure 4.9, where the vertical axis denotes index i and the horizontal axis denotes j. In the array, every forward path[6] from $(0, 0)$ through (m, n) corresponds to a match.

For describing the algorithm for finding the optimal path (optimal sequence of edit operations) of string matching, two arrays of accumulated distance $D(i, j)$ and backpointer $T(i, j)$ are defined. $D(i, j)$ records the accumulated cost of optimal path from $(0, 0)$ to (i, j). $T(i, j)$ records the coordinates $t \in \{(i - 1, j - 1), (i - 1, j), \text{ and } (i, j - 1)\}$ that $D(i, j)$ is derived from. $t = (i - 1, j - 1)$ indicates that b_j

---

[6]In a forward path, every step has an increment $\Delta i > 0$ or $\Delta j > 0$ or both.

is substituted by (matched with) a_i, $t = (i - 1, j)$ indicates that a_i is inserted, and $t = (i, j - 1)$ indicates that b_j is deleted. On updating all the values of $D(i, j)$ and $T(i, j)$, $D(m, n)$ gives the value of edit distance, and $T(m, n)T(T(m, n)) \cdots (0, 0)$ gives the backtrack path of editing. The algorithm is as follows:

Algorithm: Computing string edit distance

1. Initialization.
   - Set $D(0, 0) = 0$.
   - For $i = 1, \ldots, m$, update $D(i, 0) = D(i - 1, 0) + d(i, 0)$, $T(i, 0) = (i - 1, 0)$.
   - For $j = 1, \ldots, n$, update $D(0, j) = D(0, j - 1) + d(0, j)$, $T(0, j) = (0, j - 1)$.

2. Forward updating. For $i = 1, \ldots, m$, $j = 1, \ldots, n$,
   - Update $D(i, j)$ by

$$D(i, j) = \min \left\{ \begin{array}{l} D(i - 1, j) + d(i, 0), \\ D(i, j - 1) + d(0, j), \\ D(i - 1, j - 1) + d(i, j) \end{array} \right\}. \tag{4.133}$$

   - Corresponding to the minimum of the three rows of Eq. (4.133), update $T(i, j)$ by

$$T(i, j) = \text{select} \left\{ \begin{array}{l} (i - 1, j), \\ (i, j - 1), \\ (i - 1, j - 1) \end{array} \right\}. \tag{4.134}$$

3. Backtracking. Backtrack the editing path from (m, n) to $(0, 0)$ as $T(m, n)T(T(m, n))(T(T(m, n))) \cdots (0, 0)$.

In the algorithm, we can see that the main computation is to update the accumulated cost and the trace at each point of the array, so the complexity of computation is $O((m + 1) \cdot (n + 1)) = O(m \cdot n)$. The updating of $D(i, j)$ in Eq. (4.133) follows DP, which relies on the Bellman principle of optimality, that is, the minimum cost of all paths from the start node to the terminal node through an intermediate node equals the sum of two minimum costs, from the start to the intermediate and from the intermediate to the terminal. This is why every point in the array retains only the minimum accumulated cost and traces the point that the minimum accumulated cost is derived from.

When a character is represented as a string of sampled points, the DP matching of two strings is also referred to as dynamic time warping (DTW), as have been used in speech recognition. The substitution cost between two points can be computed from the geometric distance between them, the difference of direction of adjacent line segments, and so on. As an example, two online shapes of character "2" are shown in Figure 4.8. They are encoded in 11 and 12 sampled points, respectively. The matching

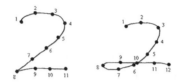

FIGURE 4.8 Two shapes of character "2" encoded as strings of sampled points.

distance between them is computed by DP as in Figure 4.9, where the editing path is shown by the thick line, and the correspondence between points is shown on the right.

For shape matching for sequence of line segments, Tsai and Yu proposed an attributed string matching algorithm with merging [88], where the editing path at node (i, j) is extended not only from $\{(i - 1, j - 1), (i - 1, j), (i, j - 1)\}$ but also from $(i - k, j - l)$ $(k > 1, l > 1)$. Tsay and Tsai have applied this kind of matching algorithm to online Chinese character recognition [89].

Another extension of string matching is the normalized edit distance [61], which is defined as the ratio between the weighted edit distance and the length of editing path (number of edit operations). The normalization of edit distance is important to overcome the variability of string length in shape recognition. As the length of editing path is unknown a priori, the minimum normalized edit distance must be computed by an algorithm of higher complexity. If we simply normalize w.r.t. the length of strings $(m, n,$ or any combination of them), the weighted edit distance is computed by DP and then divided by the length.

4.5.2 Attributed Graph Matching

A graph comprises a set of nodes (vertices) and a set of arcs (edges) for defining the relationship between nodes. An attributed relational graph (ARG) is a graph with nodes and edges described by attributes (features) and can be denoted by a triple $G = (V, E, A)$ $(V, E, A$ denote the sets of nodes, edges, and attributes, respectively). A character pattern can be represented as an ARG, generally with nodes denoting strokes and edges denoting interstroke relationships. A simple example of ARG is

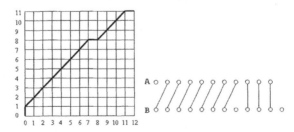

FIGURE 4.9 Editing path of the strings in Figure 4.8 and the correspondence of points.

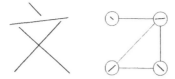

FIGURE 4.10   A character shape and its ARG representation.

shown in Figure 4.10, where a character pattern has four strokes and the closely dependent strokes are related by edges in the graph. For character recognition using ARGs, each class is modeled as one or more template graphs, and the graph of the input pattern is matched with the templates and classified to the class of the template of minimum matching distance.

Similar to string matching, the matching distance between two ARGs can be formulated as an edit distance, which is the sum of costs in transforming the input pattern graph to a model graph by node/edge deletion, insertion, and substitution [76]. However, to find the transformation of minimum graph edit distance is much more complicated than that of string edit distance. Graph matching can also be viewed as a correspondence, assignment, or consistent labeling problem. Consider a graph with m nodes $V_1 = \{u_1, \ldots, u_m\}$, to match it with another graph with n nodes $V_2 = \{v_1, \ldots, v_n\}$, the nodes in V_2 are actually assigned labels from $V_1 \cup \{\phi\}$ (ϕ is a null label) under the constraint that when two nodes v_{i_2} and v_{j_2} are assigned labels u_{i_1} and u_{j_1}, the relationships (edges) (v_{i_2}, v_{j_2}) and (u_{i_1}, u_{j_1}) should be consistent (compatible).

The 1:1 correspondence of nodes between two graphs is generally called graph isomorphism or subgraph isomorphism. The former refers to the 1:1 correspondence between all nodes of two graphs, whereas the latter refers to the correspondence between one graph and a subgraph of another. Subgraph isomorphism is also called as graph monomorphism or homomorphism. It can be generalized to the case of error-correcting or inexact isomorphism, in which approximate node/edge matching and deletion/insertion operations are allowed. Most of the practical shape or vision matching problems can be formulated as inexact subgraph isomorphism.

For character recognition using ARG matching, the most important task is to find the correspondence between the nodes of two graphs. Based on node correspondence, the distance or similarity between two graphs is computed from the between-node and between-edge distances and is then used for recognition. For example, the distance between two strokes can be computed from the geometric distance between two center points, the difference in lengths and orientations, and so on. The relationship between a pair of strokes may have the attributes of difference of orientations, the distance between two center points, and the orientation of the line connecting the center points of two strokes, and accordingly, the between-edge distance can be computed from the difference of these attributes.

The correspondence between two sets of graph nodes can be found by two methods: heuristic tree search and relaxation labeling. In the following, we describe the basic algorithms of the two methods. It is noteworthy that graph matching is still a research issue in pattern recognition and computer vision fields. The readers should consult the literature for the recent advances.

4.5.2.1 Tree Search
Graph matching is a combinatorial optimization problem, and it is natural to use a tree to represent the search space of optimization [91]. The efficiency of search depends on the order of node[7] expansion in the tree: Breadth-first search and depth-first search are brute-force (exhaustive), whereas heuristic search expands those nodes that are likely to approach the solution.

For inexact graph matching, usually an objective function of matching distance is formulated and minimized using combinatorial optimization algorithms, including tree search. Tsai and Fu [87] and Wong et al. [96] proposed to solve this problem using an efficient heuristic search algorithm, the A* algorithm [64]. In the following, we outline the method of Wong et al.

Consider two ARGs $G_1 = (V_1, E_1, A_1)$ and $G_2 = (V_2, E_2, A_2)$, with m nodes and n nodes, respectively. G_2 is to be transformed for optimally matching with G_1. Accordingly, the nodes of G_2 are assigned labels from $V_1 \cup \{\phi\}$, or the nodes of G_1 are assigned labels from $V_2 \cup \{\phi\}$. Assume that the nodes of G_1, $V_1 = \{u_1, u_2, \ldots, u_m\}$, correspond to the nodes in G_2, $\{v_{l_1}, v_{l_2}, \ldots, v_{l_m}\} \in V_2 \cup \{\phi\}$, the matching distance between these two graphs can be defined as

$$D(G_1, G_2) = \sum_{i=1}^{m} d(u_i, v_{l_i}) + \sum_{i=1}^{m} \sum_{j=1}^{i-1} d(u_i, u_j, v_{l_i}, v_{l_j}), \qquad (4.135)$$

where $d(u_i, v_{l_i})$ and $d(u_i, u_j, v_{l_i}, v_{l_j})$ denote node-matching cost and edge-matching cost, respectively; and if $v_{l_i} = \phi$ or $v_{l_j} = \phi$, they are the costs of node/edge deletion. The cost functions can be defined incorporating the attributes of nodes and edges and the compatibility between edges.

To minimize the distance of Eq. (4.135), the combinations of labels from $V_2 \cup \{\phi\}$ are represented in the search tree. The root node of the tree denotes an empty match, that is, no node of G_1 is matched with node of G_2. Then, each node in the depth-k layer (u_k, v_{l_k}) indicates that the kth node of G_1, u_k, is matched with a node v_{l_k} of G_2. When a node (u_k, v_{l_k}) is expanded, the next node of G_1, u_{k+1}, is matched with a node from $(V_2 - \{v_{l_1}, \ldots, v_{l_k}\}) \cup \{\phi\}$. The accumulated cost from the root node to (u_k, v_{l_k}) (a partial match in the search tree) is computed by

$$g(u_k, v_{l_k}) = \sum_{i=1}^{k} d(u_i, v_{l_i}) + \sum_{i=1}^{k} \sum_{j=1}^{i-1} d(u_i, u_j, v_{l_i}, v_{l_j}) = \sum_{i=1}^{k} d'(u_i, v_{l_i}), \quad (4.136)$$

---

[7]Note the context of "node": whether it is the node of a graph or the node of the search space (tree). A node of the search tree represents a state of search.

where

$$d'(u_i, v_{l_i}) = d(u_i, v_{l_i}) + \sum_{j=1}^{i-1} d(u_i, u_j, v_{l_i}, v_{l_j}).$$ (4.137)

The order of node expansion follows branch-and-bound (BAB); that is, every time the node of minimum accumulated cost is expanded to generate successors. BAB guarantees that the globally optimal solution, that is, the complete match of minimum distance, is found. However, the search efficiency of BAB is low because no heuristics about the remaining path is utilized.

In heuristic search, the node selected to be expanded is the one of minimum heuristic function

$$f(N) = g(N) + h(N),$$ (4.138)

where N is the current node of the search tree, $N = (u_k, v_{l_k})$, $g(N) = g(u_k, v_{l_k})$, and $h(N)$ is an estimate of the remaining cost of the path from N to the goal (complete match). If $h(N)$ is a lower bound of the real remaining cost, BAB search guarantees finding the globally optimal solution, and the tighter the bound is, the more efficient (fewer nodes are expanded before the goal is reached) the search is. Hence, to find a tight lower bound estimate of the remaining cost is the key to achieve optimal and efficient search.

Wong et al. compute the estimate of the remaining cost as follows. At the tree node $N = (u_k, v_{l_k})$ denotes the sets of unmatched nodes of G_1 and G_2 as M_1 and M_2, respectively. We have $M_1 = \{u_{k+1}, \ldots, u_m\}$, $M_2 = (V_2 - \{v_{l_1}, \ldots, v_{l_k}\}) \cup \{\phi\}$. For a node in M_1 matched with a label from M_2, define a cost function

$$d''(u_i, v_{l_i}) = d(u_i, v_{l_i}) + \sum_{j=1}^{k} d(u_i, u_j, v_{l_i}, v_{l_j}).$$ (4.139)

Finding for each node in M_1 an optimal label from M_2, we obtain the minimum cost

$$a(N) = \sum_{i=k+1}^{m} \min_{v_{l_i} \in M_2} d''(u_i, v_{l_i}).$$ (4.140)

Similarly, for each edge (pair of nodes) in M_1, find an optimal pair from M_2. The minimum cost is

$$b(N) = \sum_{i=k+1}^{m} \sum_{j=i+1}^{m} \min_{v_{l_i}, v_{l_j} \in M_2} d(u_i, u_j, v_{l_i}, v_{l_j}).$$ (4.141)

It can be proved that $h(N) = a(N) + b(N)$ is an estimate of the lower bound of the real remaining cost. Its use in heuristic search was shown to find optimal graph matching efficiently [96].

4.5.2.2 *Relaxation Labeling* Relaxation labeling techniques utilize contextual information to reduce the uncertainty of labels in assignment or labeling problems. They have been widely applied to image analysis since Rosenfeld et al. [74]. Yamamoto and Rosenfeld first applied relaxation labeling to Chinese character recognition and obtained impressive performance [100].

Consider a set of objects or elements (can be viewed as the nodes of a graph) $B = \{b_1, \ldots, b_n\}$ and a set of labels (can be viewed as the nodes of another graph) $\Lambda = \{1, \ldots, m\}$. The aim is to assign a distinct label to each object such that under a relational structure, each pair of objects is compatible with the labels assigned to them. In character recognition, the objects can be the strokes of the input pattern, and the labels are the strokes of a template structure. Relaxation labeling techniques solve the labels by iteratively updating the assignment probabilities until each object is unambiguously assigned a unique label.

Denote the probability (certainty) of assigning label λ to object b_i as $p_{i\lambda}$ ($0 \leq p_{i\lambda} \leq 1$). The label probabilities of an object can be represented in a vector $\mathbf{p}_i$ that satisfies $|\mathbf{p}_i| = \sum_{\lambda=1}^{m} p_{i\lambda} = 1$. The initial probability, $p_{i\lambda}^{(0)}$, is set at random or computed from the matching similarity of attributes between the object b_i and its label λ. For updating the probabilities, compatibility functions, between pairs of objects and pairs of labels, are defined: $r_{ij}(\lambda, \mu) = r(b_i, b_j, \lambda, \mu)$. In character recognition, the compatibility functions can be computed according to the differences between relative geometric features of stroke pairs.

For updating the assignment probabilities at time t, a support function for each label on an object is computed:

$$s_{i\lambda}^{(t)} = \sum_{b_j \in N(b_i)} \sum_{\mu=1}^{m} r_{ij}(\lambda, \mu) p_{j\mu}^{(t)}, \qquad (4.142)$$

where $N(b_i)$ denotes a neighborhood of object b_i. The support functions are then used to update the probabilities by

$$p_{i\lambda}^{(t+1)} = \frac{p_{i\lambda}^{(t)} s_{i\lambda}^{(t)}}{\sum_{\mu=1}^{m} p_{i\mu}^{(t)} s_{i\mu}^{(t)}}, \qquad (4.143)$$

wherein the probabilities of an object have been normalized. After a number of iterations, the probabilities are expected to converge to the state that for each object, only one label probability remains large whereas the other probabilities tend to vanish. As a result, the object is assigned the label with the maximum probability.

In a modified updating rule called radial projection [66] both the probabilities and the support values of an object are viewed as vectors in m-dimensional space. The normalized vector of support, $\mathbf{s}_i^{(t)}/|\mathbf{s}_i^{(t)}|$, is used to update the probability vector by

$$\mathbf{p}_i^{(t+1)} = \mathbf{p}_i^{(t)} + \alpha_i^{(t)} \left(\frac{\mathbf{s}_i^{(t)}}{|\mathbf{s}_i^{(t)}|} - \mathbf{p}_i^{(t)} \right), \qquad (4.144)$$

where

$$\alpha_i^{(t)} = \frac{|\mathbf{s}_i^{(t)}|}{1 + |\mathbf{s}_i^{(t)}|}.$$ (4.145)

The updated probability vector satisfies $|\mathbf{p}_i^{(t+1)}| = 1$ automatically.

The relaxation labeling algorithm has polynomial complexity. The complexity of updating assignment probabilities is $O(m \cdot n)$, and the complexity of computing support functions is $O((mn)^2)$. The convergence of relaxation, however, depends on the computing of initial probabilities and the compatibility functions. Designing proper initial probabilities and compatibility functions relies on the domain knowledge of the application problem.

Graph matching and relaxation labeling methods have been widely applied to the structural matching of character patterns. An example of graph-matching-based offline character recognition can be found in [54]. For online character recognition, many structural matching methods are reviewed in [53].

4.6 COMBINING MULTIPLE CLASSIFIERS

The Bayes decision classifier is optimal but the conditions to achieve the optimality are rather unrealistic: sufficiently large (nearly infinite) training data set, flexible enough density model, and powerful learning tool for parameter estimation from data. Practical classifiers instead learn constrained density models or discriminant functions from finite data set and make approximate decisions. Different classifiers approximate the Bayes decision in different ways and show varying performance: varying classification accuracy and speed, and different errors on concrete patterns. It is hence natural to combine the strengths of different classifiers to achieve higher performance. This is usually accomplished by combining the decisions (outputs) of multiple classifiers.

Classification with multiple classifiers (also called an ensemble of classifiers) has been an active research area since early 1990s. Many combination methods have been proposed, and the applications to practical problems have proven the advantage of ensemble over individual classifiers [84, 98, 31, 42, 83]. A recent survey [70] categorizes the methods into parallel (horizontal) and sequential (vertical, cascaded) ones. Parallel combination is more often adopted for improving the classification accuracy, whereas sequential combination is mainly used for accelerating the classification of large category set (coarse classification in Chinese character recognition [53]).

In this section, we mainly address parallel combination of classifiers, but when discussing ensemble generation, there will be a method that sequentially generates individual classifiers (boosting). All the issues concerned have been studied intensively and we summarize only some selected methods. A comprehensive treatment of combining classifiers has been given in the textbook of Kuncheva [47].

4.6.1 Problem Formulation

Assume the classification of a pattern X into M classes $\{\omega_1, \ldots, \omega_M\}$ using K individual classifiers $\{E_1, \ldots, E_K\}$, each using a feature vector $\mathbf{x}_k$, $k = 1, \ldots, K$. Each feature vector can be viewed as a subspace of the entire space occupied by the union of all features (united feature vector $\mathbf{x}$). On an input pattern, each classifier E_k outputs decision scores to all classes: $d_{ki}(\mathbf{x}_k)$, $j = 1, \ldots, M$. The classifier outputs can be categorized into three levels: abstract level (unique class), rank level (rank order of classes), and measurement level (confidence scores of classes) [98].

We can represent the outputs of a classifier at different levels uniformly in a decision vector $\mathbf{d}_k = (d_{k1}, \ldots, d_{kM})^T$. At the abstract level, $\mathbf{d}_k$ has only one nonzero element corresponding to the decided class. The rank order can be converted to class scores such that d_{ki} is the number of classes ranked below ω_i. At the measurement level, d_{ki} is the discriminant value (similarity or distance) or probability-like confidence of ω_i. The measurement-level outputs can be easily reduced to rank level and abstract level.

The combination of K classifiers assigns fused scores to M classes, which are the functions of the scores of individual classifiers:

$$y_i(X) = F_i \begin{pmatrix} d_{11}, & \cdots, & d_{1M}, \\ \vdots & \ddots & \vdots \\ d_{K1}, & \cdots, & d_{KM} \end{pmatrix}, \qquad i = 1, \ldots, M, \qquad (4.146)$$

In practice, the fused score of one class often considers the individual classifier outputs of the same class only:

$$y_i(X) = F_i(d_{1i}, \ldots, d_{Ki}), \qquad i = 1, \ldots, M. \qquad (4.147)$$

The formulas of (4.146) and (4.147) are referred to as class-indifferent combination and class-conscious combination, respectively [47].

The class-indifferent combiner can be viewed as a new classifier (called meta-classifier) in a $K \times M$-dimensional feature space. Class-conscious combination rules can be divided into fixed ones and trained ones [18]. As a meta-classifier does, a trained rule has some tunable parameters that are estimated on a training data set. Desirably, the data set for training the combiner consists of classifier outputs on samples that are not used in training the individual classifiers. If an extra sample set disjoint from the training set is available, the classifier outputs on the extra set are good for combiner training. The individual classifiers can also be trained with different sample sets. If only one sample set is available, it can be partitioned for training both individual classifiers and the combiner by a strategy called stacking [95, 86], which we outline below.

By stacking, the sample set $\mathcal{X}$ is partitioned into two subsets, by either rotation or holdout: one for temporarily training individual classifiers and the other (validation set) for collecting classifier outputs. The output data on validation samples are used to train the combiner. Yet for classifying new data, the individual classifiers are still trained with the whole sample set. Specifically, let us partition $\mathcal{X}$ into J subsets $\mathcal{X}_j$, $j = 1, \ldots, J$. We rotationally hold out $\mathcal{X}_j$ and use $\mathcal{X} - \mathcal{X}_j$ to train classifiers $E_k^{(j)}$, $k = 1, \ldots, K$, whose outputs on the samples in $\mathcal{X}_j$ form a data subset $\mathcal{D}^{(j)}$. The union $\mathcal{D}^{(1)} \cup \cdots \cup \mathcal{D}^{(J)} = \mathcal{D}$ is used for training the combiner. The classifiers $E_k^{(j)}$, $j = 1, \ldots, J$ and $k = 1, \ldots, K$, are used only for collecting output data. For classification, the classifiers E_k, $k = 1, \ldots, K$, are still trained with $\mathcal{X}$.

Holdout partition is different from rotation in that for collecting output data, the individual classifiers are trained only once with a subset $\mathcal{X}_1$ and their outputs on $\mathcal{X}_2 = \mathcal{X} - \mathcal{X}_1$ form the data for combiner training (usually, $\mathcal{X}_1$ contains more samples than $\mathcal{X}_2$). The individual classifiers are then trained with $\mathcal{X}$ for classification. It is also a practice that the classifiers trained with $\mathcal{X}_1$ are directly used for classification.

4.6.2   Combining Discrete Outputs

Some particular methods are used for combining discrete (abstract-level and rank-level) outputs of classifiers. If the discrete outputs of a classifier on a pattern is represented as a vector $\mathbf{d}_k = (d_{k1}, \ldots, d_{kM})^T$, they can also be combined in the same way as continuous scores.

4.6.2.1   *Majority Vote*   For combining abstract-level decisions, the most frequently considered rules are the majority vote and weighted majority vote. If the input pattern is classified to ω_j by classifier E_k, the class ω_j is assigned a vote $d_{kj} = 1$, and the other classes are assigned vote $d_{ki} = 0$ ($i \neq j$). The combined votes are computed by

$$y_i = \sum_{k=1}^{K} d_{ki}, \quad i = 1, \ldots, M. \tag{4.148}$$

By majority vote, the input pattern is classified to the class of $\max_{i=1,\ldots,M} y_i$ if $\max_{i=1,\ldots,M} y_i > K/2$, and rejected otherwise. According to this rule, the combination gives correct classification if at least $\lfloor K/2 \rfloor + 1$ classifiers are correct. The rule that always classifies to the class of $\max_{i=1,\ldots,M} y_i$ without rejection is also called plurality.

It is shown that by majority vote, combining K independent classifiers of the same accuracy ($p > 0.5$) can always improve the accuracy [47]. The probability that the ensemble makes correct decision (at least $\lfloor K/2 \rfloor + 1$ classifiers are correct) is

$$P_{\text{ens}} = \sum_{k=\lfloor K/2 \rfloor + 1}^{K} \binom{K}{k} p^k (1-p)^{K-k}. \tag{4.149}$$

If $p > 0.5$, P_{ens} is monotonically increasing with K, and when $K \to \infty$, $P_{\text{ens}} \to 1$. In practice, however, the individual classifiers are more or less dependent, and the accuracy of ensemble is dependent on the dependence. A more detailed analysis of the performance of majority vote can be found in [50].

In the case of imbalanced or dependent classifiers, better combination performance can be achieved by weighted majority vote (or weighted plurality):

$$y_i = \sum_{k=1}^{K} w_k d_{ki}, \quad i = 1, \dots, M, \tag{4.150}$$

where w_k is the weight for classifier E_k. The constraints $0 \le w_k \le 1$ and $\sum_{k=1}^{K} w_k = 1$ are recommended but not compulsory. It is shown that for combining K independent classifiers with accuracies $p_1, \dots, p_K$, the ensemble accuracy of weighted majority vote is maximized by assigning weights [47]

$$w_k \propto \log \frac{p_k}{1 - p_k}. \tag{4.151}$$

For combining dependent classifiers, the weights can be optimized by training with a validation data set. This is handled in the same way as combining continuous outputs. Besides, Lam and Suen have proposed to optimize the weights using a genetic algorithm [49].

4.6.2.2 Bayesian Combination

Consider the decided class of classifier E_k as a discrete variable $e_k \in \{\omega_1, \dots, \omega_M\}$, the combined decision of K classifiers can be made by the Bayes decision rule on K variables $e_1, \dots, e_K$. To do this, the a posteriori probabilities of M classes on an input pattern are computed by the Bayes formula:

$$P(\omega_i | e_1, \dots, e_K) = \frac{P(e_1, \dots, e_K | \omega_i) P(\omega_i)}{P(e_1, \dots, e_K)}, \quad i = 1, \dots, M. \tag{4.152}$$

The input pattern is classified to the class of maximum a posteriori probability, equivalently, the class of maximum likelihood:

$$\max_i P(e_1, \dots, e_K | \omega_i) P(\omega_i). \tag{4.153}$$

$P(\omega_i)$ and $P(e_1, \dots, e_K | \omega_i)$, the a priori probability and the joint conditional probability, can be estimated from a validation sample set. If the validation set contains N samples in total, $P(\omega_i)$ is approximated by N_i / N (N_i is the number of samples from class ω_i). Denote n_{ij}^k as the number of samples from ω_i that are classified to ω_j by classifier E_k, then n_{ij}^k / N_i approximates the probability $P(e_k = \omega_j | \omega_i)$. A matrix composed of the elements $P(e_k = \omega_j | \omega_i)$, $i, j = 1, \dots, M$, is called the confusion matrix of classifier E_k. Assume that the K decisions are independent, the joint conditional probability is reduced to

$$P(e_1 = \omega_{j_1}, \dots, e_K = \omega_{j_K} | \omega_i) = \prod_{k=1}^{K} P(e_k = \omega_{j_k} | \omega_i). \tag{4.154}$$

Inserting Eq. (4.154) into (4.153), this Naive Bayes formulation of combination can perform fairly well, even in combining dependent classifiers.

Bayesian combination by estimating the joint probability $P(e_1, \ldots, e_K | \omega_i)$ is also called behavior knowledge space (BKS) method [33], where the joint confusion of K classifiers is represented by a lookup table with M^{K+1} cells. A cell $n_{j_1,\ldots,j_K}(i)$ stores the number of samples from class ω_i that are classified to ω_{j_k} by classifier E_k, $k = 1, \ldots, K$. The conditional probability is then estimated by

$$P(e_1 = \omega_{j_1}, \ldots, e_K = \omega_{j_K} | \omega_i) = \frac{n_{j_1,\ldots,j_K}(i)}{N_i}. \tag{4.155}$$

The BKS method is expected to yield higher combination performance than the Naive Bayes if the joint probability $P(e_1, \ldots, e_K | \omega_i)$ is precisely estimated. In practice, however, the number of samples for estimating probabilities is limited compared to the number of cells M^{K+1}, and so the probabilities cannot be estimated with sufficient precision. A way to alleviate this problem is to approximate high-order probability ($K + 1$th order in combining K classifiers) with the product of lower order probabilities [38]. The storage complexity of this method is intermediate between the BKS method and the Naïve Bayes.

4.6.2.3 *Combination at Rank Level*   A simple rule for combining rank-level outputs is the Borda count [31]. Denote d_{ki} as the number of classes ranked below ω_i by classifier E_k, the total count of each class given by K classifiers is computed in the same way as Eq. (4.148). The input pattern is classified to the class of maximum count. For a two-class problem, the Borda count is equivalent to the sum of votes.

The Borda counts of multiple classifiers can also be combined by weighted sum, as in Eq. (4.150). The weights are estimated by optimizing an objective on a validation data set, particularly by logistic regression [31]. The same way is also used for combining abstract-level and measurement-level outputs. We will go into the details of weight estimation in the context of combining continuous outputs.

In weighted combination, the weights of classifiers are static once trained on validation data. For better performance, it is desirable to use dynamic weights because the strength of each individual classifier varies in different regions of a feature space (the features in this space can be those used by the individual classifiers, the outputs of classifiers, or any measures related to the behaviors of classifiers). As this philosophy also applies to combination at the abstract level and the measurement level, we will describe the details later in Section 4.6.4.

4.6.3   Combining Continuous Outputs

For combining continuous outputs of classifiers, we will focus on class-conscious combination rules, whereas class-indifferent combination can be accomplished using a generic classifier with parameters estimated on a validation data set. The class-conscious combination rules, especially weighted combination, perform sufficiently well in practice.

The transformation of classifier outputs is important because different classifiers often output scores with much different scales and physical meanings. The outputs should be at least normalized to the same scale, and, preferably, transformed to probability-like confidence scores (confidence transformation). On confidence scores, it is possible to achieve high performance using simple combination rules.

4.6.3.1 Confidence Transformation
Classifier outputs are commonly transformed to a posteriori probabilities using soft-max or sigmoid functions. We describe some details of sigmoid-based confidence transformation. We have shown in Section 4.3 that the a posteriori probability of a two-class Gaussian classifier is a sigmoid function. The outputs of most classifiers (for neural networks, the weighted outputs without sigmoid activation) can be viewed as linear or generalized linear discriminant functions and can be similarly transformed to a posteriori probabilities by sigmoid function.

The outputs of a generic classifier[8] can be transformed to confidence values via a scaling function [52]:

$$f_i(\mathbf{d}) = \sum_{j=1}^{M} \beta_{ij} d_j + \beta_{i0}, \quad i = 1, \ldots, M. \qquad (4.156)$$

Sigmoid scores are obtained by

$$z_i^s(\mathbf{d}) = \frac{1}{1 + e^{-f_i(\mathbf{d})}}. \qquad (4.157)$$

The sigmoid function, like that in Eqs. (4.49) and (4.130), primarily functions as a two-class a posteriori probability. In multiclass classification, $z_i^s(\mathbf{d})$ is actually the probability of ω_i from two classes ω_i and $\overline{\omega}_i$. The two-class sigmoid probabilities can be combined to multiclass probabilities according to the Dempster–Shafer theory of evidence [81, 3]. In the framework of discernment, we have $2M$ focal elements (singletons and negations) $\{\omega_1, \overline{\omega}_1, \ldots, \omega_M, \overline{\omega}_M\}$ with basic probability assignments (BPAs) $m_i(\omega_i) = z_i^s$, $m_i(\overline{\omega}_i) = 1 - z_i^s$, then the combined evidence of ω_i is

$$z_i^c = m(\omega_i) = A \cdot m_i(\omega_i) \prod_{j=1, j \neq i}^{M} m_j(\overline{\omega}_j) = A \cdot z_i^s \prod_{j=1, j \neq i}^{M} (1 - z_j^s), \qquad (4.158)$$

---

[8]We drop off the classifier index k when considering only one classifier in confidence transformation.

where

$$A^{-1} = \sum_{i=1}^{M} z_i^s \prod_{j=1, j\neq i}^{M} (1 - z_j^s) + \prod_{i=1}^{M}(1 - z_i^s).$$

The multiclass a posteriori probabilities $P(\omega_i|\mathbf{d}) = z_i^c$, $i = 1, \ldots, M$, satisfy $\sum_{i=1}^{M} z_i^c \leq 1$. The complement to one is the probability of input pattern being out of the M defined classes.

The scaling function is often reduced to the one taking only one input variable:

$$f_i(\mathbf{d}) = \beta_{i1} d_i + \beta_{i0}, \quad i = 1, \ldots, M. \tag{4.159}$$

The weight parameters can even be shared by M classes: $\beta_{i1} = \beta_1$, $\beta_{i0} = \beta_0$, $i = 1, \ldots, M$. The parameters can be estimated using two techniques: Bayesian formula assuming one-dimensional Gaussian densities on two classes (ω_i and $\overline{\omega}_i$) [80] and logistic regression [31, 67]. We introduce the two techniques in the following.

Similar to Eq. (4.49), assuming one-dimensional Gaussian densities for two classes (ω_i and $\overline{\omega}_i$) with means μ_i^+ and μ_i^- and equal variance σ_i^2, the a posteriori probability of ω_i versus $\overline{\omega}_i$ is

$$\begin{aligned} z_i^s(\mathbf{d}) &= \frac{P(\omega_i)p(d_i|\omega_i)}{P(\omega_i)p(d_i|\omega_i) + P(\overline{\omega}_i)p(d_i|\overline{\omega}_i)} \\ &= \frac{P(\omega_i) \exp\left[-\frac{(d_i-\mu_i^+)^2}{2\sigma_i^2} \right]}{P(\omega_i) \exp\left[-\frac{(d_i-\mu_i^+)^2}{2\sigma_i^2} \right] + P(\overline{\omega}_i) \exp\left[-\frac{(d_i-\mu_i^-)^2}{2\sigma_i^2} \right]} \\ &= \frac{1}{1 + e^{-\alpha[d_i-(\beta+\gamma/\alpha)]}}, \end{aligned} \tag{4.160}$$

where $\alpha = \frac{\mu_i^+ - \mu_i^-}{\sigma_i^2}$, $\beta = \frac{\mu_i^+ + \mu_i^-}{2}$, and $\gamma = \log \frac{P(\overline{\omega}_i)}{P(\omega_i)}$. The scaling function is extracted as

$$f_i(\mathbf{d}) = \alpha \left[d_i - \left(\beta + \frac{\gamma}{\alpha} \right) \right]. \tag{4.161}$$

The a priori probabilities and the parameters $\{\mu_i^+, \mu_i^-, \sigma_i^2\}$ are estimated on a validation data set by maximum likelihood (ML).

In logistic regression, the parameters of scaling functions $f_i(\mathbf{d})$, $i = 1, \ldots, M$, are optimized by gradient descent to maximize the likelihood on a validation data set $\{(\mathbf{d}^n, c^n)|n = 1, \ldots, N\}$ (c^n is the class label of sample $\mathbf{d}^n$):

$$\max L = \prod_{n=1}^{N} \prod_{i=1}^{M} [z_i^s(\mathbf{d}^n)]^{t_i^n} [1 - z_i^s(\mathbf{d}^n)]^{1-t_i^n}, \tag{4.162}$$

where the target probability is $t_i^n = 1$ for $i = c^n$ and $t_i^n = 0$ otherwise. Maximizing the likelihood is equivalent to minimizing the negative log-likelihood, which is actually the cross-entropy (CE) of Eq. (4.64). The parameters of scaling functions can also be optimized by minimizing the sum of the squared error of Eq. (4.52). Both the squared error and the cross-entropy can be regularized by weight decay, as that in Eq. (4.75).

On scaling parameter estimation by either one-dimensional Gaussian density modeling or logistic regression, the sigmoid scores are computed by Eq. (4.157) and are combined to multiclass probabilities by Eq. (4.158).

In the following description of combination rules for continuous outputs, d_{ki} is replaced by z_{ki}, $k = 1, \ldots, K$, $i = 1, \ldots, M$, which represent either transformed confidence scores or raw classifier outputs.

4.6.3.2 Fixed Rules   The confidence scores of a class given by K classifiers, z_{ki}, $k = 1, \ldots, K$, can be combined using fixed rules that have no tunable parameters to be estimated on validation data. The frequently considered rules are

- Sum rule:

$$y_i = \sum_{k=1}^{K} z_{ki}.$$

Regarding the decision of classification, the sum rule is equivalent to the so-called simple average: $y_i = \frac{1}{K} \sum_{k=1}^{K} z_{ki}$.

- Product rule:

$$y_i = \prod_{k=1}^{K} z_{ki}.$$

- Max rule:

$$y_i = \max_{k=1}^{K} z_{ki}.$$

- Min rule:

$$y_i = \min_{k=1}^{K} z_{ki}.$$

- Median rule: Reorder the K scores of class ω_i: $z_i^{(1)} \geq z_i^{(2)} \geq \cdots \geq z_i^{(K)}$, then if K is odd,

$$y_i = z_i^{(\frac{K+1}{2})},$$

and otherwise,

$$y_i = \frac{1}{2}\left(z_i^{(\frac{K}{2})} + z_i^{(\frac{K}{2}-1)}\right).$$

The input pattern is classified to the class of $\max_i y_i$.

If z_{ki} is an estimate of a posteriori probability $z_{ki} = P(\omega_i|\mathbf{x}_k)$, it is shown that the decision of input pattern to the maximum product of a posteriori probabilities is equivalent to the decision to the maximum joint a posteriori probability on K conditionally independent feature vectors [42]. Further, assume that the a posteriori probability does not deviate from the a priori probability significantly, the product rule is approximated to the sum rule. In practice, the sum rule performs fairly well even in combining dependent classifiers. It often outperforms the product rule because it is shown to be less sensitive to the estimation error of a posteriori probabilities [42].

4.6.3.3 *Weighted Combination*   In weighted combination, the fused score of a class is a weighted average of K individual scores:

$$y_i = \sum_{k=1}^{K} w_{ki} z_{ki}, \quad i = 1, \dots, M. \tag{4.163}$$

Usually, the weights are required to be non-negative, and, more strictly, sum to one: $\sum_{k=1}^{K} w_{ki} = 1$. Class-independent weights can be used instead of class-dependent ones: $w_{ki} = w_k$.

To show that weighted average (WA) is superior to simple average (SA), let us review some theoretical implications of Fumera and Roli [24]. Assume that the classifier outputs or transformed scores are the estimates of a posteriori probabilities with random prediction error:

$$z_{ki}(\mathbf{x}) = P(\omega_i|\mathbf{x}) + \epsilon_{ki}(\mathbf{x}), \tag{4.164}$$

where $\epsilon_{ki}(\mathbf{x})$ is a random variable with mean μ_{ki} and variance σ_{ki}^2. $\epsilon_{ki}(\mathbf{x})$ and $\epsilon_{lj}(\mathbf{x})$ ($i \neq j$) are assumed to be uncorrelated. Using class-independent weights, the weighted average of a class is

$$y_i^{\text{WA}} = \sum_{k=1}^{K} w_k z_{ki}(\mathbf{x}) = P(\omega_i|\mathbf{x}) + \epsilon_{ki}^{\text{WA}}(\mathbf{x}),$$

where $\epsilon_{ki}^{\text{WA}}(\mathbf{x}) = \sum_{k=1}^{K} w_k \epsilon_{ki}(\mathbf{x})$.

Compared to the Bayes error rate, the expected added error of WA, $E_{\text{add}}^{\text{WA}}$, is shown to depend on the a posteriori probability at the decision boundary, μ_{ki}, σ_{ki}^2, and the correlation coefficient $\rho_i(k, l)$ between $\epsilon_{ki}(\mathbf{x})$ and $\epsilon_{li}(\mathbf{x})$. Smaller correlation coefficients lead to smaller added error of ensemble. With optimal weights that minimize $E_{\text{add}}^{\text{WA}}$, the overall error rate of WA is not higher than the best individual classifier, whereas the SA is guaranteed only to be better than the worst individual classifier.

Assume unbiased and uncorrelated errors $\mu_{ki} = 0$, $\rho_i(k, l) = 0$, $\forall k, l, i$, the optimal weights can be solved analytically:

$$w_k = \frac{1}{E_{\text{add}}(k)} \left(\sum_{l=1}^{K} \frac{1}{E_{\text{add}}(l)} \right)^{-1}, \qquad (4.165)$$

where $E_{\text{add}}(k)$ is the expected added error of individual classifier k. It is easy to see that on K classifiers of equal added error, the optimal weights are equal. Under optimal weights, the difference of added error between WA and SA ensembles, $\Delta E = E_{\text{add}}^{\text{SA}} - \Delta E_{\text{add}}^{\text{WA}} \geq 0$ and increases when the added errors of individual classifiers are imbalanced.

It is also shown in [24] that assuming class-independent variances and correlations for $\epsilon_{ki}(\mathbf{x})$, ΔE is greater than zero even when the added errors of individual classifiers are equal, but the correlation coefficients are imbalanced.

In practice, to compute the optimal weights is not trivial. Even the formula (4.165) is not practical because the Bayes error rate is unknown. Empirically, the weights are estimated on a validation data set to optimize an objective. In combining neural networks for function approximation, Hashem solves the optimal weights by regression under the mean squared error (MSE) criterion [27]. This can be applied to classification as well by adding sigmoid function to the combined scores. Ueda optimizes the weights under the minimum classification error (MCE) criterion [90], where the combined scores need not be transformed by sigmoid.

Solving optimal weights to minimize the CE criterion (Eq. (4.64)) is generally called as logistic regression, where the combined class score is a sigmoid function of weighted average with a bias:

$$y_i = \frac{1}{1 + e^{-(\sum_{k=1}^{K} w_{ki} z_{ki} + w_{0i})}}. \qquad (4.166)$$

In this formulation, the weights can also be estimated under the MSE criterion. Either the MCE and CE or MSE criterion can be regularized by weight decay and can be optimized to update the weights by gradient descent.

The MCE method for weight estimation deserves some details as it had shown superior performance in [90]. Given N validation samples $(\mathbf{z}^n, c^n)$, $n = 1, \ldots, N$, where $\mathbf{z}^n$ is the vector of classifier outputs or confidence values, and c^n is the class label. On each sample, the combined class scores are computed as in Eq. (4.163). The empirical loss on the validation set is similar to that for learning vector quantization

(LVQ) as in Eq. (4.107), on which we add a regularization term:

$$
\begin{aligned}
L_1 &= \frac{1}{N}\sum_{n=1}^{N} l_{c^n}(\mathbf{z}^n) + \frac{\beta}{2}\sum_{k=1}^{K}\sum_{i=1}^{M} w_{ki}^2 \\
&= \frac{1}{N}\sum_{n=1}^{N}\left[l_{c^n}(\mathbf{z}^n) + \frac{\beta}{2N}\sum_{k=1}^{K}\sum_{i=1}^{M} w_{ki}^2 \right] \\
&= \frac{1}{N}\sum_{n=1}^{N} E(\mathbf{z}^n),
\end{aligned}
\tag{4.167}
$$

where $l_c(\mathbf{z}) = l_c(\mu_c) = \frac{1}{1+e^{-\mu_c}}$, $\mu_c = y_r - y_c$, and $y_r = \max_{i \neq c} y_i$ (c is the class label of pattern $\mathbf{z}$).

By stochastic gradient descent, the weights are iteratively updated on each input pattern. Specifically, the weights are moved along the negative direction of the gradient of $E(\mathbf{z}^n)$: $\Delta w_{ki} = -\eta \frac{\partial E(\mathbf{z}^n)}{\partial w_{ki}}$, where η is the learning rate, and

$$
\begin{aligned}
\frac{\partial E(\mathbf{z})}{\partial w_{ki}} &= l_c(1 - l_c)\frac{\partial(y_r - y_c)}{\partial w_{ki}} + \frac{\beta}{N} w_{ki} \\
&= \begin{cases}
-l_c(1 - l_c)z_{ki} + \dfrac{\beta}{N} w_{ki}, & \text{if } i = c, \\[2mm]
l_c(1 - l_c)z_{ki} + \dfrac{\beta}{N} w_{ki}, & \text{if } i = r, \\[2mm]
\dfrac{\beta}{N} w_{ki}, & \text{otherwise.}
\end{cases}
\end{aligned}
$$

In the case of class-independent weights, $y_i = \sum_{k=1}^{K} w_k z_{ki}$, $i = 1, \ldots, M$, and

$$
\frac{\partial E(\mathbf{z})}{\partial w_k} = l_c(1 - l_c)(z_{kr} - z_{kc}) + \frac{\beta}{N} w_k, \quad k = 1, \ldots, K.
$$

If the weights are constrained to be non-negative, each weight is checked on updating: if it is negative, it is modified to 0.

As to the effects of non-negativity constraint of weights and the bias, the experiments of Ting and Witten, using logistic regression, show that non-negative weights and free weights yield comparable performance [86]. On the contrary, the experiments of Ueda show that class-dependent weights outperform class-independent ones [90]. In practice, it is recommended to try the choices of both with and without non-negativity constraint, with and without bias, and both class-dependent and class-independent weights.

4.6.4   Dynamic Classifier Selection

Assume that the imbalance of individual classifier performances and the correlation between prediction errors remain constant in the feature space occupied by the union of features of the ensemble or the decision outputs of individual classifiers, the weighted combination scheme, using data-independent weights, is optimal in all regions of the feature space. This assumption, however, does not hold true in practice. The strengths of individual classifiers are generally variable in the feature space. It is hence desirable to use different (data-dependent) weights in different regions and select active individual classifier or combiner dynamically. This strategy, called dynamic classifier selection or dynamic combiner selection, has shown superior combination performance over constant combination in previous works [31, 97, 93].

In a dynamic classifier selection scheme proposed by Woods et al. [97], the local accuracies of individual classifiers are estimated in a region of the pattern feature space surrounding the unknown test sample $\mathbf{x}$. The local region is defined by the k-nearest neighbors of training samples to $\mathbf{x}$, and the local accuracies of individual classifiers are the correct rates on the nearest neighbors. The local accuracy is dichotomized into overall local accuracy and local class accuracy. The local class accuracy of a classifier E_i is estimated as the fraction of correctly classified training samples in the neighborhood whose assigned class labels by E_i are the same as $\mathbf{x}$. The decision of the individual classifier of the highest local accuracy is assigned to the test sample. Ties of multiple equally accurate classifiers are broken by majority vote.

For dynamic combiner selection, the feature space of pattern features or classifier outputs are more often partitioned into local regions by clustering the training or validation samples. Each region is represented by a cluster center, and in classification, the region of the test sample is decided by the nearest distance to the cluster center. In the decided region, the decision of the most accurate individual classifier is taken, or region-dependent weighted combination or meta-classifier is activated. Region partitioning by clustering is computationally much less expensive in operation than the k-nearest neighbor partitioning, as only the cluster centers are to be stored.

In combining rank-level decisions, Ho et al. proposed a dynamic combiner selection approach [31], where the training samples are partitioned into subsets (regions) according to the agreement on top choices of individual classifiers. Specifically, for K classifiers, there are $\binom{K}{K} + \binom{K}{K-1} + \cdots + \binom{K}{2} + \binom{K}{0}$ states (and, accordingly, regions) of agreement on the top choices from "all agree" to "all disagree." The test sample is first assigned to a region according to the agreement of individual decisions on it, and a locally optimal individual classifier is selected or the region-dependent logistic regression model is activated.

4.6.5   Ensemble Generation

The performance of multiple classifier systems not only depend on the combination scheme, but also rely on the complementariness (also referred to as independence or diversity) of the individual classifiers. Compared to the best individual classifier, the combination of independent classifiers can give maximum performance gain, but it is hard to make the individual classifiers totally independent. In practice, complementary classifiers can be generated by training with different sample sets, using different features, classifier structures, learning algorithms, and so on, for the individual classifiers. In character recognition, combining classifiers based on different pre-processing techniques is also effective.

Some approaches have been proposed to generate multiple classifiers systematically with the aim of optimizing the ensemble performance. We will describe some representatives of these: overproduce-and-select, bagging, random subspace, and boosting. The first approach selects a subset of individual classifiers from a large set of candidate classifiers. Bagging and random subspace methods generate classifiers in parallel, by randomly selecting training data and features, respectively. By boosting, the individual classifiers are generated sequentially, with each successive classifier designed to optimally complement the previous ones. Bagging, random subspace, and boosting methods are usually used to generate classifiers of the same structure on the same pattern features, wherein the underlying classifier structure is called base classifier or base learner.

4.6.5.1 *Overproduce and Select*   For character recognition, a very large number of classifiers can be generated by varying the techniques of preprocessing, feature extraction, classifier structure and learning algorithm, and so on. To combine a large number (say, 100) of classifiers is neither efficient in computation nor effective in performance. As the candidate classifier set may contain many correlated and redundant classifiers, it is possible to achieve a higher combination performance by selecting a subset from overly produced classifiers.

To select classifiers that are independent, one method is to partition the candidate classifiers into clusters such that the classifiers within a cluster are more dependent than those in different clusters. By defining a dependence measure (like a similarity) between two classifiers, the classifiers can be partitioned using a hierarchical agglomerative clustering algorithm [35]. On clustering, the ensemble is formed by drawing one classifier from each cluster [25]. The dependence measure is the rate of double fault, also called compound error rate, which is defined as the percentage of samples that are misclassified by both the two classifiers. A low compound error rate indicates that two classifiers complement well. The classifiers drawn from different clusters thus form an ensemble of good complementariness.

The problem of classifier subset selection is very similar to that of feature selection. On defining an evaluation criterion (called diversity [73]) of classifier subsets, the search methods for feature selection in Chapter 3, like forward search, backward search, genetic algorithm, and so on, can be used to find the optimal or near-optimal subset of classifiers. The diversity of a classifier subset is measured on a validation data set. The accuracy of classifier combination (e.g., by majority vote or confidence average) is immediately a diversity measure, but the selected classifiers maximizing the ensemble accuracy on validation data do not necessarily generalize well to unseen test data. A well-performing diversity measure is the compound diversity (CD), which is actually the complement of the compound error rate: CD = 1 − Prob(both the classifiers fail). The diversity of multiple classifiers can be measured as the average CD of all pairs of classifiers.

4.6.5.2 *Bagging and Random Subspace*   The term *bagging* is the acronym of *b*ootstrap *agg*regat*ing* [5]. In this method, each individual classifier is trained with a bootstrap replicate of the original training data set. Assume that the training data set contains N vectors, a bootstrap replicate is formed by randomly sampling the vectors with replacement for N times. The classifiers trained with different replicates of training data are combined by plurality vote or simple average.

Bagging can improve the accuracy over the single classifier trained with the original training data set when the base classifier is unstable: A small change of training data causes a large change to the output of prediction. Examples of unstable classifiers include neural networks and decision trees, whereas parametric Gaussian classifiers and nonparametric k-NN classifiers are fairly stable. From the statistical viewpoint, unstable classifiers have high variance of prediction error, and the average of multiple bagged predictors can reduce the variance.

The random subspace method generates multiple classifiers in parallel by sampling features instead of training vectors. Each individual classifier is trained with all the training samples, each represented by a randomly selected subset of features. A subset of features spans a subspace of the original feature space occupied by all the features. The random subspace method was experimented with a decision tree classifier in [30], but it is applicable to other classifier structures as well. According to the observations therein, combining random subspace ensemble is beneficial when the data set has a large number of features and samples and is not good when there are very few features coupled with a small number of samples. It is expected that the random subspace method is good when there is certain redundancy in the data set, especially for the collection of features.

4.6.5.3 *Boosting*   Boosting is a general method for improving the classification performance by combining multiple classifiers generated sequentially. The theoretical work of Schapire [78] shows that by boosting it is possible to convert a learning machine with error rate less than 0.5 to an ensemble with arbitrarily low error rate. The underlying learning machine (base classifier) is called a weak learner or weak classifier. On training the first classifier with the original training data set, each successive classifier is trained with reweighted training data (misclassified samples by the preceding classifier are assigned larger weights than correct ones) and is used to update the weights. After a number of iterations, the trained classifiers are combined by weighted plurality vote (can be extended to confidence-based combination).

The first successful implementation of boosting for pattern recognition was reported by Drucker et al. [16]. For handwritten digit recognition, they used a convolutional neural network as the base classifier and selected some deformed images to form the second and third training sets. Using only three classifiers, they could achieve a very high recognition accuracy.

Freund and Schapire proposed the AdaBoost (adaptive boosting) algorithm [20], which is the first principled algorithm and has triggered the multitude of theoretical and algorithmic research since the mid-1990s. The AdaBoost assumes that the base classifier gives binary hypothesis. Two extensions of AdaBoost, called AdaBoost.M1 and AdaBoost.M2, were proposed for training multiclass classifiers [19]. The AdaBoost.M1, covering AdaBoost as a special case, is applicable to a wide range of base classifiers. Assume that a weak classifier $f_t(\mathbf{x}) \in \{1, \ldots, M\}$ assigns an input pattern to one of M classes. Given a training data set $\{(\mathbf{x}_i, y_i)|i = 1, \ldots, N\}, y_i \in \{1, \ldots, M\}$, AdaBoost.M1 generates weak classifiers successively in a procedure, as described below.

Algorithm: AdaBoost.M1

1. Initialize data weights $D_1(i) = 1/N, i = 1, \ldots, N$.
2. Iterate for $t = 1, \ldots, T$,
    - Train weak classifier $f_t(\mathbf{x})$ with data distribution D_t.
    - Calculate the weighted error rate

$$\epsilon_t = \sum_{i=1}^{m} D_t(i) I[f_t(\mathbf{x}_i) \neq y_i].$$

   If $\epsilon_t > 0.5$, set $T = t - 1$ and terminate iteration.
    - Choose $\beta_t = \frac{\epsilon_t}{1-\epsilon_t}$.
    - Update weights:

$$D_{t+1}(i) = \frac{D_t(i)}{Z_t} \cdot \begin{cases} \beta_t & \text{if } f_t(\mathbf{x}_i) = y_i, \\ 1 & \text{otherwise}, \end{cases}$$

   where Z_t is used to normalize the weights to unity sum.
3. Output the final hypothesis by weighted plurality vote:

$$F(\mathbf{x}) = \arg \max_{y=1,\ldots,M} \sum_{t=1}^{T} \alpha_t I[f_t(\mathbf{x}) = y],$$

where $\alpha_t = \log \frac{1}{\beta_t}$.

On a data distribution D_t, the weak classifier $h_t(\mathbf{x})$ can be trained in two ways. One way is to directly incorporate the weights into classifier parameter estimation. This is possible for parametric statistical classifiers and the Naive Bayes classifier that approximate the density of each dimension with a histogram. Another way is to generate a new data set of N samples by resampling with replacement according to the distribution (the weight of each sample serves the probability of being drawn). This is applicable to any classifier structure. In practice, the two ways do not differ significantly in classification performance.

4.7 A CONCRETE EXAMPLE

To demonstrate the effects of some classifiers and combination schemes described in this chapter, we show some experimental results in a problem of handwritten digit recognition on a publicly available database. The database, called USPS, can be downloaded from the kernel machines Web site [39]. The digit images in the database were originally segmented from the envelop images of US Postal Service and were normalized to 16×16 images with 256-level gray values. The images were partitioned into two sets: 7291 images for training and 2007 images for testing. Some examples of the test images are shown in Figure 4.11.

We trained five individual classifiers, all using the 256 pixel values of a digit image as features. The individual classifiers are the MLP with one hidden layer of 50 units, the RBF network with 120 hidden units, the polynomial network classifier (PNC) with 60 principal components of features, the LVQ classifier with four prototypes per class trained under the MCE criterion, the MQDF classifier with 30 eigenvectors per class, and the minor eigenvalue estimated by fivefold cross-validation. The coefficient of weight decay was set equal to 0.05, 0.02, 0.1 for MLP, RBF, and PNC, respectively.

The validation data for confidence transformation and combiner training was generated by stacking with fivefold partitioning of training data. The outputs of neural classifiers are linear outputs without sigmoidal squashing. The output score of LVQ classifier is the nearest square Euclidean distance to each class, and the class score of MQDF is as in Eq. (4.18).

For combining the decisions of five classifiers at abstract level, we tested the plurality vote (PV), dynamic classifier selection with overall local accuracy (DCS-ov) and local class accuracy (DCS-lc) in three neighboring training points, nearest mean (NMean, also called as decision template in [48]), linear SVM, and weighted average (WA) with the weights estimated under the MCE criterion. The WA has four

FIGURE 4.11 Examples of handwritten digits in USPS database.

TABLE 4.1   Test accuracies (%) of individual classifiers, oracle, and DCS.

Classifier	MLP	RBF	PNC	LVQ	MQDF	Oracle	DCS-ov	DCS-lc
Top one	93.52	94.17	*95.12*	94.22	94.72	97.31	95.27	95.12
Top two	96.81	97.26	98.01	97.71	97.26			

variants: WA00, WA01, WA10, and WA11 (the first "1" denotes class-independent weights, and the second "1" denotes non-negative constraint). The weight decay coefficient in MCE-based estimation was set equal to 0.5.

For measurement-level combination, the classifier outputs are transformed to sigmoid scores, where the class-independent scaling parameters (Eq. (4.159)) were estimated by logistic regression with the coefficient of weight decay set equal to 1.0.

The accuracies of individual classifiers on the test digits are shown in Table 4.1, which also includes the accuracies of Oracle (at least one classifier assigns correct label) and DCS. The last row of the table shows the cumulative accuracies of top two ranks. We can see that the PNC yields the highest accuracy, 95.12%. The accuracy of Oracle indicates that 2.69% of test samples are misclassified by all the individual classifiers. Combining the decisions of five classifiers by DCS-ov (overall local accuracy) yields a higher accuracy (95.27%) than the best individual classifier.

The accuracies of abstract-level combination by plurality vote (PV) and trained rules are shown in Table 4.2. The PV yields an accuracy of 94.92%, which is lower than the PNC but higher than the other individual classifiers. The class-indifferent combiners, NMean and SVM, yield accuracies comparable to the best individual classifiers. The comparable accuracies of four variants of WA indicate that the sharing of class weights and the non-negative constraint are not influential. The highest accuracy is obtained by WA00, with free class-specific weights.

The accuracies of measurement-level combination by fixed rules (sum, product, and median) and trained rules (NMean, SVM, and four variants of WA) are shown in Table 4.3. By fusing confidence measures, the fixed rules yield lower accuracies than the abstract-level combination (PV). However, it is evident that measurement-level combination (except the NMean) maintains a higher cumulative accuracy of two ranks than abstract-level combination. The trained combiners, SVM and WA, yield higher accuracies than the best individual classifier, and again the four variants of WA do not differ significantly. The highest accuracy, 95.32%, is obtained by WA00.

It is interesting to look into the classifier weights of WA. On normalizing the non-negative weights of WA11 into unity sum, the weights corresponding to five classifiers

TABLE 4.2   Accuracies (%) of combination at abstract level: plurality vote and trained rules.

Method	PV	NMean	SVM	WA00	WA01	WA10	WA11
Top one	94.92	95.02	95.07	*95.22*	95.12	95.07	95.17
Top two	97.16	97.36	96.96	97.01	96.91	97.16	97.21

TABLE 4.3 Accuracies (%) of combination at measurement level: fixed and trained rules.

Method	Sum	Prod	Median	NMean	SVM	WA00	WA01	WA10	WA11
Top one	94.62	94.62	94.52	94.57	94.97	95.32	95.17	95.17	95.17
Top two	97.71	97.51	97.46	97.31	97.26	97.71	97.56	97.81	97.91

are $\{0.23, 0, 0.43, 0.12, 0.22\}$ for the abstract level and $\{0.06, 0, 0.42, 0, 0.53\}$ for the measurement level. The best two individual classifiers, PNC and MQDF, are assigned large weights. The RBF classifier, though performing better than MLP, is totally ignored in combination. The decision of measurement-level combination is almost made by only two classifiers: PNC and MQDF.

In another experiment, we combined only two classifiers, PNC and MQDF. The test accuracies of plurality vote and sum rule (average of sigmoidal confidence) are 95.17 and 95.37%, respectively. The higher accuracy of averaging two classifiers than the weighted combination of a larger set of five classifiers can be explained that the validation data set is not representative enough for estimating the combining weights. We observed that by confidence averaging of two classifiers, eight test digits misclassified by PNC were corrected, as shown in Figure 4.12.

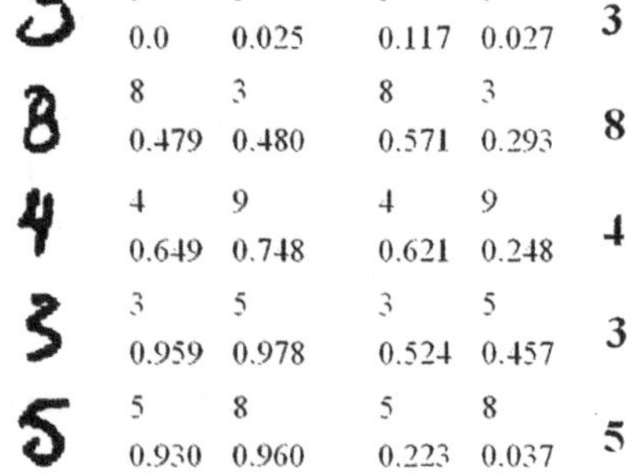

FIGURE 4.12 Test digits corrected by combining PNC and MQDF. The second and third columns show the confidence values of two classes by PNC, the fourth and fifth columns show the confidence values of two classes by MQDF, and the rightmost column shows the final decision.

4.8 CHAPTER SUMMARY

Oriented to character recognition applications, this chapter first gives an overview of pattern classification methods, and then goes into details for some representative methods, including those that have been long proven effective and some emerging ones that are still under research. Some classifiers and multiple classifier combination schemes are tested in an experiment of handwritten digit recognition.

Statistical pattern recognition is based on the Bayes decision theory and is instantiated by classifiers based on parametric and nonparametric density estimation. Its principles are also important for better understanding and implementing neural networks, support vector machines (SVMs), and multiple classifier systems. Unlike statistical methods that are based on classwise density estimation, neural networks, SVMs, and boosting are based on discriminative learning, that is, their parameters are estimated with the aim of optimizing a classification objective. Discriminative classifiers can yield higher generalization accuracies when trained with a large number of samples.

For structural pattern recognition, we describe only two methods that have been widely used, especially in online character recognition: attributed string matching and attributed graph matching. Despite that the automatic learning of structural models from samples is not well solved, structural recognition methods have some advantages over statistical methods and neural networks: They interpret the structure of characters, store less parameters, and are sometimes more accurate.

Neural networks are considered to be pragmatic and somewhat obsolete compared to SVMs and boosting, but, actually, they yield competitive performance at much lower (training and operation) complexity. Nevertheless, for neural classifiers to achieve good performance, skilled implementation of model selection and nonlinear optimization are required. Potentially higher accuracies can be obtained by SVMs and multiple classifier methods.

Though the classifiers described in this chapter assume single character recognition, they are all applicable to integrated segmentation and recognition of character strings. For string recognition, the objective is no longer the accuracy of single characters, but the holistic accuracy of string segmentation-recognition. Accordingly, some classifier structures and learning algorithms need to be customized. Classifier design strategies oriented to string recognition have been addressed in Chapter 5.

REFERENCES

1. N. E. Ayat, M. Cheriet, and C. Y. Suen. Automatic model selection for the optimization of SVM kernels. *Pattern Recognition.* **38**(10), 1733–1745, 2005.
2. G. H. Bakir, L. Bottou, and J. Weston. Breaking SVM complexity with cross-training. In *Advances in Neural Information Processing Systems 18*. Vancouver, Canada, 2004.
3. J. A. Barnett. Computational methods for a mathematical theory of evidence. In *Proceedings of the 7th International Joint Conference on Artificial Intelligence*. Vancouver, Canada, 1981, pp. 868–875.

4. C. M. Bishop. *Neural Networks for Pattern Recognition.* Oxford University Press, 1995.

5. L. Breiman. Bagging predictors. *Machine Learning.* **24**(2), 123–140, 1996.

6. C. J. C. Burges. Simplified support vector decision rules. In *Proceedings of the 13th International Conference on Machine Learning.* Bari, Italy, 1996, pp. 71–77.

7. C. J. C. Burges. A tutorial on support vector machines for pattern recognition. *Knowledge Discovery and Data Mining.* **2**(2), 1–43, 1998.

8. C. C. Chang and C. J. Lin. *LIBSVM: A library for support vector machines.* Technical Report, Department of Computer Science and Information Engineering, National Taiwan University, 2003.

9. O. Chapelle, V. Vapnik, O. Bousquet, and S. Mukherjee. Choosing multiple parameters for support vector machines. *Machine Learning.* **46**(1–3), 131–159, 2002.

10. N. Cristianini and J. Shawe-Taylor. *An Introduction to Support Vector Machines and Other Kernel-Based Learning Methods.* Cambridge University Press, 2000.

11. A. P. Dempster, N. M. Laird, and D. B. Rubin. Maximum likelihood from incomplete data via the EM algorithm. *Journal of Royal Statistical Society B.* **39**(1), 1–38, 1977.

12. P. A. Devijver and J. Kittler. *Pattern Recognition: A Statistical Approach.* Prentice Hall, Englewood Cliffs, NJ, 1982.

13. J.-X. Dong, A. Krzyzak, and C. Y. Suen. Fast SVM training algorithm with decomposition on very large data sets. *IEEE Transactions on Pattern Analysis and Machine Intelligence.* **27**(4), 603–618, 2005.

14. T. Downs. Exact simplification of support vector solutions. *Journal of Machine Learning Research.* **2**, 293–297, 2001.

15. P. M. L. Drezet and R. F. Harrison. A new method for sparsity control in support vector classification and regression. *Pattern Recognition.* **34**(1), 111–125, 2001.

16. H. Drucker, R. Scahpire, and P. Simard. Boosting performance in neural networks. *International Journal on Pattern Recognition and Artificial Intelligence.* **7**(4), 705–719, 1993.

17. R. O. Duda, P. E. Hart, and D. G. Stork. *Pattern Classification.* 2nd edition, Wiley Interscience, New York, 2001.

18. R. P. W. Duin. The combining classifiers: To train or not to train. In *Proceedings of the 16th International Conference on Pattern Recognition.* Quebec, Canada, 2002, Vol. 2, pp. 765–770.

19. Y. Freund and R. E. Schapire. Experiments with a new boosting algorithm, In *Proceedings of the 13th International Conference on Machine Learning.* Bari, Italy, 1996, pp. 148–156.

20. Y. Freund and R. E. Schapire. A decision-theoretic generalization of on-line learning and an application to boosting. *Journal of Computer and System Sciences.* **55**(1), 119–139, 1997.

21. H. Friedman. Regularized discriminant analysis. *Journal of the American Statistical Association.* **84**(405), 166–175, 1989.

22. K. S. Fu and B. K. Bhargava. *Syntactic Methods for Pattern Recognition.* Academic Press, 1974.

23. K. Fukunaga. *Introduction to Statistical Pattern Recognition.* 2nd edition, Academic Press, 1990.

24. G. Fumera and F. Roli. A theoretical and experimental analysis of linear combiners for multiple classifier systems. *IEEE Transactions on Pattern Analysis and Machine Intelligence.* **27**(6), 942–856, 2005.

25. G. Giacinto and F. Roli. An approach to the automatic design of multiple classifier systems. *Pattern Recognition Letters.* **22**(1), 25–33, 2001.

26. R. M. Gray. Vector quantization. *IEEE ASSP Magazine.* **1**(2), 4–29, 1984.

27. S. Hashem. Optimal linear combinations of neural networks. *Neural Networks.* **10**(4), 599–614, 1997.

28. T. Hastie and R. Tibshirani. Classification by pairwise coupling. *The Annals of Statistics.* **26**(2), 451–471, 1998.

29. S. Haykin. *Neural Networks: A Comprehensive Foundation.* Prentice-Hall, Inc., 1999.

30. T. K. Ho. The random subspace method for constructing decision forests. *IEEE Transactions on Pattern Analysis on Machine Intelligence.* **20**(8), 832–844, 1998.

31. T. K. Ho, J. Hull, and S. N. Srihari. Decision combination in multiple classifier systems. *IEEE Transactions on Pattern Analysis on Machine Intelligence.* **16**(1), 66–75, 1994.

32. L. Holmström, P. Koistinen, J. Laaksonen, and E. Oja. Neural and statistical classifiers— taxonomy and two case studies. *IEEE Transactions on Neural Networks.* **8**(1), 5–17, 1997.

33. Y. S. Huang and C. Y. Suen. A method of combining multiple experts for the recognition of unconstrained handwritten numerals. *IEEE Transactions on Pattern Analysis on Machine Intelligence.* **17**(1), 90–94, 1995.

34. A. K. Jain, R. P. W. Duin, and J. Mao. Statistical pattern recognition: A review. *IEEE Transactions on Pattern Analysis on Machine Intelligence.* **22**(1), 4–37, 2000.

35. A. K. Jain, M. N. Murty, and P. J. Flynn. Data clustering: A review. *ACM Computing Survey.* **31**(3), 264–323, 1999.

36. T. Joachims. Making large-scale support vector machine learning practical. In B. Schölkopf, C. J. C. Burges, and A. J. Smola, editors, *Advances in Kernel Methods: Support Vector Learning.* MIT Press, 1999, pp. 169–184.

37. B.-H. Juang and S. Katagiri. Discriminative learning for minimum error classification. *IEEE Transactions on Signal Processing.* **40**(12), 3043–3054, 1992.

38. H.-J. Kang, K. Kim, and J. H. Kim. Optimal approximation of discrete probability distribution with kth-order dependency and its application to combining multiple classifiers. *Pattern Recognition Letters.* **18**(6), 515–523, 1997.

39. Kernel machines web site. http://www.kernel-machines.org/

40. F. Kimura, S. Inoue, T. Wakabayashi, S. Tsuruoka, and Y. Miyake. Handwritten numeral recognition using autoassociative neural networks. In *Proceedings of the 14th International Conference on Pattern Recognition,* Brisbane, Australia, 1998, Vol. 1, pp. 166–171.

41. F. Kimura, K. Takashina, S. Tsuruoka, and Y. Miyake. Modified quadratic discriminant functions and the application to Chinese character recognition. *IEEE Transactions on Pattern Analysis on Machine Intelligence.* **9**(1), 149–153, 1987.

42. J. Kittler, M. Hatef, R. P. W. Duin, and J. Matas. On combining classifiers. *IEEE Transactions on Pattern Analysis on Machine Intelligence.* **20**(3), 226–239, 1998.

43. T. Kohonen. The self-organizing map. *Proceedings of the IEEE.* **78**(9), 1464–1480, 1990.

44. T. Kohonen. Improved versions of learning vector quantization. In *Proceedings of the 1990 International Joint Conference on Neural Networks*. San Diego, CA, Vol. 1, pp. 545–550.

45. U. Kreßel. Pairwise classification and support vector machines. In B. Schölkopf, C. J. C. Burges, and A. J. Smola, editors, *Advances in Kernel Methods: Support Vector Learning*. MIT Press, 1999, pp. 255–268.

46. U. Kreßel and J. Schürmann. Pattern classification techniques based on function approximation. In H. Bunke and P. S. P. Wang, editors, *Handbook of Character Recognition and Document Image Analysis*. World Scientific, Singapore, 1997, pp. 49–78.

47. L. I. Kuncheva. *Combining Pattern Classifiers: Methods and Algorithms*. Wiley Interscience, 2004.

48. L. I. Kuncheva, J.C. Bezdek, and R.P.W. Duin. Decision templates for multiple classifier fusion: an experimental comparison. *Pattern Recognition*. **34**(2), 299–314, 2001.

49. L. Lam and C. Y. Suen. Optimal combinations of pattern classifiers. *Pattern Recognition Letters*. **16**, 945–954, 1995.

50. L. Lam and C. Y. Suen. Application of majority voting to pattern recognition: an analysis of its behavior and performance. *IEEE Transactions on System Man Cybernet. Part A*. **27**(5), 553–568, 1997.

51. Y. LeCun, L. Bottou, Y. Bengio, and P. Haffner. Gradient-based learning applied to document recognition. *Proceedings of the IEEE*. **86**(11), 2278–2324, 1998.

52. C.-L. Liu. Classifier combination based on confidence transformation. *Pattern Recognition*. **38**(1), 11–28, 2005.

53. C.-L. Liu, S. Jaeger, and M. Nakagawa. Online handwritten Chinese character recognition: The state of the art. *IEEE Transactions on Pattern Analysis on Machine Intelligence*. **26**(2), 198–213, 2004.

54. C.-L. Liu, I.-J. Kim, and J. H. Kim. Model-based stroke extraction and matching for handwritten Chinese character recognition. *Pattern Recognition*. **34**(12), 2339–2352, 2001.

55. C.-L. Liu and M. Nakagawa. Evaluation of prototype learning algorithms for nearest neighbor classifier in application to handwritten character recognition. *Pattern Recognition*. **34**(3), 601–615, 2001.

56. C.-L. Liu, K. Nakashima, H. Sako, and H. Fujisawa. Handwritten digit recognition: Benchmarking of state-of-the-art techniques. *Pattern Recognition*. **36**(10), 2271–2285, 2003.

57. C.-L. Liu, H. Sako, and H. Fujisawa. Discriminative learning quadratic discriminant function for handwriting recognition. *IEEE Transactions on Neural Networks*. **15**(2), 430–444, 2004.

58. O. L. Mangasarian and D. R. Musicant. Successive overrelaxation for support vector machines. *IEEE Transactions on Neural Networks*. **10**(5), 1032–1037, 1999.

59. S. Mannor, D. Peleg, and R. Y. Rubinstein. The cross entropy method for classification. In L. De Raedt and S. Wrobel, editors, *Proceedings of the 22nd International Conference on Machine Learning*. Bonn, Germany, ACM Press, 2005.

60. T. M. Martinetz, S. G. Berkovich, and K. J. Schulten. "Neural-gas" network for vector quantization and its application to time-series prediction. *IEEE Transactions on Neural Networks*. **4**(4), 558–569, 1993.

61. A. Marzal and E. Vidal. Computation of normalized edit distance and applications. *IEEE Transactions on Pattern Analysis on Machine Intelligence.* **15**(9), 926–932, 1993.

62. W. S. McCulloch and W. Pitts. A logical calculus of the ideas immanent in nervous activity. *Bulletin of Mathematical Biophysics.* **5**, 115–133, 1943.

63. B. Moghaddam and A. Pentland. Probabilistic visual learning for object representation. *IEEE Transactions on Pattern Analysis on Machine Intelligence.* **19**(7), 696–710, 1997.

64. N. J. Nilsson. *Principles of Artificial Intelligence.* Springer-Verlag, 1980.

65. I.-S. Oh and C. Y. Suen. A class-modular feedforward neural network for handwriting recognition. *Pattern Recognition.* **35**(1), 229–244, 2002.

66. P. Parent and S. W. Zucher. Radial projection: an efficient update rule for relaxation labeling. *IEEE Transactions on Pattern Analysis on Machine Intelligence.* **11**(8), 886–889, 1989.

67. J.C. Platt. Fast training of support vector machines using sequential minimal optimization. In B. Schölkopf, C. J. C. Burges, and A. J. Smola, editors, *Advances in Kernel Methods: Support Vector Learning.* MIT Press, 1999, pp. 185–208.

68. J. Platt. Probabilistic outputs for support vector machines and comparisons to regularized likelihood methods. In A. J. Smola, P. Bartlett, D. Scholkopf, and D. Schuurmanns, editors, *Advances in Large Margin Classifiers.* MIT Press, 1999.

69. J. Platt, N. Cristianini, and J. Shawe-Taylor. Large margin DAGs for multiclass classification. In *Advances in Neural Information Processing Systems 12.* MIT Press, 2000, pp. 547–553.

70. A. F. R. Rahman and M. C. Fairhurst. Multiple classifier decision combination strategies for character recognition: A review. *International Journal on Document Analysis and Recognition.* **5**(4), 166–194, 2003.

71. M. D. Richard, R. P. Lippmann. Neural network classifiers estimate Bayesian a posteriori probabilities. *Neural Computation.* **4**, 461–483, 1991.

72. H. Robbins and S. Monro. A stochastic approximation method. *Annals of Mathematical Statistics.* **22**, 400–407, 1951.

73. F. Roli, G. Giacinto, and G. Vernazza. Methods for designing multiple classifiers systems. In J. Kittler and F. Roli, editors, *Multiple Classifier Systems.* LNCS Vol. 2096, Springer, 2001, pp. 78–87.

74. A. Rosenfeld, R. A. Hummel, and S. W. Zucker. Scene labeling by relaxation operations. *IEEE Transactions on System, Man, and Cybernetics.* **6**(6), 420–433, 1976.

75. D. E. Rumelhart, G. E. Hinton, and R. J. Williams. Learning internal representations by error propagation. In D. E. Rumelhart, J. L. McClelland, and the PDP Research Group, editors, *Parallel Distributed Processing: Explorations in the Microstructure of Cognition.* MIT Press, Cambridge, MA, 1986, Vol. 1, pp. 318–362.

76. A. Sanfeliu and K.-S. Fu. A distance measure between attributed relational graphs for pattern recognition. *IEEE Transactions on System, Man, and Cybernetics.* **13**(3), 353–362, 1983.

77. A. Sato and K. Yamada. A formulation of learning vector quantization using a new misclassification measure. In *Proceedings of the 14th International Conference on Pattern Recognition.* Brisbane, 1998, Vol. I, pp. 322–325.

78. R. E. Schapire. The strength of weak learnability. *Machine Learning.* **5**(2), 197–227, 1990.

79. B. Schölkopf and A. J. Smola. *Learning with Kernels.* MIT Press, Cambridge, MA, 2002.

80. J. Schürmann. *Pattern Classification: A Unified View of Statistical and Neural Approaches.* Wiley Interscience, New York, 1996.

81. G. Shafer. *A Mathematical Theory of Evidence.* Princeton University Press, 1976.

82. Y. Shin and J. Ghosh. Ridge polynomial networks. *IEEE Transactions on Neural Networks.* **6**(3), 610–622, 1995.

83. C. Y. Suen and L. Lam. Multiple classifier combination methodologies for different output levels. In J. Kittler and F. Roli, editors, *Multiple Classifier Systems.* LNCS Vol. 1857, Springer, 2000, pp. 52–66.

84. C. Y. Suen, C. Nadal, R. Legault, T. A. Mai, and L. Lam. Computer recognition of unconstrained handwritten numerals. *Proceedings of the IEEE.* **80**(7), 1162–1180, 1992.

85. C. C. Tappert, C. Y. Suen, and T. Wakahara. The state of the art in on-line handwriting recognition. *IEEE Transactions on Pattern Analysis on Machine Intelligence.* **12**(8), 787–808, 1990.

86. K. M. Ting and I. H. Witten. Issues in stacked generalization. *Journal on Artificial Intelligence Research.* **10**, 271–289, 1999.

87. W.-H. Tsai and K.-S. Fu. Subgraph error-correcting isomorphisms for syntactic pattern recognition. *IEEE Transactions on System, Man, and Cybernetics.* **13**(1), 48–62, 1983.

88. W.-H. Tsai and S.-S. Yu. Attributed string matching with merging for shape recognition. *IEEE Transactions on Pattern Analysis on Machine Intelligence.* **7**(4), 453–462, 1985.

89. Y.-T. Tsay and W.-H. Tsai. Attributed string matching by split-and-merge for on-line Chinese character recognition. *IEEE Transactions on Pattern Analysis on Machine Intelligence.* **15**(2), 180–185, 1993.

90. N. Ueda. Optimal linear combination of neural networks for improving classification performance. *IEEE Transactions on Pattern Analysis on Machine Intelligence.* **22**(2), 207–215, 2000.

91. J. R. Ullmann. An algorithm for subgraph isomorphism. *Journal of ACM.* **23**(1), 31–42, 1976.

92. V. Vapnik. *The Nature of Statistical Learning Theory.* Springer, New Work, 1995.

93. A. Verikas, A. Lipnickas, K. Malmqvist, M. Bacauskiene, and A. Gelzinis. Soft combination of neural classifiers: A comparative study. *Pattern Recognition Letters.* **20**(4), 429–444, 1999.

94. R. A. Wagner and M. J. Fischer. The string-to-string correction problem. *Journal of ACM.* **21**(1), 168–173, 1974.

95. D. H. Wolpert. Stacked generalization. *Neural Networks.* **5**, 241–259, 1992.

96. A. K. C. Wong, M. You, and S. C. Chan. An algorithm for graph optimal monomorphism, *IEEE Transactions on System, Man, and Cybernetics.* **20**(3), 628–636, 1990.

97. K. Woods, W. P. Kegelmeyer, and K. Bowyer. Combination of multiple classifiers using local accuracy estimates. *IEEE Transactions on Pattern Analysis on Machine Intelligence.* **19**(4), 405–410, 1997.

98. L. Xu, A. Krzyzak, and C. Y. Suen. Methods of combining multiple classifiers and their applications to handwriting recognition. *IEEE Transactions on System, Man, and Cybernetics.* **22**(3), 418–435, 1992.

99. E. Yair, K. Zeger, and A. Gersho. Competitive learning and soft competition vector quantizer design. *IEEE Transactions on Signal Processing.* **40**, 294–309, 1992.

100. K. Yamamoto and A. Rosenfeld. Recognition of handprinted Kanji characters by a relaxation method. In *Proceedings of the 6th International Conference on Pattern Recognition.* Munich, Germany, 1982, pp. 395–398.

CHAPTER 5

WORD AND STRING RECOGNITION

Practical documents consist of words or character strings rather than isolated characters as elementary units. Very often, the characters composing a word or string cannot be reliably segmented before they are recognized. Although the character feature extraction and classification techniques described in the previous chapters are integral for word/string recognition, special strategies are required to integrate segmentation and classification. This chapter addresses such kind of strategies, including explicit segmentation, implicit segmentation, classifier design, path search, lexicon organization, holistic recognition, and so on. It is organized in an analytical-to-holistic fashion: starts from oversegmentation, then describes the techniques involved in classification-based (character-model-based) recognition, hidden Markov models (HMMs)-based recognition, and ends with holistic recognition.

5.1 INTRODUCTION

The techniques of feature extraction described in Chapter 3 and those of classification in Chapter 4 are primarily aimed for isolated character recognition. In practical document images and online handwritten texts, however, characters are not in isolation. In English texts, words are separated by apparent space, but the letters within a word are not well separated, so words can be considered as natural units to recognize. In Chinese and Japanese texts, a sentence separated by punctuation marks is an integral unit because there is no difference between the interword and intercharacter

Character Recognition Systems: A Guide for Students and Practitioner, by M. Cheriet, N. Kharma, C.-L. Liu and C. Y. Suen Copyright © 2007 John Wiley & Sons, Inc.

gaps. Word or sentence recognition, or generally, character string recognition, faces the difficulty of character segmentation: the constituent characters cannot be reliably segmented before they are recognized (Sayre's paradox [82]). Viewing a whole word/string as a single pattern circumvents segmentation, but then, the number of pattern classes will be huge!

This chapter describes strategies and techniques for word/string recognition. As a recognition method, especially segmentation-based one, can apply to both word and string recognition, we will refer to word and string interchangeably. The context of language, available in most environments, can greatly reduce the ambiguity in the segmentation and recognition of characters. The language context can be defined by a set of legal words, called a lexicon, or a statistical form (often, the n-gram [85]) in an open vocabulary. It can be applied while segmentation and recognition are being performed or in a postprocessing stage after lexicon-free recognition.

Numerous word/string recognition methods have been developed since the 1970s, and a major progress was made in the 1990s [1, 75, 83, 89]. Many recognition systems have reported fairly high recognition accuracies, sometimes for very large vocabularies with tens of thousands of words [51]. According to the underlying techniques of segmentation, character recognition, and linguistic processing, we categorize the string recognition methods as in Figure 5.1, which is similar to the hierarchy of segmentation methods of Casey and Lecolinet [13].

Depending on whether the characters are segmented or not, the word recognition methods can be categorized into two major groups: analytical (segmentation-based) and holistic (segmentation-free). Holistic recognition, where each lexicon entry (word class) should be built in a model, generally applies to small and fixed lexicons. Segmentation-based methods, which form string classes by concatenating character classes, apply to both lexicon-driven and lexicon-free recognition and can accommodate a very large vocabulary. Lexicon-free recognition is typically applied to numeral strings, where little linguistic context is available, or to word/string recognition followed by linguistic postprocessing.

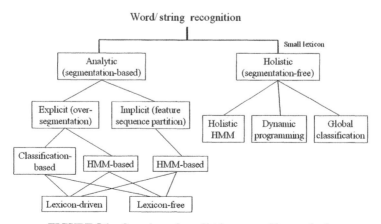

FIGURE 5.1 Overview of word/string recognition methods.

Segmentation methods can be generally divided into explicit ones and implicit ones. Explicit segmentation tries to separate the string image at character boundaries and usually results in an over list of candidate segmentation points. So, explicit segmentation is also called oversegmentation. Implicit segmentation blindly slices the string image into frames of equal size (width). The frames, each represented by a feature vector, are grouped into characters by recognition. Explicit oversegmentation can be similarly performed by slicing (merging slices into candidate character patterns) or windowing (e.g., [68]). The process of oversegmentation and candidate pattern formation is also called presegmentation.

The character recognizer for segmentation-based string recognition can be any classifier or a multiple classifier system as described in Chapter 4. A mathematical tool, hidden Markov models (HMM) [76], is often used with implicit segmentation, but can be used for modeling character classes in explicit segmentation and modeling word classes in holistic recognition as well. Character-model HMM for segmentation-based recognition is also called path discriminant, whereas holistic word-model HMM is also called model discriminant. Holistic recognition can also be performed by dynamic programming (DP) for word template matching and classification based on vector representation of global features.

The preprocessing techniques for word recognition (image smoothing, deskewing, slant correction, baseline detection, etc.) have been discussed in Chapter 2. In the remainder of this chapter, we first describe the segmentation techniques, focusing on presegmentation. We then describe explicit segmentation-based methods under a statistical string classification model, which unifies character classification and context analysis based on oversegmentation. We introduce HMM-based recognition in a separate section due to the special characteristics of HMMs compared to the other classification tools. Holistic recognition methods are described with emphasis on global feature extraction and word template matching.

We assume that the object to recognize is an offline word/string image, but except preprocessing and presegmentation, most strategies are common to online and offline handwriting recognition.

5.2 CHARACTER SEGMENTATION

Machine-printed or hand-drawn scripts can have various font types or writing styles. The writing styles can be roughly categorized into discrete style (handprint or boxed style), continuous style (cursive style), and mixed style. Figure 5.2 shows some character string images of different languages and various styles. We can see that the ambiguity of character segmentation has three major sources: (1) variability of character size and intercharacter space; (2) confusion between intercharacter and within-character space (some characters have multiple components); and (3) touching between characters. To solve the ambiguity, segmentation is often integrated with character recognition: character patterns are hypothesized and verified by recognition. Linguistic context is effective to disambiguate both the classes and the boundaries of

由最近距离确定检验字符的书写人

举杯邀明月 对影成三人

新泽市東狄山纽2-2004

Please print the following text

People of the United

Justice promote

SS174184 78874 12286

504692 815AG 3009

FIGURE 5.2 Character strings of different languages and styles (English words and numeral strings are separated by white space).

characters (it is the case that a composite character can be split into multiple legal characters).

The problem of segmentation has been attacked by many works, among which are many techniques of image dissection, integrated segmentation and recognition, and lexicon manipulation [13, 62]. As string recognition methods will be specially addressed later, in this section we describe only presegmentation (oversegmentation) techniques, which decompose a string image into subunits (primitive segments).

Slicing and windowing techniques do not consider the shape features in string image. To guarantee that all the characters are separated apart, the number of candidate character patterns will be very large. This brings heavy burden to the succeeding character recognition and lexical matching procedures, and may deteriorate the string recognition accuracy because of the imperfection of character verification and recognition. In the following, we consider only the dissection techniques that take into account shape cues to detect character-like components. We first give an overview of dissection techniques, and then describe an approach specially for segmenting handwritten digits.

5.2.1 Overview of Dissection Techniques

In oversegmentation, a character can be split into multiple primitive segments, but different characters should be contained in different segments. Thus, the candidate patterns formed by merging successive segments will contain the true characters, which are to be detected by recognition. The ideal case that each character is contained in a unique segment is hard to achieve, but it is hoped that the number of segments is as small as possible.

Projection analysis (PA) and connected component analysis (CCA) are two popular segmentation techniques. If the characters in a string image are not connected, all the characters can be formed by merging connected components (CCs). PA succeeds if each pair of neighboring characters can be separated by a white space (a straight line in background). For touching characters (multiple characters contained in one connected component), special techniques are needed to split at the touching points.

(a)

print   following Mid zone

Threshold   Threshold

(b)   (c)

FIGURE 5.3   Character segmentation by projection analysis. (a) Characters separable by vertical projection; (b) separation at points of small projection; (c) separation by projection in the middle zone.

In the following, we briefly describe PA, CCA, and a ligature analysis technique for segmenting cursive words.

5.2.1.1 *Projection Analysis*

If the characters in a string image of horizontal writing are separated by vertical straight lines, the image can be decomposed into segments at the breaks of horizontal projection profile (vertical projection) (Fig. 5.3(a)). If the characters can be separated by slanted lines but not vertical lines, then projection at an angle can solve, but this can be converted to vertically separable case by slant correction. Further, for string image separable by projection, if the within-character gap is apparently smaller than the intercharacter gap, then the characters can be unambiguously segmented before recognition. Nevertheless, this only applies to very limited cases.

PA can also separate touching characters in some special touching situations (usually for machine-printed characters). If the border between two characters has only one touching point and the projection at this point is small, the string image can be oversegmented at points of small projection below a threshold (Fig. 5.3 (b)). For some scripts, characters cannot be separated by projecting the whole string image, but are separable in a zone of the image (Fig. 5.3 (c)).

5.2.1.2 *Connected Component Analysis*

CCs are generally considered in binary images. Two pixels are said to be 8-connected (4-connected) if they are connected by a chain of 8-connected (4-connected) pixels. A CC is a set of pixels in which each pixel is connected to the rest. Usually, CCs in foreground (pixels valued 1) are defined under 8-connectivity and those in background (pixels valued 0) are defined under 4-connectivity. In document images, the stroke pixels of the same character are connected more often than those between different characters. If there are no touching characters, then all the characters can be separated by CCA. Touching characters that have stroke pixels in a common CC should then be split at the points of touching.

Character segmentation by CCA is accomplished in several steps: CC labeling, overlapping components merging, and touching character splitting. By CC labeling, the pixels of different components are stored in different sets or labeled with different pixel values. There have been many effective algorithms for labeling CCs, which can be roughly divided into two categories: raster scan [37, 86] and contour tracing [14].

FIGURE 5.4   Character segmentation by connected component analysis. Upper: connected components (left) and segments after merging horizontally overlapping components.

By raster scan, all the CCs can be found in two passes (forward and background) or a forward scan with local backtracking. By contour tracing, the pixels enclosed by different outer contours (outer and inner contours are traced in different directions) belong to different CCs.

For multicomponent characters like Chinese characters, the components of the same character largely overlap on the axis of writing direction. The heavily overlapping components can be safely merged because the components of different characters overlap less frequently and more slightly. The degree of overlapping can be quantitatively measured (e.g., [56, 60]). Figure 5.4 shows some examples of segmentation by CCA. Each component in Figure 5.4 is enclosed within a rectangular box. The five characters of the upper string are decomposed into 18 CCs, which are merged into eight components (primitive segments). The lower left string has a character disconnected into two CCs, and the lower right one has a CC containing two touching characters. All the three string images have overlapping components between different characters. This can be easily separated by CCA but cannot be separated by projection.

5.2.1.3 Ligature Analysis

5.2.1.3 Ligature Analysis   Handwritten characters in cursive scripts are connected frequently. Very often, the letters of a word form one single connected component (see Figure 5.5). The stroke connecting two neighboring characters is called a ligature. Generally, the ligature turns out to be a horizontal sequence of vertical black runs in binary images. In the region of ligature, most columns of the image have only one black run, and the height of the run is small (corresponding to a horizontal or nearly horizontal stroke). These columns are called single-run columns, and the corresponding ligature is "visible" from both the top and the bottom of the image. To separate the word image into primitive segments at visible ligatures, first the sequences of single-run columns are found. For each sequence, if the upper or lower contour has an apparent valley (the lowest point significantly lower than the other points), the word image is separated at the valley, otherwise separated at the middle of the horizontal sequence of runs.

After separating the word image at visible ligatures, if any primitive segment is wide enough (compared to the height of middle zone of word image), it is a potential touching pattern, in which two or more characters are connected by "invisible" ligatures. Invisible ligatures can be detected in two cases. In one case, the lower contour of word image has an apparent valley, at which the image can be separated forcedly (left word of Fig. 5.5). In another case, the ligature can be traced on the outer contour.

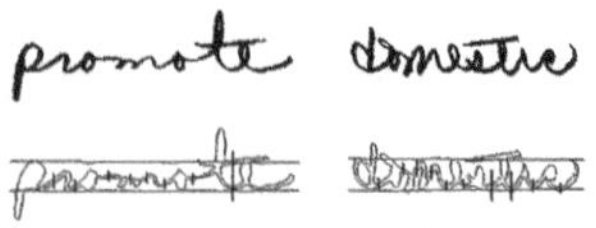

promote domestic

FIGURE 5.5   Cursive word segmentation at visible ligatures (denoted by short vertical lines) and invisible ligatures (denoted by long vertical lines).

On detecting the ligature in this case, the word image is separated at the valley or the middle of the ligature (right word of Fig. 5.5).

5.2.2   Segmentation of Handwritten Digits

The segmentation and recognition of unconstrained handwritten numeral strings is one of the most difficult problems in the area of optical character recognition (OCR). It has been a popular topic of research for many years and has many potential applications such as postal code recognition, bank check processing, tax form reading, and reading of many other special forms. Generally, numeral strings contain isolated (with a lot of variations), touched, overlapped, and noisy or broken digits. Some examples of these cases are shown in Figure 5.6. The segmentation of touching digits is one of the main challenges in handwritten numeral recognition systems. Even though handwritten digits are not so frequently touched as cursive words, they show different features of touching points, and hence need special techniques to separate.

Many algorithms have been proposed for touching digits segmentation, and generally they can be classified into three categories. The first uses foreground features extracted from the black pixels in the image, frequently, by contour shape analysis [84]. The second uses background features extracted from the white pixels in the image [20, 61], and the third category uses a combination of both foreground and background features [19, 80]. Normally, approaches of the third category yield better results. For numeral string recognition, the segmentation module first provides a list of candidate character patterns, and each of them is evaluated by the character

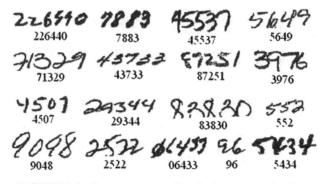

FIGURE 5.6   Some examples of handwritten numeral strings.

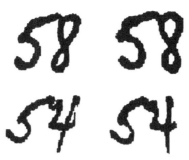

FIGURE 5.7   Numeral string image preprocessing (smoothing and slant correction). Left: original; right: preprocessed.

recognition module. The character recognition scores are postprocessed to find the best partitioning of character patterns.

Before segmentation and recognition, the numeral string image is preprocessed using the techniques introduced in Chapter 2. First, the string image is smoothed in order to remove small spots of noise and to smooth the contour of characters. Slant correction is then performed to make the segmentation task easier and the segmentation paths as straight as possible. Some results of preprocessing operations are shown in Figure 5.7.

If the digits in a string image are not touched, the digits can be easily separated by CCA. Touching digits in one connected component need to be separated at the character boundary by shape analysis. In the following, we focus on touching digits separation utilizing foreground and background features.

5.2.2.1   *Feature Points' Extraction*   To find feature points on the foreground (black pixels) of each connected component, an algorithm based on skeleton tracing is introduced. First, by using the Zhang–Suen thinning algorithm introduced in Chapter 2, the skeleton of each connected component is extracted (see Figure 5.8.) On the skeleton, a starting point and an ending point (denoted by S and E, respectively) are found. The starting point is the leftmost skeleton point (if there are multiple skeleton points in the leftmost column, the top point), and the ending point is the rightmost skeleton point (top point in the rightmost column). From the starting point, the skeleton is traced to the ending point in two different directions: clockwise and counterclockwise. In Figure 5.8(c), the trace in clockwise direction is denoted as top skeleton, and the trace in counterclockwise direction is denoted as bottom skeleton.

In clockwise and counterclockwise tracing, the intersection points (IPs) on the skeleton are detected. The IPs are the points that have more than two 8-connected branches on the skeleton. Some intersection points on the top/bottom skeleton may be visited more than once. Corresponding to each visit of an intersection point at the skeleton, the bisector of the angle formed by two branches of the skeleton intersects with the outer contour of the connected component. This intersection point at the outer contour is called a foreground feature point (denoted by in □ Figure 5.8(d).

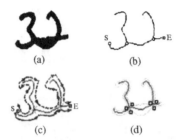

FIGURE 5.8 Foreground feature points' extraction. (a) Original image; (b) skeleton after thinning, starting point (S) and ending point (E) are denoted by □; (c) the skeleton is traversed from left to right in two directions: clockwise and counterclockwise; (d) mapping intersection points to the foreground features on the outer contour (denoted by □).

These feature points can be divided into two subsets: top-foreground features corresponding to the top skeleton and bottom-foreground features corresponding to the bottom skeleton. These feature points carry important information about the location of touching strokes and segmentation regions.

For background feature extraction, first the vertical top and bottom projection profiles of the connected component image are found, as seen in Figure 5.9 (b) and (c). On thinning these profile images (black regions), the skeletons, as seen in Figure 5.9 (e) and (f)), are called top-background skeleton and bottom-background skeleton, respectively. The branch end points of these skeletons sometimes do not reach the segmentation regions (an example is shown in Figure 5.10(a)). As shown in Figure 5.10(b), all parts of the top-background skeleton that are below the middle line (the line divides the image by two equal halves) and all parts of bottom-background skeleton above the middle line are removed. Then, the end points on the top- and bottom-background skeletons (denoted by □ in Figure 5.10(b)) are found for later use in segmentation path construction (wherein the first and last end points on each skeleton are not considered).

5.2.2.2 Constructing Segmentation Paths The feature points extracted from foreground and background are divided into four subsets: top- and bottom-

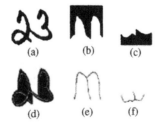

FIGURE 5.9 Background skeleton extraction. (a) Preprocessed connected component image; (b) top projection profile; (c) bottom projection profile; (d) background region (white pixels outside the black object); (e) top-background skeleton; (f) bottom-background skeleton.

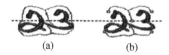

(a)   (b)

FIGURE 5.10   Background feature points' extraction. (a) Original top/bottom-background skeletons; (b) after removing the parts crossing the middle line, the end points are background features.

foreground features, and top- and bottom-background features. From these points, two paths of search are conducted: upward search and downward search. If two feature points in two different subsets match each other, they are connected to construct a segmentation path. Two feature points, A and B are said to match each other if

$$|x_A - x_B| \leq \alpha \cdot H, \quad \alpha \in [0.25, 0.5],$$

where x_A and x_B are the horizontal coordinates of A and B, respectively. The constant α is taken to be 0.4, and H is the vertical height of the string image. The detailed procedure of segmentation path construction is depicted in the flowchart of Figure 5.11, and some examples are shown in Figure 5.12.

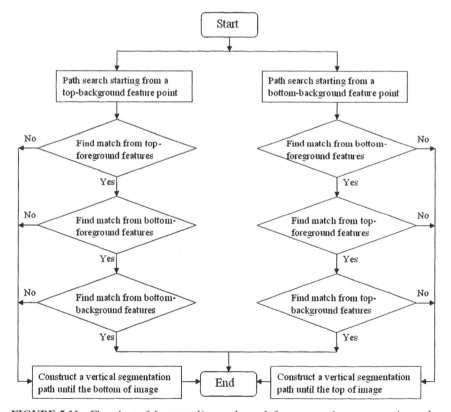

FIGURE 5.11   Flowchart of downward/upward search for constructing segmentation paths.

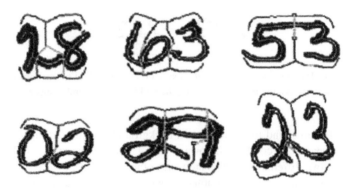

FIGURE 5.12 Feature points on the background and foreground (from the top and bottom) are matched, and connected to construct segmentation paths.

5.3 CLASSIFICATION-BASED STRING RECOGNITION

We call a string recognition method based on explicit segmentation and character recognition as classification-based, as a character classifier is underlying for verifying and classifying the candidate character patterns hypothesized in presegmentation. It is also called character-model-based or, in the case of word recognition, letter-model-based. By the character classifier, the candidate patterns are assigned confidence (probability or similarity/dissimilarity) scores to defined classes (characters in an alphabet), which are combined to measure the confidence of segmentation hypothesis and string class. A string class is a sequence of character classes assigned to the segmented character patterns. The combination of a segmentation hypothesis (or called segmentation candidate) and its string class is also called a segmentation-recognition hypothesis (or segmentation-recognition candidate).

If segmentation is stressed, classification-based string recognition is also called classification-based or recognition-based segmentation. On segmentation, the segmented character patterns can be reclassified using a more accurate classifier (some classifiers are more suitable for classification than for segmentation) or combining multiple classifiers.

The final goal of string recognition is to classify the string image (or online string pattern) to a string class. Character segmentation, character recognition, and linguistic processing can be formulated into a unified string classification model. Under this model, the optimal string class (associated with a segmentation hypothesis) of maximum confidence is selected. How to determine the optimal string class involves sophisticated operations of presegmentation, character classification, linguistics representation, path search, and so on. In the following, we first describe the unified string classification model, and then go into some details of the operations.

5.3.1 String Classification Model

A string image can be partitioned into character patterns in many ways, and each segmented pattern can be assigned to different character classes. Figure 5.13 shows

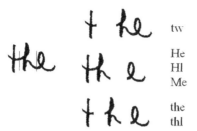

FIGURE 5.13   A word image (left), its segmented pattern sequences (middle), and string classes (right).

the segmentation of a simple word image into different pattern sequences (segmentation hypotheses) and the string classes assigned to the pattern sequences. The task of string recognition is to find the optimal segmentation and its optimal string class, incorporating character recognition scores, character likeliness (geometric context), and optionally, linguistic context. To do this, the segmentation candidate and the string class should be first evaluated by a string confidence measure.

Many works have tried to score segmentation candidate and string class probabilistically [10, 60, 72, 88]. Viewing string segmentation-recognition as a string classification problem, that is, classifying the string image to one of all string classes, according to the Bayes decision rule (see Chapter 4), the minimum classification error rate will be achieved by classifying the string image to the string class of maximum a posteriori (MAP) probability. The computation of a posteriori probabilities (or other types of scores) of string classes on a string image encounters two difficulties. First, the number of string classes is very large. In lexicon-free recognition, the number of string classes is as many as $\sum_{n=1}^{L} M^n$ (L is the maximum string length and M is the number of character classes). Even in lexicon-driven recognition, it is hard to compute the scores of all string classes in the lexicon. Second, for scoring a string class, the string image needs to be segmented into characters, wherein segmentation and character recognition interact. The number of segmentation candidates can be fairly large.

5.3.1.1 Probabilistic Models   An early string class probability model can be found in [10]. Assume that a string image X is segmented into n candidate patterns $S = s_1, \ldots, s_n$ (note that there are many segmentation candidates even with the same string length) and is assigned to a string class $W = w_1 \cdots w_n$ (character w_i is assigned to s_i), the a posteriori probability of the string class is

$$P(W|X) = \sum_{S} P(W, S|X). \tag{5.1}$$

The segmentation candidate is constrained to have the same length as W: $|S| = |W| = n$, and the candidate patterns are represented by feature vectors[1] $X = \mathbf{x}_1, \ldots, \mathbf{x}_n$. To avoid summing over multiple segmentation candidates in Eq. (5.1), the optimal string class can instead be decided by

$$W^* = \arg \max_W \max_S P(W, S|X). \tag{5.2}$$

This is to search for the optimal segmentation candidate S for each string class. $P(W, S|X)$ is decomposed into

$$P(W, S|X) = \frac{P(X|W, S)P(W, S)}{P(X)} = \frac{P(X|W, S)P(S|W)P(W)}{P(X)}, \tag{5.3}$$

and by assuming context independence of character shapes, can be approximated as

$$P(W, S|X) \approx P(W) \prod_{i=1}^{n} \frac{p(\mathbf{x}_i|w_i, s_i)P(s_i|w_i)}{p(\mathbf{x}_i)} = P(W) \prod_{i=1}^{n} \frac{P(w_i, s_i|\mathbf{x}_i)}{P(w_i)}. \tag{5.4}$$

The a priori probability of string class $P(W)$ represents the linguistic context. It is assumed to be a constant for lexicon-free recognition. For lexicon-driven recognition, $P(W)$ can be assumed to be equal among the string classes in the lexicon. For an open vocabulary, it is often approximated by second-order statistics (bigram):

$$P(W) = P(w_1) \prod_{i=2}^{n} P(w_i|w_{i-1}). \tag{5.5}$$

$P(w_i, s_i|\mathbf{x}_i)$ in Eq. (5.4) can be decomposed into

$$P(w_i, s_i|\mathbf{x}_i) = P(w_i|s_i, \mathbf{x}_i)P(s_i|\mathbf{x}_i). \tag{5.6}$$

Breuel used a multilayer neural network to directly output the a posteriori probability $P(w_i|\mathbf{x}_i)$ (approximating $P(w_i|s_i, \mathbf{x}_i)$) but did not specify how to estimate $P(s_i|\mathbf{x}_i)$. We will see that $P(s_i|\mathbf{x}_i)$ measures the plausibility of character in geometric context (character likeliness considering the relative width, height, and position in string, and the relationship with neighboring characters). Ignoring the geometric context, $P(s_i|\mathbf{x}_i)$ can be viewed as a constant, and the string class probability of Eq. (5.4) is approximately

$$P(W, S|X) \approx P(W) \prod_{i=1}^{n} \frac{P(w_i|s_i, \mathbf{x}_i)}{P(w_i)} = P(W) \prod_{i=1}^{n} \frac{P(w_i|\mathbf{x}_i)}{P(w_i)}. \tag{5.7}$$

---

[1]For describing character patterns in feature vectors, please refer to Chapter 3.

In the context of Japanese online handwriting recognition, Nakagawa et al. proposed a string class probability model incorporating the geometry of intercharacter gap [72]. The candidate pattern sequence is denoted by $S = s_1 g_1 s_2 g_2 \cdots s_n g_n$, where s_i represents the geometric features of the ith pattern, including the width and height of bounding box, the within-character gaps, and so on, and g_i represents the between-character gap, and possibly, the relative size and position between characters. In the formula of (5.3), $P(X)$ is omitted because it is independent of string class. Hence, the pattern sequence is classified to the string class of max $P(X|W, S)P(S|W)P(W)$. $P(W)$ is estimated using a bigram as in Eq. (5.5). Assuming context-independent character shapes, $P(X|W, S)$ is approximated by

$$P(X|W, S) \approx \prod_{i=1}^{n} p(\mathbf{x}_i|w_i, s_i) \approx \prod_{i=1}^{n} p(\mathbf{x}_i|w_i), \qquad (5.8)$$

where $p(\mathbf{x}_i|w_i)$ is the likelihood of pattern $\mathbf{x}_i$ with respect to class w_i, which will be estimated by a character classifier.

The likelihood of geometric context, $P(S|W)$, plays an important role in string segmentation. Different characters or characters versus noncharacters differ in the distribution of geometric features (size and position in the string, within-character gap), and two patterns of classes $w_i w_{i+1}$ in the same string observe special distributions of relative size/position and gap between them. These features can be incorporated into the probability $P(S|W)$ (character likeliness or compatibility score) to improve string segmentation and recognition.

In [72], $P(S|W)$ is estimated by

$$P(S|W) \approx \prod_{i=1}^{n} P(s_i|w_i)P(g_i|w_i w_{i+1}), \qquad (5.9)$$

where $P(s_i|w_i)$ and $P(g_i|w_i w_{i+1})$ can be seen as character likeliness and between-character compatibility, respectively.

On extracting single-character and between-character geometric features, the scores of likeliness and compatibility can be estimated by parametric density functions or artificial neural networks, both trained with character samples segmented from strings. The samples can be labeled as some superclasses instead of character classes, such as character versus noncharacter (two classes), core (C), ascender (A), and descender (D) (three classes [32]). The compatibility between two patterns can similarly be labeled as "compatible" and "incompatible" (two classes) or the combinations of character superclasses.

5.3.1.2 *Simplified and Heuristic Models*

The geometric context is often ignored in integrated segmentation-recognition of strings. Rather, it is used in presegmentation to preclude candidate character patterns, or in postprocessing to compare a few selected segmentation candidates. Even without geometric context, if the character classifier is trained to be resistant to noncharacters (all defined classes are assigned low confidence values on noncharacter patterns), it can still give high string recognition accuracy [60]. Without geometric context score, the string class probability model is simplified to

$$P(W|X) = \frac{P(X|W)P(W)}{P(X)} \approx P(W) \prod_{i=1}^{n} \frac{p(\mathbf{x}_i|w_i)}{p(\mathbf{x}_i)}. \tag{5.10}$$

The right-hand side of Eq. (5.10) is equivalent to that of Eq. (5.7). By omitting the string-class-independent term $P(X)$, the string pattern is classified to

$$W^* = \arg\max_W P(W)P(X|W) = \arg\max_W P(W) \prod_{i=1}^{n} p(\mathbf{x}_i|w_i). \tag{5.11}$$

When $P(W)$ is equal (lexicon-free or closed-lexicon-driven), the classification rule is further simplified to

$$W^* = \arg\max_W P(X|W) = \arg\max_W \prod_{i=1}^{n} p(\mathbf{x}_i|w_i). \tag{5.12}$$

Care should be taken when using the rule of (5.12) for string recognition. A string pattern can be segmented into character pattern sequences of variable lengths. For any pattern $\mathbf{x}_i$ and character class w_i, the likelihood $p(\mathbf{x}_i|w_i) < 1$ mostly, and the score $\prod_{i=1}^{n} p(\mathbf{x}_i|w_i)$ decreases as the string length n increases. Thus, when comparing the scores of string classes of different lengths, the scores of shorter strings will be favored. This will raise the segmentation error of merging multiple characters into one pattern. To overcome this bias, Tulyakov and Govindaraju proposed a normalized string probability score:

$$W^* = \arg\max_W \left(\prod_{i=1}^{n} p(\mathbf{x}_i|w_i) \right)^{\frac{1}{n}}. \tag{5.13}$$

To explain the poor performance of the simplified rules (5.11) and (5.12) in string recognition, let us examine whether omitting $P(X)$ is reasonable or not. When the segmented pattern sequence $\mathbf{x}_1 \cdots \mathbf{x}_n$ is fixed, $P(X) = P(\mathbf{x}_1 \cdots \mathbf{x}_n)$ is really string class independent and the omission is reasonable. When comparing two string classes W_1 and W_2 on two different sequences X_1 and X_2 (segmented from the same string pattern), however, $P(X_1)$ and $P(X_2)$ are different and should not be omitted. On the contrary, the decision rule of Eq. (5.13) performs fairly well in practice.

In contrast to Eq. (5.12), the bias to short strings does not happen to the right-hand formula of (5.7). Though $P(w_i|\mathbf{x}_i)$ is proportional to $p(\mathbf{x}_i|w_i)$ ($P(w_i|\mathbf{x}_i)$ is obtained by normalizing $p(\mathbf{x}_i|w_i)p(\mathbf{x}_i)$ to unity sum for M classes) and $P(w_i|\mathbf{x}_i) \leq 1$, the denominator $P(w_i)$ prevents the product from shrinking with increased n. String classification according to (5.7), however, is sensitive to the precision of a posteriori probabilities estimation, whereas the estimation of conditional probability density in (5.13) can be circumvented by the proportional relationship between classifier outputs and log-likelihood [60] as follows.

For parametric statistical classifiers, such as linear discriminant function (LDF) and quadratic discriminant function (QDF), the discriminant score of a class ω_i equals $\log P(\omega_i)p(\mathbf{x}|\omega_i) + \mathrm{CI}$ (CI is a class-independent term), and when assuming equal a priori probabilities, is equivalent to $\log p(\mathbf{x}|\omega_i) + \mathrm{CI}$. LDF and QDF are derived from the logarithm of multivariate Gaussian density function, and the negative of QDF can be viewed as a distance metric (a special case is the square Euclidean distance). For neural network classifiers, considering the analogy of output layer to LDF, the output score (weighted sum without nonlinear squashing) can be regarded to be proportional to the log-likelihood:

$$y_i(\mathbf{x}) = y(\mathbf{x}, \omega_i) \propto \log p(\mathbf{x}|\omega_i). \qquad (5.14)$$

For distance-based classifiers, taking the analogy of distance metric with the negative of QDF, the class distance can be regarded to be proportional to the minus log-likelihood:

$$d(\mathbf{x}, \omega_i) \propto -\log p(\mathbf{x}|\omega_i).$$

Viewing $d(\mathbf{x}, \omega_k) = -y_k(\mathbf{x})$, the output scores of all classifiers can be related to conditional probabilities as (5.14). The string classification rule of (5.12) is then equivalent to

$$W^* = \arg\max_W \sum_{i=1}^{n} y(\mathbf{x}_i, w_i) = \arg\min_W \sum_{i=1}^{n} d(\mathbf{x}_i, w_i),$$

and (5.13) is equivalent to

$$W^* = \arg\max_W \frac{1}{n}\sum_{i=1}^{n} y(\mathbf{x}_i, w_i) = \arg\min_W \frac{1}{n}\sum_{i=1}^{n} d(\mathbf{x}_i, w_i). \qquad (5.15)$$

By this formulation, the classifier outputs, regarded as generalized similarity or distance measures, can be directly used to score string classes.

The geometric context likelihood in (5.9) can be similarly converted to

$$\log(P(S|W))^{\frac{1}{n}} = \frac{1}{n}\sum_{i=1}^{n} [\log P(s_i|w_i) + \log P(g_i|w_i w_{i+1})], \qquad (5.16)$$

where $\log P(s_i|w_i)$ and $\log P(g_i|w_i w_{i+1})$ are, respectively, proportional to $y(s_i, w_i)$ and $y(g_i, w_i w_{i+1})$, the output scores of classifiers on geometric features.

The transformed heuristic scores of character classification, geometric context and linguistic context can be combined to a normalized string class score as a weighted sum:

$$\text{NL}(X, W) = \frac{1}{n} \sum_{i=1}^{n} [y(\mathbf{x}_i, w_i) + \beta_1 y(s_i, w_i) + \beta_2 y(g_i, w_i w_{i+1}) + \gamma \log P(W)],$$
(5.17)

or equivalently, a distance measure of string class:

$$\text{ND}(X, W) = \frac{1}{n} \sum_{i=1}^{n} [d(\mathbf{x}_i, w_i) + \beta_1 d(s_i, w_i) + \beta_2 d(g_i, w_i w_{i+1}) - \gamma \log P(W)].$$
(5.18)

Assuming equal a priori string class probabilities, the term $\log P(W)$ can be omitted. The weights $\{\beta_1, \beta_2, \gamma\}$ can be adjusted empirically for customizing to different applications.

In (5.17) and (5.18), $y(s_i, w_i)$ and $d(s_i, w_i)$ measure the geometric context plausibility of single characters, and $y(g_i, w_i w_{i+1})$ and $d(g_i, w_i w_{i+1})$ measure the plausibility of pairs of characters, usually pairs of neighboring characters only.

Some researchers have utilized the geometric context in string recognition in different ways, but the obtained scores can be generally incorporated by the framework of (5.17) or (5.18). Xue and Govindaraju compute for a character pattern the distance between its geometric features and the class expectation [92]. Gader et al. train a neural network (called intercharacter compatibility network) to output the probabilities of a number of superclasses on a pair of neighboring characters [32].

The segmentation statistics of Kim and Govindaraju [47] can be viewed as a geometric context measure. Based on oversegmentation, the distribution of number of primitive segments (upper bounded by a defined maximum number) for each character class is estimated on training string samples. In string recognition, the probability of the number of segments of a candidate pattern for a character class is combined with the classification score for evaluating the string class.

5.3.2 Classifier Design for String Recognition

In Eq. (5.17) or (5.18), $y(\mathbf{x}_i, w_i)$ or $d(\mathbf{x}_i, w_i)$ is given by a character classifier, to represent the measure of similarity or distance of the candidate character pattern (represented by a feature vector $\mathbf{x}_i$) to a character class w_i. The classifier structure can be selected from those described in Chapter 4 and the HMM. On estimating the classifier parameters on a set of character samples (character-level training), the classifier can then be inserted into the string recognition system to perform candidate pattern classification.

A classifier trained on segmented character samples does not necessarily perform well in string recognition. This is because the candidate patterns generated in pre-segmentation include both characters and noncharacters. The character class scores on noncharacter patterns are somewhat arbitrary and can mislead the comparison of segmentation candidates: It is probable that a segmentation with noncharacters has a higher confidence than the one composed of correctly segmented characters. This can be overcome by two ways: enhancing the noncharacter resistance in character-level training or training the classifier directly on string samples.

5.3.2.1 Character-Level Training

Most classifiers are trained to separate the patterns of M defined classes regardless of outliers (patterns that do not belong to the defined classes). Typically, a classifier partitions the feature space into disjoint decision regions corresponding to different classes. If the whole feature space is partitioned into regions of the defined classes, an outlier falling in a region and distant from decision boundaries will be classified to the corresponding class with high confidence.

Rejection has been considered in pattern recognition: When it is not sure enough to which class the input pattern belongs, the pattern is rejected. There are two types of rejects: ambiguity reject and distance reject [25]. A pattern is ambiguous when its membership degrees to multiple classes are comparable. Normalizing the a posteriori probabilities of M defined classes to unity sum, ambiguity can be judged when the maximum a posteriori probability is below a threshold [22]. Distance reject is the case that the input pattern does not belong to any of the defined classes; that is, it is an outlier. It can be judged by the mixture probability density or the maximum conditional probability density below a threshold.

In character-level classifier training, noncharacter resistance can be obtained by three methods: density-based classifier design, hybrid generative–discriminative training, and training with noncharacter samples [11, 60]. In the following, we briefly discuss the mathematical consideration of outlier rejection, and then describe the classifier design methods.

Consider classification in a feature space, the patterns of M defined classes inhabit only a portion of the space, and the complement is inhabited by outliers, which we include into an "outlier" class ω_0. The closed world assumption that the a posteriori probabilities sum up to unity is now replaced by an open class world:

$$\sum_{i=0}^{M} P(\omega_i|\mathbf{x}) = 1.$$

Apparently, $\sum_{i=1}^{M} P(\omega_i|\mathbf{x}) \leq 1$, and the complement is the probability of outlier. The a posteriori probabilities are computed according to the Bayes formula:

$$P(\omega_i|\mathbf{x}) = \frac{P(\omega_i)p(\mathbf{x}|\omega_i)}{p(\mathbf{x})} = \frac{P(\omega_i)p(\mathbf{x}|\omega_i)}{\sum_{j=0}^{M} P(\omega_j)p(\mathbf{x}|\omega_j)}.$$

As $p(\mathbf{x}|\omega_0)$ is hard to estimate due to the absence or insufficiency of outlier samples, outlier can be judged according to $\sum_{i=1}^{M} P(\omega_i)p(\mathbf{x}|\omega_i)$ or $\max_{i}^{M} p(\mathbf{x}|\omega_i)$ below a threshold.

When using the string class score of (5.7) or (5.13) for string recognition, the a posteriori probability $P(\omega_i|\mathbf{x})$ and the conditional probability $p(\mathbf{x}|\omega_i)$ are required to be estimated in the open class world.

Assume that the regions of M-class patterns and outliers are disjoint or overlap slightly in the feature space, classifiers that approximate the class-conditional probability density or the log-likelihood are inherently resistant to outliers, because in the region of outliers, the density of all the defined classes is low. The density functions of defined classes can be estimated from training samples by parametric methods (like the Gaussian density), semiparameter methods (like mixture of Gaussians), or non-parametric methods (like the Parzen window). In character recognition, the QDF and the modified QDF (MQDF) [50], derived from multivariate Gaussian density, are both accurate in classification and resistant to outliers, and have shown good performance in string segmentation and recognition [48, 49, 81].

Probability density functions can be viewed as generative models, which include HMMs and structural descriptions as well. In contrast, discriminative classifiers, like neural networks and support vector machines (SVMs), have discriminant functions connected to the a posteriori probabilities more than to the probability density. The parameters of discriminant functions are adjusted such that the training samples of different classes are maximally separated without regarding the boundary between in-class patterns and outliers. For obtaining outlier resistance, the classifier can be trained discriminatively with outlier (noncharacter) samples. Noncharacter samples can be collected from the candidate patterns generated in presegmentation of string images.

In training with outlier samples, the classifier can have either M or $M + 1$ output units (discriminant functions), corresponding to M-class or $(M + 1)$-class classification. When outliers are treated in an extra class, the classifier is trained in the same way as an M-class classifier is trained. After training the $(M + 1)$-class classifier, the class output of outliers is abandoned in string recognition. That is to say, only the output scores of M defined classes (have been trained to be low in the region of outliers) are used in scoring string classes.

For training an M-class classifier with outlier samples, if the classifier is a neural network with weights estimated under the mean squared error (MSE) or cross-entropy (CE) criterion, the target values for M classes are all set to zero on an outlier sample, unlike that on an in-class sample where one of target values is set to 1. For a one-versus-all support vector classifier, the outlier samples are added to the negative samples of each class such that each class is trained to be separated from outliers. For classifiers trained under the minimum classification error (MCE) criterion [43, 44], a threshold is set to be the discriminant function of the "outlier" class and is adjusted together with the classifier parameters [60]. The threshold is abandoned after training.

In a hybrid generative–discriminative classifier, the discriminant functions are tied with probability density functions or structural descriptions. The parameters are initially estimated to optimize a maximum likelihood (ML)-like criterion, and then adjusted discriminatively to further improve the classification accuracy. Two classifier structures of this kind have been tested in [60]: regularized learning vector quantization (LVQ) and regularized discriminative learning QDF (DLQDF) [59]. The general learning method is as follows.

The class output (discriminant score) of either LVQ or DLQDF classifier is a distance measure. The class prototypes of LVQ classifier and the class parameters (mean, eigenvalues, and eigenvectors) of DLQDF classifier are initialized by ML estimation. In MCE training [43, 44], the misclassification loss on a sample is computed based on the difference of discriminant score between the genuine class and competing classes, and is transformed to soft (sigmoid) 0–1 loss $l_c(\mathbf{x})$. The empirical loss on the training sample set is regularized to

$$L_1 = \frac{1}{N} \sum_{n=1}^{N} [l_c(\mathbf{x}^n) + \alpha d(\mathbf{x}^n, \omega_c)], \qquad (5.19)$$

where N is the number of training samples and ω_c denotes the genuine class of $\mathbf{x}^n$. The regularization term $\alpha d(\mathbf{x}^n, \omega_c)$ is effective to attract the classifier parameters to the ML estimate and hence helps preserve the resistance to outliers. The coefficient α is empirically set to compromise between classification accuracy and outlier resistance. The classifier parameters are updated by stochastic gradient descent to optimize the regularized MCE criterion (5.19).

5.3.2.2 *String-Level Training*

String-level training aims to optimize the performance of string recognition. A performance criterion involving character classification scores is optimized on a string sample set, wherein the classifier parameters are adjusted. Among various criteria, the MCE criterion of Juang et al. [43, 44], which was originally proposed in speech recognition [21, 46], is popularly used. The MCE criterion applies to arbitrary forms of discriminant functions, unlike the maximum mutual information (MMI) criterion (which is usually applied to HMM models, e.g., [87]) that requires the class probability estimates. The application of MCE-based string-level training to handwriting recognition has reported superior performance [7, 8, 16, 57].

Classification-based string recognition involves two levels of classifiers, namely a character classifier and a string classifier. The string classifier combines character classification scores to form string class scores and computes explicitly the scores of a small portion of string classes only. Assume that a character pattern (represented by a feature vector $\mathbf{x}$) is classified into M defined classes $\{\omega_1, \ldots, \omega_M\}$, the string class of a sequence of candidate patterns $X = \mathbf{x}_1 \cdots \mathbf{x}_n$ is denoted by a sequence of character classes $W = w_1 \cdots w_n = \omega_{i_1} \cdots \omega_{i_n}$. The parameters Θ of the embedded character classifier include a subset of shared parameters θ_0 and M subsets of class-specific parameters $\theta_i, i = 1, \ldots, M$. The classifier assigns similarity/dissimilarity scores to the defined classes:

$$y_i(\mathbf{x}) = -d(\mathbf{x}, \omega_i) = f(\mathbf{x}, \theta_0, \theta_i).$$

In string-level training, the classifier parameters are estimated on a data set of string samples $D_X = \{X^n, W^n)|n = 1, \ldots, N_X\}$ (W^n denotes the string class of sample X^n)

by optimizing an objective function, for example, the log-likelihood:

$$\max \mathrm{LL}_X = \sum_{n=1}^{N_X} \log p(X^n | W^n).$$

In ML estimation of classifier parameters, the string samples are dynamically segmented into character patterns $X^n = \mathbf{x}_1^n \cdots \mathbf{x}_m^n$, which best match the ground-truth string class $W^n = w_1^n \cdots w_m^n$ on the current parameters. The parameters are iteratively updated on segmented characters with the aim of maximizing the log-likelihood, in a way similar to expectation maximization (EM) [23].

In discriminative string-level training, the string sample is matched with the genuine string class and a number of competing classes. The contrast between the scores of genuine string and competing strings corresponds to the string recognition error. The MCE method transforms this contrast of scores into a probabilistic loss and updates the classifier parameters with the aim of minimizing the empirical string recognition error.

The MCE criterion can be applied to either character-level training or string-level training. Assume to classify an object X into M classes $\{C_1, \ldots, C_M\}$ (regardless of character or string classes), each class has a discriminant function (class score) $g_i(X, \Theta)$. A sample set $D_X = \{X^n, c^n) | n = 1, \ldots, N_X\}$ (c^n denotes the class label of sample X^n) is used to estimate the parameters Θ.

Following Juang et al. [43], the misclassification measure on a pattern from class C_c is defined by

$$h_c(X, \Theta) = -g_c(X, \Theta) + \log \left[\frac{1}{M-1} \sum_{i \neq c} e^{\eta g_i(X, \Theta)} \right]^{1/\eta}, \qquad (5.20)$$

where η is a positive number. When $\eta \to \infty$, the misclassification measure becomes

$$h_c(X, \Theta) = -g_c(X, \Theta) + g_r(X, \Theta), \qquad (5.21)$$

where $g_r(X, \Theta)$ is the discriminant score of the closest rival class:

$$g_r(X, \Theta) = \max_{i \neq c} g_i(X, \Theta).$$

This simplification significantly saves the computation of training by stochastic gradient descent (also called generalized probabilistic descent (GPD) [46]), where only the parameters involved in the loss function are updated on a training pattern.

The misclassification measure is transformed to loss function by

$$l_c(X, \Theta) = l_c(h_c) = \frac{1}{1 + e^{-\xi h_c}}, \qquad (5.22)$$

where ξ is a parameter that controls the hardness of sigmoid nonlinearity. On the training sample set, the empirical loss is computed by

$$
\begin{aligned}
L_0 &= \frac{1}{N_X} \sum_{n=1}^{N_X} \sum_{i=1}^{M} l_i(X^n, \Theta) I(X^n \in C_i) \\
&= \frac{1}{N_X} \sum_{n=1}^{N_X} l_c(X^n, \Theta),
\end{aligned}
\tag{5.23}
$$

where $I(\cdot)$ is the indicator function.

In minimizing L_0 by stochastic gradient descent, the training patterns are fed into the classifier repeatedly. On a training pattern, the classifier parameters are updated by

$$
\Theta(t+1) = \Theta(t) - \epsilon(t) U \nabla l_c(X, \Theta)|_{\Theta=\Theta(t)},
\tag{5.24}
$$

where U is a positive definite matrix and $\epsilon(t)$ is the learning step. U is related to the inverse of Hessian matrix and is usually approximated to a diagonal matrix. In this case, the diagonal elements U are absorbed into the learning step, which is thus parameter dependent. The parameters converge to a local minimum of L_0 (where $\nabla l_c(X, \Theta) = 0$) under the following conditions:

$$
\begin{cases}
\lim_{t \to \infty} \epsilon(t) = 0, \\
\sum_{t=1}^{\infty} \epsilon(t) = \infty, \\
\sum_{t=1}^{\infty} \epsilon^2(t) < \infty.
\end{cases}
$$

In practice, the parameters are updated in finite iterations, and setting the learning rate as a gradually vanishing sequence leads to convergence approximately.

To specify the MCE training method to string-level classifier training, we use the string class score of Eq. (5.15). If geometric context and linguistic context are incorporated in training, then the score of (5.17) or (5.18) is used instead.

In MCE training, each training sample is segmented by candidate pattern classification and path search using the current classifier parameters $\Theta(t)$. By path search, we obtain a number of string classes matched with the string sample with high scores, each corresponding to a segmented pattern sequence. The search techniques will be described in the next subsection. Denote the genuine string class by $W_c = \omega_{i_1} \ldots \omega_{i_n}$ and the closest competing string by $W_r = \omega_{j_1} \ldots \omega_{j_m}$, corresponding to two sequences of segmented patterns $\mathbf{x}_1^c \ldots \mathbf{x}_n^c$ and $\mathbf{x}_1^r \ldots \mathbf{x}_m^r$, respectively. The scores of two string classes are computed by

$$
NL(X, W_c) = \frac{1}{n} \sum_{k=1}^{n} y(\mathbf{x}_k^c, \omega_{i_k})
$$

and

$$\text{NL}(X, W_r) = \frac{1}{m} \sum_{k=1}^{m} y(\mathbf{x}_k^r, \omega_{j_k}),$$

respectively. The misclassification measure is then

$$h_c(X, \Theta) = \frac{1}{m} \sum_{k=1}^{m} y(\mathbf{x}_k^r, \omega_{j_k}) - \frac{1}{n} \sum_{k=1}^{n} y(\mathbf{x}_k^c, \omega_{i_k}), \qquad (5.25)$$

which is transformed to loss by (5.22).

By stochastic gradient descent, the classifier parameters are updated on a string sample at time step t by

$$\Theta(t + 1) = \Theta(t) - \epsilon(t) \frac{\partial l_c(X, \Theta)}{\partial \Theta} \bigg|_{\Theta=\Theta(t)}$$

$$= \Theta(t) - \epsilon(t) \xi l_c (1 - l_c) \frac{\partial h_c(X, \Theta)}{\partial \Theta} \bigg|_{\Theta=\Theta(t)}$$

$$= \Theta(t) - \epsilon(t) \xi l_c (1 - l_c) \left[\frac{1}{m} \sum_{k=1}^{m} \frac{\partial y(\mathbf{x}_k^r, \omega_{j_k})}{\partial \Theta} - \frac{1}{n} \sum_{k=1}^{n} \frac{\partial y(\mathbf{x}_k^c, \omega_{i_k})}{\partial \Theta} \right] \bigg|_{\Theta=\Theta(t)}.$$

$$(5.26)$$

In practice, the pattern sequence of genuine string and the sequence of competing string have many shared patterns. This can largely simplify the parameter updating of (5.26).

We partition the candidate patterns in two sequences (corresponding to W_c and W_r) into three disjoint subsets: $P_c \cup P_r \cup P_s$. P_s contains the patterns shared by the two string classes, P_c contains the patterns of W_c that are not shared, and P_r contains the patterns of W_r that are not shared. P_s is further divided into two subsets: $P_s = P_{s1} \cup P_{s2}$. The patterns in P_{s1} are assigned different character classes, whereas the patterns in P_{s2} are assigned the same class in the two string classes. The partial derivative of the second line of (5.26) is specified to

$$\frac{\partial h_c(X, \Theta)}{\partial \Theta} = \frac{1}{m} \sum_{\mathbf{x}_k^r \in P_r} \frac{\partial y(\mathbf{x}_k^r, \omega_{j_k})}{\partial \Theta} - \frac{1}{n} \sum_{\mathbf{x}_k^c \in P_c} \frac{\partial y(\mathbf{x}_k^c, \omega_{i_k})}{\partial \Theta}$$

$$+ \sum_{\mathbf{x}_k \in P_{s1}} \left[\frac{1}{m} \frac{\partial y(\mathbf{x}_k, \omega_{j_k})}{\partial \Theta} - \frac{1}{n} \frac{\partial y(\mathbf{x}_k, \omega_{i_k})}{\partial \Theta} \right] \qquad (5.27)$$

$$+ \left(\frac{1}{m} - \frac{1}{n} \right) \sum_{\mathbf{x}_k \in P_{s2}} \frac{\partial y(\mathbf{x}_k, \omega_{i_{j_k}})}{\partial \Theta},$$

where $\omega_{i_{j_k}}$ denotes the common class of a shared pattern in P_{s2}. When $n = m$, we can see that the partial derivative on a pattern in P_{s2} vanishes, and consequently, the classifier parameters are not updated on such patterns.

The learning procedure is customized to various classifier structures by specializing $\frac{\partial y(\mathbf{x}, \omega_i)}{\partial \Theta}$. By partitioning the classifier parameters into shared parameters θ_0 and class-specific parameters θ_i, $i = 1, \ldots, M$, the partial derivative of a class output $y_i(\mathbf{x})$ or $d(\mathbf{x}, \omega_i)$ is computed only with respect to θ_0 and θ_i.

The above learning algorithm was applied to lexicon-free numeral string recognition [57], but is applicable to word recognition as well. If a closed lexicon is used in training, the competing string class on a string sample is selected from the in-lexicon entries, and the classifier parameters are adjusted to optimize the accuracy of in-lexicon string classification. For an open vocabulary, the a priori probability of string class from the n-gram model is incorporated into the string class score in matching the string image with genuine string class and competing classes.

For lexicon-driven word recognition, Chen and Gader [16] and Biem [7, 8] have used different string class scoring functions in the same MCE training framework and have investigated the effect of lexicon on training. Lexicon-driven training may suffer from the sparsity and imbalance (both imbalanced string classes and imbalanced character classes) of string samples and generalize poorly to off-lexicon string classes. Chen and Gader generated an artificial lexicon for training by randomly replacing characters in the lexicon entries. The experimental results of Biem show that lexicon-free training yields higher generalized word recognition accuracy than lexicon-driven training.

5.3.3 Search Strategies

On defining a matching score between the string image and string classes, the task of string recognition is to search for the optimal string class of maximum score or minimum cost (distance). The string image can be segmented into many candidate pattern sequences, and each can be assigned different string classes. The exhaustive search strategy that computes the scores of all segmentation-recognition candidates and then selects the optimal is computationally expensive.

Heuristic search algorithms that evaluate only a portion of segmentation-recognition candidates have been commonly used in string recognition. According to the manner of incorporating lexicon, the search techniques can be divided into two groups: matching string image with one lexicon entry and matching with all string classes simultaneously. We will describe two important search algorithms, dynamic programing and beam search, for accomplishing the two kinds of matching, respectively.

5.3.3.1 Segmentation-Recognition Candidates' Representation The candidates of string segmentation can be represented by a hypothesis network, called segmentation candidate lattice [71]. An example is shown in Figure 5.14. Each node in the lattice corresponds to a separation point. One or more consecutive segments form candidate patterns, which are denoted by edges. All the candidate patterns can be imposed constraints like the maximum number of segments, least and maximum width, height, vertical position in string image, and so on. The candidate patterns that do not meet the constraints are abandoned before being recognized. The maximum

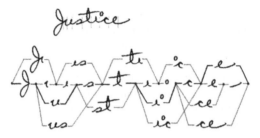

FIGURE 5.14   Segmentation candidate lattice of a word image. The optimal segmentation candidate is denoted by thick lines.

number of segments is usually set to a constant, typically from 3 to 5, but a variable number specific to character class is beneficial [47].

Each candidate pattern is recognized by a character classifier and assigned scores to (all or selected) character classes. On pairing a candidate pattern with character classes, the corresponding edge in the segmentation lattice is expanded into multiple edges, each corresponding to a pattern-class pair. The expanded candidate lattice is called a segmentation-recognition lattice (see Fig. 5.15). In this lattice, each path (segmentation-recognition path or segmentation-recognition candidate) from the leftmost node to the rightmost node denotes a sequence of pattern-class pairs $((\mathbf{x}_1, w_1), \ldots, (\mathbf{x}_n, w_n))$ or a pair of pattern sequence and string class $((\mathbf{x}_1 \cdots \mathbf{x}_n), (w_1 \cdots w_n)) = (X, W)$.

The segmentation-recognition lattice can be formed either before string recognition or dynamically during string recognition. In the former way, the string image is oversegmented and candidate patterns are formed and recognized to assign character classes without considering linguistic context. In lexicon-driven recognition, the candidate patterns and their corresponding classes are dependent on the context. For example, if a candidate pattern $\mathbf{x}_i$ has been assigned to a character class w_i and the character following w_i is constrained to be from two classes $\{v_{i1}, v_{i2}\}$, the candidate pattern succeeding $\mathbf{x}_i$ is then only paired with two class v_{i1} and v_{i2} and is subject to geometric constraints specific to class v_{i1} or v_{i2}.

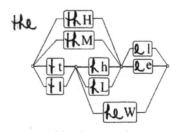

FIGURE 5.15   Segmental   nage of Figure 5.13. Each edge denotes a candidate pattern and its character class.

The segmentation lattice can be stored in a two-dimensional array $S(j, k)$, with j denoting the index of node (separation point) at the right boundary of candidate pattern and k denoting the number of primitive segments in the candidate pattern. Assume that the string image is oversegmented into n primitive segments $t_1 \cdots t_n$, then the element $S(j, k)$ denotes a candidate pattern composed of k primitive segments $\langle t_{j-k+1} \cdots t_j \rangle$.

The segmentation-recognition lattice inherits the array $S(j, k)$ of segmentation lattice, and each element is enriched with a list of character classes and their scores: $S(j, k, c), c \in \{\omega_1, \ldots, \omega_M\}$. Hence, the segmentation-recognition lattice is stored in a three-dimensional array $S(j, k, c)$.

For describing the search algorithms, we assume that a segmentation-recognition path is measured by a distance (cost of path) like that in Eq. (5.18), between a sequence of primitive segments $X = t_1 \cdots t_n$ and a string class $W = w_1 \cdots w_m$. We re-formulate the normalized distance as

$$\text{ND}(X, W) = \frac{1}{m} \sum_{i=1}^{m} [\alpha d_1(s_i, w_i) + \beta d_2(s_i, w_{i-1} w_i)], \quad (5.28)$$

where s_i denotes a candidate pattern $\langle t_{e_i - k_i + 1} \cdots t_{e_i} \rangle$. $d_1(s_i, w_i)$ covers the character classification score and single-character geometric plausibility, and $d_2(s_i, w_{i-1} w_i)$ covers the character-pair geometric compatibility and linguistic bigram. A partial segmentation-recognition path is scored by a partial cost $(i < m)$

$$
\begin{aligned}
\text{ND}(s_1 \cdots s_i, w_1 \cdots w_i) &= \frac{1}{i} \sum_{j=1}^{i} [\alpha d_1(s_j, w_j) + \beta d_2(s_j, w_{j-1} w_j)] \\
&= \frac{1}{i} D(s_1 \cdots s_i, w_1 \cdots w_i),
\end{aligned} \quad (5.29)
$$

where $D(s_1 \cdots s_i, w_1 \cdots w_i)$ denotes the accumulated cost, which can be reformulated as

$$
\begin{aligned}
D(s_1 \cdots s_i, w_1 \cdots w_i) &= \sum_{j=1}^{i} [\alpha d_1(s_j, w_j) + \beta d_2(s_j, w_{j-1} w_j)] \\
&= D(s_1 \cdots s_{i-1}, w_1 \cdots w_{i-1}) + d_c(s_i, w_i),
\end{aligned} \quad (5.30)
$$

where $d_c(s_i, w_i) = \alpha d_1(s_i, w_i) + \beta d_2(s_i, w_{i-1} w_i)$ is the added cost when matching a new pattern s_i with a character w_i. The partial cost is normalized with respect to the substring length i because the true string length m is unknown before string recognition is completed.

5.3.3.2 Matching with One Lexicon Entry

When the number of string classes (words or phrases) is not very large (say, less than 1000), string recognition can be performed by matching the string image with each lexicon entry one by one. To match with a string class, the string image is segmented into character patterns that best match the characters. The match between the string image (represented as a sequence of primitive segments) and a string class is similar to the dynamic time warping (DTW) in speech recognition and is commonly solved by DP.

For matching with a given string class, as the string length m is known, the segmentation-recognition path and partial paths can be evaluated using the accumulated cost. Denote the number of primitive segments in candidate pattern s_i as k_i, $s_i = \langle t_{e_i - k_i + 1} \cdots t_{e_i} \rangle$, then $s_1 \cdots s_i = t_1 \cdots t_{e_i}$, and the accumulated partial cost is

$$D(t_1 \cdots t_{e_i}, w_1 \cdots w_i) = D(t_1 \cdots t_{e_i - k_i}, w_1 \cdots w_{i-1})$$
$$+ d_c(\langle t_{e_i - k_i + 1} \cdots t_{e_i} \rangle, w_i). \tag{5.31}$$

Thus, the full path cost $D(t_1 \cdots t_n, w_1 \cdots w_m)$ can be minimized stagewise: the optimal sequence of candidate patterns ending with $s_i = \langle t_{e_i - k_i + 1} \cdots t_{e_i} \rangle$ corresponding to w_i, $\arg \min_{e_1 \cdots e_{i-1}} D(t_1 \cdots t_{e_i}, w_1 \cdots w_i)$, comprises a subsequence of $\arg \min_{e_1 \cdots e_{i-2}} D(t_1 \cdots t_{e_i - k_i}, w_1 \cdots w_{i-1})$.

To search for the optimal match of $\min_{e_1, \dots, e_{m-1}} D(t_1 \cdots t_n, w_1 \cdots w_m)$ by DP, an array $D(j, i)$ is used to record the minimum partial cost up to a primitive segment t_j: $\min_{e_1, \dots, e_{i-1}} D(t_1 \cdots t_j, w_1 \cdots w_i)$, and a backpointer array $T(j, i)$ records the index of primitive segment $j - k_i$ that the candidate pattern s_{i-1} ends with:

$$D(j, i) = \min_{k_i} [D(j - k_i, i - 1) + d_c(\langle t_{j - k_i + 1} \cdots t_j \rangle, w_i)]. \tag{5.32}$$

Based on the above formulations, the DP algorithm for matching a string image with a string class is as below.

Algorithm: DP for matching with a string class

1. *Initialization.* Set $D(0, 0) = 0$, $D(j, 0) = \infty$ and $D(0, i) = \infty$, $j = 1, \dots, n$, $i = 1, \dots, m$.

2. *Forward updating.* For $j = 1, \dots, n$, $i = 1, \dots, m$, update $D(j, i)$ by Eq. (5.32), denote the resulting number of primitive segments as $\arg \min_{k_i}$, and update $T(j, i) = j - \arg \min_{k_i}$.

3. *Backtracking.* The sequence of primitive segments is partitioned into characters at the separation points $J_1 \cdots J_{m-1}$, where $J_{m-1} = T(n, m)$, $J_{i-1} = T(J_i, i)$, $i = m, \dots, 2$.

Figure 5.16 shows an example of DP matching: the string image of Figure 5.14 is matched with two word classes separately. The search space is represented as a two-dimensional array $D(j, i)$, and the optimal segmentation is defined by the separation points $J_0 J_1 \cdots J_{m-1} J_m$ ($J_0 = 0$, $J_m = n$) and is denoted by the diagonal lines in the array.

In string matching by the above DP algorithm, the primitive segments of string image are forced to match with the characters, so the matching path corresponds to diagonal lines in the array of search space. If, more generally, not only primitive segments merging but also segment deletion (character insertion) and segment insertion (character deletion) are considered in matching, the partial cost of (5.32) is minimized

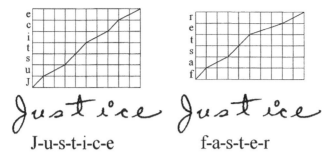

J-u-s-t-i-c-e   f-a-s-t-e-r

FIGURE 5.16   Match the string image of Figure 5.14 with two word classes. Upper: matching path in search space (array); lower: segmented character patterns.

with respect to operations of deleting segment t_j and deleting character w_i as well as merging k_i segments, and the matching path comprises of vertical and horizontal lines as well as diagonal lines.

5.3.3.3 *Matching with All String Classes*   In string recognition with large vocabulary, as the objective is to search for only one or a few string classes with maximum scores, many computation efforts on the low-score string classes can be saved. To do this, all the string classes are considered simultaneously in matching with the string image and those classes that are unlikely to give the maximum score are abandoned before the matching is completed.

In lexicon-free recognition, each candidate pattern $s_i = \langle t_{e_i - k_i + 1} \cdots t_{e_i} \rangle$ is assigned to all character classes $\{\omega_1, \ldots, \omega_M\}$. On rejecting some characters according to the classification scores, s_i is paired with a subset of characters, and the corresponding edge in the segmentation lattice is expanded into multiple segmentation-recognition edges accordingly. For lexicon-driven recognition, the lexicon is often represented in a tree structure called trie (to be described in the next subsection). When a candidate pattern ending with segment t_j is paired with a character w_i, a candidate pattern starting from segment t_{j+1} will be paired with the characters that are the offspring of w_i in the lexicon tree.

A segmentation-recognition solution is a pair of candidate pattern sequence $s_1 \cdots s_m$ (partitioned by separation points $e_1 \cdots e_{m-1}$) and its corresponding string class $W = w_1 \cdots w_m$. A partial solution, corresponding to a partial path in the segmentation-recognition lattice, is a pair of pattern sequence $s_1 \cdots s_i$ and a substring $w_1 \cdots w_i$. To avoid evaluating all the full paths, the optimal full path is searched heuristically via selectively extending partial paths.

A partial segmentation-recognition path $(s_1 \cdots s_i, w_1 \cdots w_i)$ is viewed as a state in a search space and is represented as a node in a tree. The root of the search tree denotes an empty match, an intermediate node corresponds to a partial match, and a terminal node gives a segmentation-recognition result. Each edge in the segmentation-recognition lattice is converted to one or more nodes in the search tree. Specifically, an edge $S(j, k, c)$, which denotes a candidate pattern $\langle t_{j-k+1} \cdots t_j \rangle$ paired with character

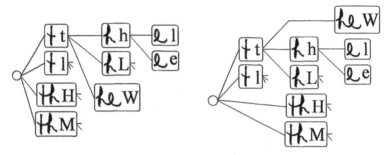

FIGURE 5.17 Character-synchronous (left) and frame-synchronous (right) node generation in search tree. The nodes in a column are generated simultaneously.

c, is converted to a search node SN(j, k, c, p), where p denotes the pointer to the parent node of SN.

Search algorithms vary in the order of node generation. In character-synchronous search, the offspring of a node, which have the same depth, are generated simultaneously. The candidate patterns of these nodes start from the same primitive segment but may end with different primitive segments. In frame-synchronous search, the nodes whose candidate patterns end with the same primitive segment are generated simultaneously, and each node is linked to the parent node whose candidate pattern immediately precedes the pattern of the new node. If there are multiple such parent nodes, the new node is generated with multiple copies, each linked to a respective parent node. Figure 5.17 shows the order of node generation for the segmentation-recognition lattice in Figure 5.15.

In either character-synchronous or frame-synchronous search, the nodes are generated by pairing a candidate pattern with character classes via classification and rejection. On computing the class similarity measure $y(\mathbf{x}, \omega_i)$ or distance $d(\mathbf{x}, \omega_i)$, some classes can be rejected by setting an abstract threshold T_1 and a relative threshold T_2 [58]. Specifically, if $y(\mathbf{x}, \omega_i) < T_1$ or $d(\mathbf{x}, \omega_i) > T_1$, ω_i is rejected; denote the top-rank class as ω_m, if $y(\mathbf{x}, \omega_m) - y(\mathbf{x}, \omega_i) > T_2$ or $d(\mathbf{x}, \omega_i) - d(\mathbf{x}, \omega_m) > T_2$, and ω_i is rejected. [2]

In matching with multiple string classes, as the length of the optimal string class is unknown a priori, the search nodes (corresponding to partial segmentation-recognition paths) should be evaluated using a normalized score like that in Eq. (5.29). The partial cost is now nonmonotonic in the sense that the cost of a longer path may be smaller than the cost of its partial path.

In the following, we first describe the algorithms for character-synchronous search, including best-first search [9, 15] and beam search [56, 66, 78], and then a beam search algorithm for frame-synchronous search [60].

In character-synchronous search, the operation of generating the offspring of a node in the search tree is called expansion. The search algorithm is variable depending on

---

[2]These simple heuristic rejection rules perform fairly well. More accurate rejection can be achieved by estimating probabilistic confidence precisely.

the order of node expansion [74, 91]. By breadth-first search, the nodes of a same depth are expanded synchronously, whereas by depth-first search, the deepest node is expanded first. By best-first search, the node of minimum partial cost is expanded recursively until the minimum cost node is a terminal node. Best-first search is still not efficient enough in the sense that the optimal terminal node cannot be found before many intermediate nodes are expanded. The heuristic search, A* algorithm, accelerates best-first search by adding an underestimate of remaining path cost. The remaining cost, however, is not trivial to estimate.

In the case of nonmonotonic path cost, the first found terminal node does not necessarily give an optimal string match though its cost is minimum compared to the nonterminal nodes because the nonterminal nodes can achieve even lower cost after expansion. For better solution, node expansion should continue until multiple terminal nodes are found. Because many nonterminal nodes have comparable partial costs, a large number of nodes are expanded before terminal nodes are reached.

A good trade-off between the optimality of solution and the efficiency of search can be achieved using a beam search strategy: It improves the optimality by expanding multiple nodes each time and saves computation by discarding intermediate nodes of low scores. For character-synchronous search, the beam search expands nodes in breadth-first manner: selected nodes of the same depth (corresponding to substrings of the same length) are expanded synchronously. To describe the search algorithm formally, we first formulate the operation of node expansion.

Denote a node in the search tree as a quadruple $SN(j, k_i, w_i, p_i)$, which corresponds to a sequence of i candidate patterns with the last one $s_i = \langle t_{j-k_i+1} \cdots t_j \rangle$ paired with a character w_i. Each node is evaluated by a normalized partial path cost as in (5.29). To expand $SN(j, k_i, w_i, p_i)$, the primitive segments starting with t_{j+1} are merged into candidate patterns, each paired with character classes following w_i (either lexicon-free or lexicon-driven). Each pattern-character pair gives a new node SN $(j + k_{i+1}, k_{i+1}, w_{i+1}, p_{i+1})$, where p_{i+1} links to its parent node $SN(j, k_i, w_i, p_i)$. When $j + k_{i+1} = n$ (n is the total number of primitive segments), the new node is a terminal node. In this case, if w_{i+1} is also the last character of a string class, the terminal node gives a complete string match; otherwise, the string match is incomplete and abandoned.

The root node of the search tree is $SN(0, 0, 0, 0)$, with accumulated path cost being zero. When a node $SN(j, k_i, w_i, p_i)$ is expanded to generate a new node $SN(j + k_{i+1}, k_{i+1}, w_{i+1}, p_{i+1})$, the accumulated path cost of the new node is updated according to (5.30) and is normalized with respect to the substring length $i + 1$. As SN corresponds to a partial segmentation-recognition path, the accumulated cost and normalized cost are also referred to as $D(SN)$ and $ND(SN)$, respectively.

The search nodes are stored in two lists. All the newly generated nodes (not yet expanded) are stored in OPEN, and those expanded are stored in CLOSED. Initially, the OPEN list contains the root node and the CLOSED list is empty. The number of nodes in OPEN is upper bounded by a number BN. If there are more than BN new nodes, the nodes with high ND(SN) are abandoned. The nodes in OPEN can be further pruned by comparing ND(SN) with an abstract threshold and comparing the difference of ND(SN) from the minimum with a relative threshold, as for character rejection in candidate pattern classification. The nodes in OPEN have the same depth and are expanded synchronously to generate new OPEN nodes. Node expansion proceeds until one or more terminal nodes are found or there is no OPEN nodes to expand. The search algorithm is described below.

Algorithm: Character-synchronous beam search

1. *Initialization.* Store root node $SN(0, 0, 0, 0)$ in OPEN.
2. *Node expansion.* Move all the OPEN nodes into CLOSED and expand them to generate at most BN new OPEN nodes. Repeat until one or more terminal nodes are found or there is no OPEN node to expand.
3. *Backtracking.* Corresponding to a terminal node $SN(n, k_m, w_m, p_m)$ (match with a string of m characters), the ending segment indices $e_1 \cdots e_{m-1}$ that partition the string image into m character patterns and the character classes are backtracked from the parent node pointers: $e_{i-1} = j_{i-1}$ and w_{i-1} are defined by the parent node of $SN(j_i, k_i, w_i, p_i)$.

Node expansion is not available in frame-synchronous search because the offspring of a node are not generated synchronously, but rather the nodes with corresponding candidate patterns ending with the same primitive segment t_j are generated simultaneously. The nodes are not differentiated into OPEN and CLOSED, but stored in a single list. Formally, for $j = 1, \ldots, n$, each candidate pattern ending with t_j, referred to as $S(j, k)$ in segmentation lattice, is linked to all the existing nodes (which serve the parents of new nodes) with candidate patterns ending with t_{j-k}. The candidate pattern $S(j, k)$ is paired with some character classes (either lexicon-free or lexicon-driven), and each pattern-character pair is linked to a parent to generate a new node. By beam search, all the new nodes at j are sorted to store at most BN of them. The nodes generated at $j = n$ are terminal nodes and those of complete match are backtracked to give segmentation-recognition results.

5.3.4 Strategies for Large Vocabulary

The number of string classes in string recognition is often very large. In English word recognition, for example, the number of daily used words is tens of thousands. In a Japanese address phrase recognition system, the number of phrases considered is as many as 111,349 [56]. Obviously, matching a string image with a large number of string classes one by one is computationally expensive, and the recognition speed is not acceptable in practical applications. To speed up string recognition, the problem of large lexicon can be dealt with in two respects: lexicon reduction and structured lexicon organization.

5.3.4.1 Lexicon Reduction   Lexicon reduction refers to the selection of a small-size subset from a large number of string classes. The string image to recognize is then matched only with the string classes in the subset, such that the computation of matching with the other string classes is saved. Various lexicon reduction techniques have been reviewed in [51, 89]. The techniques can be roughly grouped into three categories: context knowledge, string length estimation, and shape feature matching. We briefly outline these techniques, whereas their details can be found in the literature.

For specific applications, string recognition is often considered in a dynamic small-size lexicon defined by related knowledge source. For example, in postal mail address reading, the lexicon of address phrases can be restricted by the result of postal code recognition. In word recognition of ordinary text, the lexicon for recognizing a word image can be reduced by linguistic context represented as between-word dependency statistics or a grammar.

String recognition can be speeded up by matching with the string classes of a specified length or several lengths only. String length or a range of lengths can be estimated from the width and height of string image, the number of primitive segments, the number of near-vertical strokes, the number of contour peak/valley points, and so on.

Some holistic string recognition methods using global shape features can quickly match a string image with a large number of string prototypes, based on which the string classes of low scores are excluded from detailed matching. Some string shape features, like the contour envelope of string image, letter ascenders/descenders, prominent strokes and contour points, and so on, can be easily extracted and can filter out a fraction of string classes that are unlikely to match the string image.

5.3.4.2 *Lexicon Organization*
Lexicon-driven string recognition can be performed more efficiently by matching with all string classes simultaneously than by matching one by one. In matching with all string classes, the strings with low-score partial match are abandoned because they are unlikely to give the optimal complete match. The search strategy, like character-synchronous beam search, usually requires the lexicon to be organized in a lexical tree (trie) or graph structure such that common substrings are matched with the string image only once.

The trie structure of lexicon has been widely used in word and string recognition [8, 9, 15, 24, 56, 62, 66]. In the trie, the strings with a common prefix of substring, such as {ab-normal, ab-stract, ab-use}, have a common path of parent nodes for the prefix and the remaining substrings in a subtree. Each path from the root node to a terminal node corresponds to a string. So, the number of terminal nodes equals the number of lexicon entries. This structure can save both the storage of lexicon and the computation in string recognition because the common prefix is stored only once and matched only once in search. Figure 5.18 shows a trie for 12 month words, which have 74 characters in total and 69 nodes in the trie.

Assume that the average string length is L and the average branching factor of nonterminal nodes is b, then the number of nodes in the trie is approximately $(\sum_{l=0}^{L} b^l = \frac{b^{L+1}-1}{b-1} \approx \frac{b}{b-1}b^L)$, compared to $L \cdot b^L$, the total number of characters. This implies, when $\frac{b}{b-1} < L$, the number of trie nodes is smaller than the total number of characters in the lexicon.

In string recognition by character-synchronous search, a search node is associated with a node in the trie, which corresponds to the character paired with the candidate

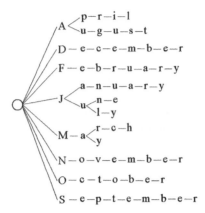

FIGURE 5.18 Trie for a small lexicon of 12 month words.

pattern. When the search node is expanded, a succeeding candidate pattern is paired with the characters given by the offspring of the trie node.

Compared to lexicon-free recognition, trie structure of lexicon also helps improve the accuracy of character recognition. This is because each candidate pattern is classified only in a restrictive subset (defined by the offspring of a trie node) of character classes, which is mostly very small compared to the whole character set. Classification in a small character set is both more efficient and more accurate than in a large character set.

The trie structure, though facilitates character-synchronous search, is not economical enough in storage. If there are many strings sharing middle substrings or suffix but differing in prefix, the trie structure cannot save much nodes compared to the flat lexicon because the common substrings and suffix are stored in different paths for multiple times. For example, in the trie of Figure 5.18, there are four words sharing suffix "-ber" and two words sharing suffix "-uary." The common substrings can be better explored using a graph structure called directed acyclic word graph (DAWG) [54]. In DAWG, a common nonprefix substring is stored in a unique path of nodes and linked to multiple parent nodes corresponding to multiple strings. Figure 5.19 shows a DAWG for four month words with common suffix.

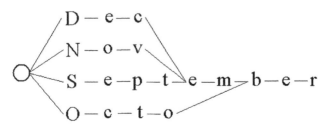

FIGURE 5.19 Directed acyclic word graph for four words.

Ikeda et al. proposed to represent a large variety of strings using a context-free grammar [40]. This structure is especially useful for the situation where a large number of strings differ only in a local substring. This is the case for Japanese address phrases, where a standard phrase is often written in variations differing in substrings. For lexicon-driven string recognition, the context-free grammar can be transformed to a directed acyclic graph for being used in search.

For string recognition with graph-structured lexicon, character-synchronous search proceeds in the same way as for trie lexicon: the succeeding characters of a trie/graph node are given by its offspring. In this sense, graph structure of lexicon does not save search time compared to trie structure, but is much more efficient in saving storage. The generation, storage, and maintenance of graph lexicon, however, is complicated.

5.4 HMM-BASED RECOGNITION

HMMs [38, 41, 76, 77] have become a mainstream technique and a ubiquitous component in current systems developed for recognizing machine-print, online, and offline handwriting data. One of the main features of these data is that, owing to noise and diversity of writing styles, the same transcription word usually takes on variable-length signal forms. This fact has motivated the use, first, of the DP paradigm whose main strength is the ability to match signals of variable lengths. Although DP techniques have proven to be effective for simple applications with small vocabularies, they turned out to be less efficient for complex applications involving large vocabularies. The reason is twofold: First, DP robustness requires a quite large number of prototypes or templates per modeling unit in order to accommodate the increasing variability in large vocabulary applications. Second, DP suffers from the lack of efficient algorithms to automatically estimate the prototype parameters. By overcoming these two limitations and featuring other desirable properties, HMMs become the natural successors of DP and have been adopted as the predominant approach for recognizing discrete sequential variable signals. These models derive single compact (average) models explaining the variability of data, and besides, automatically estimate the associated parameters using efficient algorithms.

5.4.1 Introduction to HMMs

HMMs are an extension of Markov chains, for which symbol (or observation) production along states (symbol generation along transitions is another variant that is widely used) is no longer deterministic but occurs according to an output probabilistic function, hence their description as a double stochastic process. Such a scheme, while avoiding a significant increase in the complexity of the Markov chain, offers much more freedom to the random process, enabling it to capture the underlying variability of discrete sequences. The Markov chain serves as an abstract representation of structural constraints on data causality. Such constraints are usually derived from our knowledge about the problem and the data as well as from taking into account the way data (1D signals such as online handwriting data or images such as offline handwriting data) are transformed into sequential strings.

The output probabilistic functions embody the second stochastic process and model the inherent variability of characters or of any basic unit of the language we are dealing with. The variability results from various distortions inherent to image acquisition, binarization, intra- and interscriptor diversity of writing styles, and so on.

The underlying distribution of these functions defines the HMM type. If it is discrete (nonparametric), the model is called discrete HMM. The discrete alphabet (codebook), in this case, is usually obtained through vector quantization [35, 64]. Continuous HMMs [55] model the output function by a continuous probability density function, usually a mixture of Gaussians. Semicontinuous HMMs [39] can be thought of as a compromise of these two extremes and have the output functions sharing the same set of Gaussians (but model Gaussian coefficients as state-dependent parameters).

One of the main limitations of HMMs is the assumption that every observation does not depend on the previous ones but only on the current state. Such an assumption clearly does not handle the correlation between contiguous observations, which is inherent to the diversity of writing styles. However, this very limitation is the main reason behind the success of HMMs. As will be shown in the next subsection, the conditional independence assumption is behind the existence of efficient algorithms for automatic training of the model parameters and decoding of the data.

For HMM-based text recognition, there are two main approaches: the first relies on an implicit segmentation [26, 33, 36, 45, 90], where the handwriting data are sampled into a sequence of tiny frames (overlapped or not). The second uses a more sophisticated explicit segmentation technique [17, 28, 31, 87] to cut the text into more meaningful units or graphemes, which are larger than the frames. In either case, the ultimate segmentation of the words into letters is postponed and obtained as a by-product of recognition itself. This capability is one of the most powerful features of HMMs, which make them the preferred method for word recognition.

5.4.2 Theory and Implementation

In the discrete case, an HMM is formally defined by the following parameters:

- $A = \{a_{ij}\}$, the Markov chain matrix, where a_{ij} is the probability of transition from state i to state j, with $i, j \in \{1, 2, \ldots, N\}$, N being the number of states.
- $B = \{b_j(k)\}$, the output distribution matrix, where $b_j(k)$ is the probability of producing symbol k, when the Markov process is in state j. $k \in \{1, 2, \ldots, M\}$, M being the size of the symbol alphabet.
- $\Pi = \{\pi_i\}$, the probability that the Markov process starts in state i. Without loss of generality, it will be assumed in the remaining of this section that state 1 is the only initial state. Thus, $\pi_1 = 1$ and $\pi_i = 0$ for $i \neq 1$. In the same way, we assume that N is the only terminating state. $\lambda = (A, B)$ is a compact representation of the HMM.

Before HMMs can be used in actual applications, the development of a hidden Markov model approach entails addressing three problems. First, given an observation sequence O and a model λ, how do we compute $P(O|\lambda)$? This is known as the evaluation problem. Second, given an observation sequence O and a model λ, what is the optimal state sequence in λ accounting for O (decoding problem)? Third, given a set of observation sequences, how do we estimate the parameters of the model (learning problem)?

5.4.2.1   *The Evaluation Problem*   Given an observation sequence $O = o_1 \cdots o_T$, the computation of $P(O|\lambda)$ is straightforward:

$$P(O|\lambda) = \sum_S P(O|S, \lambda)P(S|\lambda), \qquad (5.33)$$

where the sum runs over all state sequences S.

By using HMM properties regarding conditional independence of observations and first-order Markov chains, this sum becomes

$$P(O|\lambda) = \sum_S a_{s_0 s_1} b_{s_1}(o_1) a_{s_1 s_2} b_{s_2}(o_2) \cdots a_{s_{T-1} s_T} b_{s_T}(o_T). \qquad (5.34)$$

As there are N^T possible state sequences, this computation is clearly intractable. Nonetheless, because of the properties mentioned above, it is easy to see that the paths shared across state sequences S need to be computed only once. Hence, $P(O|\lambda)$ can be inductively obtained by introducing a forward probability $\alpha_t(i) : \alpha_t(i) = P(o_1, \ldots, o_t, s_t = i|\lambda)$, which leads to the Forward algorithm below.

Algorithm: Forward procedure

1. *Initialization*

$$\alpha_1(1) = 1.$$

2. *Recursion*

$$\alpha_t(j) = \left[\sum_{i=1}^N \alpha_{t-1}(i) a_{ij} \right] b_j(o_t).$$

3. *Termination*

$$P(O|\lambda) = \alpha_T(N).$$

5.4.2.2   *The Decoding Problem*   Sometimes, we are interested, for many reasons, in finding the optimal state sequence accounting for an observation sequence O. In the maximum likelihood sense, this amounts to searching for a state sequence

S^* such that

$$S^* = \arg\max_S P(O, S|\lambda). \qquad (5.35)$$

Using the properties of HMMs discussed above, this turns out to be a straightforward task, if we define the partial Viterbi probability in this way:

$$\delta_t(i) = \max_{s_1,\ldots,s_{t-1}} P(o_1, \ldots, o_t, s_1, \ldots, s_t = i|\lambda). \qquad (5.36)$$

This is the probability of the best partial state sequence, generating the t first observations and leading to state i at time t. This leads to the following Viterbi algorithm.
Algorithm: Viterbi decoding

1. *Initialization*

$$\delta_1(1) = 1,$$
$$B_1(1) = 1.$$

2. *Recursion*

$$\delta_t(j) = \max_{1 \le i \le N}[\delta_{t-1}(i)a_{ij}]b_j(o_t),$$
$$B_t(j) = \arg\max_{1 \le i \le N}[\delta_{t-1}(i)a_{ij}].$$

3. *Termination*

$$P(S^*) = \delta_T(N),$$
$$B_T(N) = \arg\max_{1 \le i \le N}[\delta_{t-1}(i)a_{iN}],$$

   where S^* is the sequence associated with the best path.

4. *Backtracking*

$$s_t^* = B_{t+1}(s_{t+1}^*).$$

$B_t(i)$ is a backpointer keeping track of the best partial path leading to state i at time t. Once reaching the end state N, we recursively apply the backpointer B to retrieve the entire optimal sequence S^*.

5.4.2.3 *The Training Problem*

Before using an HMM in evaluation or decoding, we first need to estimate its parameters. That is, given a set of training data, we want to derive a model that is able to perform reasonably well on future unseen data. Unfortunately, no satisfactory and efficient method has been devised so far to simultaneously optimize the model structure (topology) and the model parameters. The usual technique, to overcome this limitation, is to make use of a priori knowledge of the problem to design a suitable model topology and then to optimize the parameters given this topology.

(1) *ML estimation*

Training HMM parameters is a typical example of unsupervised learning where data are incomplete. The incompleteness here stems from the availability of the observation sequences (observable) but not of their generating state sequences (hidden). The most widely used technique to solve this problem is the ML-based Baum–Welch algorithm, which is an implementation of the EM algorithm [23, 69] in the case of HMMs. The objective of ML is to seek the model λ that maximizes $P(O|\lambda)$. The Baum–Welch (EM) algorithm achieves this optimization in an elegant way: After guessing an initial estimate for the model (possibly randomly), the first step consists of replacing $P(O|\lambda)$ by the Q function $Q(\lambda, \lambda')$:

$$Q(\lambda, \lambda') = \sum_S P(O, S|\lambda) \log P(O, S|\lambda'), \qquad (5.37)$$

which is essentially the expectation of $\log P(O, S|\lambda)$. The rationale behind this transformation is that

$$Q(\lambda, \lambda') \geq Q(\lambda, \lambda) \Rightarrow P(O|\lambda') \geq P(O|\lambda). \qquad (5.38)$$

Thus, maximizing $Q(\lambda, \lambda')$ over λ is equivalent to maximizing $P(O|\lambda)$. This maximization constitutes the second step. The algorithm by iteratively running through the two steps guarantees the convergence toward a local maximum of $P(O|\lambda)$. In spite of the nonglobal character of the optimization, the solutions reached are usually satisfactory, if the assumptions about the model are suitable for the problem at hand.

For real tasks, an HMM is usually trained from a set of Y independent observation sequences by looking for the model λ that maximizes the likelihood

$$L = \prod_{y=1}^{Y} P(O^y|\lambda), \qquad (5.39)$$

where O^y is the yth observation sequence. Fortunately, because of the assumption of independence between observation sequences as stated above, this leads to essentially the same calculations. By using the definition of $\alpha_t(i)$ and introducing a backward probability variable $\beta_t(i)$, $\beta_t(i) = P(o_{t+1}, \ldots, o_T | s_t = i, \lambda)$, we eventually obtain these quite intuitive reestimation formulas:

$$\overline{a_{ij}} = \frac{\sum_y \frac{1}{P(O^y|\lambda)} \sum_{t=1}^T \sum_{i=1}^N \alpha_{t-1}^y(i) a_{ij} b_j(o_t) \beta_t^y(j)}{\sum_y \frac{1}{P(O^y|\lambda)} \sum_{t=1}^T \sum_{i=1}^N \alpha_t^y(i) \beta_t^y(i)}, \qquad (5.40)$$

$$\overline{b_j(k)} = \frac{\sum_y^Y \frac{1}{P(O^y|\lambda)} \sum_{t=1}^T \sum_{i=1}^N \delta(o_t^y = v_k) \alpha_{t-1}^y(i) a_{ij} b_j(o_t) \beta_t^y(j)}{\sum_y^Y \frac{1}{P(O^y|\lambda)} \sum_{t=1}^T \sum_{i=1}^N \alpha_{t-1}^y(i) a_{ij} b_j(o_t) \beta_t^y(j)}, \tag{5.41}$$

where

$$\delta(a, b) = \begin{cases} 1, & \text{if } a = b, \\ 0, & \text{otherwise.} \end{cases}$$

$\overline{a_{ij}}$ can be interpreted as the expected number of transitions from state i to state j over the expected number of times the Markov chain is in state i. In the same way, $\overline{b_j(k)}$ is the expected number of times the state j produces observation symbol k over the expected number of times the Markov chain is in state j.

For continuous HMMs, $b_j(o_t)$ is no longer represented by a nonparametric discrete distribution $b_j(k)$ but by (usually) a mixture of Gaussians:

$$b_j(o_t) = \sum_{m=1}^M c_{jm} N_{jm}(o_t, \mu_{jm}, \textstyle\sum_{jm}) = \sum_{m=1}^M c_{jm} b_{jm}(o_t), \tag{5.42}$$

where M is the number of Gaussians under consideration, c_{jm}, μ_{jm}, and $\sum_{jm}$ are, respectively, the weight, the mean, and the covariance matrix associated with state j and mixture m. Likewise, the reestimation formulas for these parameters are intuitive and similar to the discrete case:

$$\overline{c_{jm}} = \frac{\sum_{y=1}^Y \frac{1}{P(O^y|\lambda)} \sum_{t=1}^T \sum_{i=1}^N \alpha_{t-1}^y(i) a_{ij} c_{jm} b_{jm}(o_t) \beta_t^y(j)}{\sum_y^Y \frac{1}{P(O^y|\lambda)} \sum_{t=1}^T \sum_{i=1}^N \sum_{m=1}^M \alpha_{t-1}^y(i) a_{ij} c_{jm} b_{jm}(o_t) \beta_t^y(j)}, \tag{5.43}$$

$$\overline{\mu_{jm}} = \frac{\sum_{y=1}^Y \frac{1}{P(O^y|\lambda)} \sum_{t=1}^T \sum_{i=1}^N \alpha_{t-1}^y(i) a_{ij} c_{jm} b_{jm}(o_t) \beta_t^y(j) o_t}{\sum_y^Y \frac{1}{P(O^y|\lambda)} \sum_{t=1}^T \sum_{i=1}^N \sum_{m=1}^M \alpha_{t-1}^y(i) a_{ij} c_{jm} b_{jm}(o_t) \beta_t^y(j)}, \tag{5.44}$$

$$\overline{\textstyle\sum_{jm}} = \frac{\sum_{y=1}^Y \frac{1}{P(O^y|\lambda)} \sum_{t=1}^T \sum_{i=1}^N \alpha_{t-1}^y(i) a_{ij} c_{jm} b_{jm}(o_t) \beta_t^y(j) E_{jmt}}{\sum_y^Y \frac{1}{P(O^y|\lambda)} \sum_{t=1}^T \sum_{i=1}^N \sum_{m=1}^M \alpha_{t-1}^y(i) a_{ij} c_{jm} b_{jm}(o_t) \beta_t^y(j)}, \tag{5.45}$$

where,

$$E_{jmt} = (o_t - \overline{\mu_{jm}})(o_t - \overline{\mu_{jm}})^T.$$

(2) Other approaches

One of the requirements for the success of EM training is the correctness of the model. This is hardly met in practice because of some HMM weaknesses, such as the conditional independence assumption and the Markov chain assumption, which are clearly inaccurate when studying the production of complex signals like handwriting (or speech) (actually, the underlying Markov chain may be able to take into account some dependencies between the observations, but only partially). Therefore, other alternatives have been proposed as a substitute to EM training. One of the most popular techniques is the MMI estimation [2], which not only tries to increase the probabilities of the parameters of the correct model (corresponding to current label) but also aims at decreasing the probabilities of the parameters of competing models. This optimization usually involves the use of gradient descent techniques. Although MMI and other discriminative approaches have sound theoretical foundations, EM is still one of the most widely used estimation approaches as the difference in recognition performance has not been, in general, significant in practice [38].

5.4.3 Application of HMMs to Text Recognition

HMMs are typically used in the framework of the pattern recognition approach to text recognition [4]. Given an observation sequence O, our aim is to seek the class W^* (W may correspond to characters, words, sentences, etc.) that maximizes the a posteriori probability $P(W|O)$. Using the Bayes rule,

$$W^* = \arg\max_W P(W|O) = \arg\max_W \frac{P(O|W)P(W)}{P(O)} = \arg\max_W P(O|W)P(W),$$

(5.46)

as $P(O)$ is independent of W. $P(W)$ is the a priori probability of W, and depending on the application, it might be either dropped if no a priori knowledge is available or rather introduced to guide recognition. For example, for sentence recognition, n-grams are usually used to estimate $P(W) = P(W_1 W_2 \cdots W_n)$, where W_i are the words of the sentence [67, 90].

$P(O|W)$ is the probability of O (likelihood of the data) given class W. In the case of HMMs, W is implicitly represented by its underlying model: $P(O|W) = P(O|\lambda)$. $P(O|\lambda)$ can be efficiently estimated in several ways depending on the task at hand. For applications involving small vocabularies (for instance, digit recognition [12], words belonging to literal amounts of checks [34, 36, 87], etc.), we can build an HMM model λ_i for each class and use the forward algorithm to estimate $P(O|\lambda_i)$ for $i = 1, \ldots, N_w$, where N_w is the number of classes. In such a case, the lexicon is said to be linear or flat.

For large vocabularies (for instance, city or street name recognition and sentence or continuous word recognition [17, 27, 28, 31, 34, 65, 90]), however, this approach is no longer feasible. The reason is threefold. First, considering one model for each class means that running the forward algorithm for each class separately could take a huge amount of time, which might not be affordable. Second, such a scheme may require a large amount of memory to store as much models as the size of the vocabulary. For vocabularies of thousands of entries this can be a serious issue. Third, parameter estimation might get seriously degraded as a consequence of the lack of sufficient training data to adequately represent the richness of variations inherent to such a large number of classes.

Coping with these weaknesses usually involves two steps. The first tries to find subclasses or the natural basic units of the language in consideration, which can be shared by the classes of interest. For instance, when we are dealing with words, the natural basic units are characters. Then by building characters HMMs, HMMs at the word level can be easily obtained by concatenating the elementary HMMs associated with the word transcription. One of the most important strengths of HMMs is that no segmentation of words into characters is needed in order to train the character models. HMMs are able to train the models while at the same time optimizing the segmentation of the word into its characters. The second step has to do with the structure of the lexicon. Instead of using a flat structure, the lexicon could be organized as a trie (or a lexical tree) by sharing prefixes across the words of the vocabulary. In this way, a prefix shared by K words will have its HMM evaluated only once instead of K times. By avoiding this kind of redundant computation, a significant speed up could be gained, although this depends on the compression ratio that the vocabulary of the application allows for.

Considering a tree-structured lexicon and the associated HMMs means that to retrieve the optimal class, the forward algorithm can no longer be used because it cannot give any clue as to the optimal sequence of characters. The Viterbi algorithm becomes the natural solution in this case as the optimal path it generates can be directly associated with the optimal character sequence, and hence the optimal word. However, one should keep in mind that Viterbi calculation is only an approximation as it computes the most probable sequence of states rather than the most probable sequence of characters, though this is usually satisfactory in practice. The Viterbi algorithm, moreover, makes it possible to use pruning heuristics to discard nonpromising paths. One popular technique for this purpose is the Viterbi beam search procedure [73].

There exists, however, another decoding scheme, making use of a trie, for which the forward algorithm can be used. In such a scheme, a segmentation graph is first built out of the observation sequence to generate all plausible segmentations of the sequence into character subsequences. Every character segmentation hypothesis is then submitted to the recognizer, and the accumulated scores are propagated forward using a DP-based procedure. As every character segmentation hypothesis is first fed to the HMM, the latter can be seen here as a static recognizer. Hence, one is free in this case to use the forward algorithm rather than the Viterbi algorithm. Such an approach does impede the retrieval of the most likely state sequence, but it provides the most likely character sequence. A similar technique has been used in [18], although here the "static" recognizer was a neural network.

5.4.4   Implementation Issues

In implementing HMMs for text recognition, some practical issues should be considered: the topology of HMM models, initial estimates of parameters, smoothing of parameters, and the underflow of probabilities.

5.4.4.1 Model Topology   As stated in Section 5.4.2, no straightforward method is available for training both HMM topology and parameters. Therefore, it is important

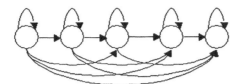

FIGURE 5.20   A left-to-right HMM.

to devise a suitable topology before the training is carried out. This is particularly important if we bear in mind that the EM method assumes that the model is correct. The topology of the model is usually built by using our a priori knowledge on the data and the way we are processing them. For instance, even though the offline handwriting or machine-print signal is two dimensional, it evolves, nonetheless, from left to right, as speech evolves over time. This is the reason why the so-called left-to-right HMMs are used for speech and text recognition. For such models, no back transitions from right to left are allowed. Figure 5.20 shows an example of a left-to-right HMM.

When an implicit segmentation approach is considered, the text signal is usually segmented into tiny frames and observation features (either discrete or continuous) are extracted from each frame. For such frame widths, observations will evolve slowly from left to right when spanning the same stroke and will abruptly change when there is a transition to another stroke. Under this process, states with self-loops can naturally represent the quasi-stationary evolution through the same stroke whereas a transition out of this state will model a transition to the following stroke. Thus, an HMM for modeling this process will have the simple topology shown in Figure 5.21 Note that for implicit segmentation, the number of states can be significantly lower than the average number of observations, as most of contiguous observations would be generated through self-loops.

In practice, this kind of topology might not be sufficient to adequately represent the variations of the data. For instance, we might consider one model for both uppercase and lowercase letters. Besides, each of these two categories usually consists of many prototypes or allographs, for instance, letter l (or h or k) with a loop versus that without a loop. For a better categorization of these well-defined shape variations, it is more convenient to have a topology with multiple branches as shown in Figure 5.22(a). Such a topology, moreover, may relax the heavy constraint of observations' conditional independence, as each parallel branch (or path) implicitly model, to some extent, the dependence between contiguous observations belonging to the same prototype.

When explicit segmentation is considered, the frames usually consist of graphemes and the sequence of graphemes or the observations extracted from can no longer be quasi-stationary. In such a case, the topology shown in Figure 5.22(a) might be used

FIGURE 5.21   A simple left-to-right HMM with no skipping transitions.

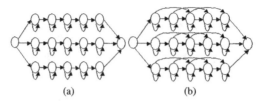

(a)   (b)

FIGURE 5.22 (a) A multiple-branch simple left-to-right HMM; (b) a multiple-branch left-to-right HMM with the possibility of skipping transitions.

without loops. (It might work with loops as well if the number of branches and the number of states are adequately chosen. In such a case, the probability of loops' transitions should be negligible after training.) Note that for explicit segmentation, the number of states is usually about the average number of observations.

One of the most popular topologies of text recognition is the one augmenting that of Figure 5.22(a) with transitions skipping one state. The rationale behind such a topology is that, in practice, the variations of the data and the imperfection of the feature extraction process might lead to missing observations. This phenomenon is conveniently modeled through transitions skipping one state (Fig. 5.22(b)). One might suggest skipping more than one state, but we should keep in mind that increasing the number of transitions leads to the increase in the number of parameters to estimate. Besides, having a sparse Markov chain rather than a dense one often leads to much more efficient calculations.

5.4.4.2 Initial Estimates Initial estimation of the parameters serves two purposes. The first is to set HMM topology by zeroing the probabilities of the transitions that are not allowed in the model. The second is to initialize the other HMM parameters with suitable values with the requirement of avoiding null values as the nature of the reestimation formulas clearly shows that once a parameter is null, it will remain so until the end of the training process. Transition probabilities can be initialized randomly or uniformly with little effect on the overall performance. The same goes for discrete output probabilities. For continuous output distributions, however, a special care should be given to initialization as Baum–Welch training heavily depends on the initial values. One suitable method is to run the EM algorithm for Gaussian mixture distribution on the nonlabeled observation vectors and initialize the distribution of all the states with the same set of trained Gaussians. A more sophisticated method is to use the k-segmental or (Viterbi training) algorithm to get a different estimation for every state according to the implicit segmentation produced by the Viterbi algorithm [77].

5.4.4.3 Parameter Smoothing Independently of the algorithm used for training, a robust estimation of HMM parameters requires the availability of a large training data set so that all events (observations) can be sufficiently represented for each state. Such amount of data, however, is rarely available and we usually end up with having some parameters poorly estimated. This constitutes a serious issue, which, if not correctly handled, could lead to a serious mismatch between the behavior of the HMM on training data on the one hand and on testing data on the other hand, with a significant drop in performance in the latter case. There are various strategies that have been adopted to alleviate this problem:

- *Flooring*: The simplest strategy is to floor all HMM parameters with some predefined threshold. The threshold can be parameter dependent (it may depend on whether we are flooring transition probabilities a_{ij}, discrete output probabilities $b_j(k)$, mixture coefficients c_{jm}, or covariance matrix elements $\sum_{mij}$).
- *Tying*: A more sophisticated strategy is to tie states that are used to model rather rare events. Parameter tying consists of sharing some parameters across different states that are believed to model similar events. Parameter tying offers, in general, a good trade-off between discrimination and reliable parameter estimation. The choice of the states to tie can be obtained either by a priori knowledge or by theoretical methods like decision trees [93].
- *Interpolation*: Another strategy is to interpolate two models with the same structure, one that is more detailed and the other that is less detailed. For instance, one can obtain a less detailed model from a detailed model by extensive parameter tying. The first model can be very discriminative but may have some of its parameters poorly estimated. The second will be less discriminative but will have its parameters better estimated. The interpolated model assigns higher weights to states of the first model when they are correctly estimated (meaning sufficient training data were available for that state) and lower weights when they are not. Implementation details for the interpolation model are straightforward and may be found in [42], and an application for handwriting recognition is reported in [29].

5.4.4.4   *Underflow*   One serious consequence of using high sampling rates in implicit-based segmentation approaches or of using multiple alphabet sets (see Section 5.4.5) is that the probabilities involved in the DP calculations ($\alpha_t(i)$, $\beta_t(i)$) may quickly reach the null value. This is because today's machines are unable to represent the range spanned by probabilities obtained by recursive multiplications. The usual solution to overcome this problem is to multiply $\alpha_t(i)$ and $\beta_t(i)$ by a scaling factor (usually this factor is $1/\sum \alpha_t(i)$) so that these parameters remain sufficiently large for adequate representation in the computer [76]. However, if the Viterbi algorithm is used, no such burden is necessary as the calculations required can be done in the log domain by adding log probabilities rather than multiplying probabilities.

5.4.5   Techniques for Improving HMMs' Performance

In the following are some strategies for improving the recognition performance of HMMs in practice.

5.4.5.1 *Multiple Codebooks*   So far, we have considered the standard HMM framework where one observation is output each time a transition is taken. This is by no means an HMM-specific constraint. One is free to consider the multiple-codebook scheme where the HMM outputs more than one observation at a time. The need for such an approach arises when we want to augment our observation features with additional ones. The latter might consist of the derivatives of the former or might be even independent new features extracted by a new feature extractor. Using this strategy, more discrimination could be obtained locally at each transition. Moreover, experiments showed that combining features inside the same HMM can be much more robust and efficient (in terms of both accuracy and speed) than combining different HMMs, each associated with a different feature set. In other words, combining evidence earlier proves more robust that combing it later.

The multiple-codebook paradigm is usually used by assuming that the different feature sets are independent. This allows us to express $b_j(o_t)$ as the product of the output probabilities for every feature set: $b_j(o_t) = \prod_{f=1}^{F} b_j^f(o_t)$, where $b_j^f(o_t)$ is the output probability associated with the fth feature set and F is the number of feature sets under consideration.

The independence assumption also makes it possible to express the Q function as a summation over independent terms, each associated with a different feature set (in addition to the term associated with transition probabilities). We end up, in this way, with reestimation formulas [30, 38] that are similar to the standard case of one feature set. For discrete HMM, the reestimation formula for $b_j^f(o_t)$ is expressed in the following way:

$$\overline{b_j^f(k)} = \frac{\sum_y \frac{1}{P(O^y|\lambda)} \sum_{t=1}^{T} \sum_{i=1}^{N} \delta((o_t^y)^f = v_k^f)\alpha_{t-1}^y(i)a_{ij}b_j(o_t)\beta_t^y(j)}{\sum_y \frac{1}{P(O^y|\lambda)} \sum_{t=1}^{T} \sum_{i=1}^{N} \alpha_{t-1}^y(i)a_{ij}b_j(o_t)\beta_t^y(i)}, \quad (5.47)$$

where $(o_t^y)^f$ is observation at time t, associated with observation sequence of index y and the fth feature set, v_k^f is the kth symbol associated with the fth alphabet, and F is the number of alphabets (number of feature sets). Similar formulas can be obtained for continuous HMM parameters.

5.4.5.2 *Null Transitions and Noise Models*   In some applications, building accurate models for characters might not be sufficient. For instance, a character pertaining to some word may not be observed at all, either because the word was misspelled or because of the imperfections of the preprocessing or the segmentation stages. On more complicated tasks, much more variability can be manifested. For real-life address recognition tasks, for instance, the location of the bloc to be recognized is not perfect and may cause the input text to be either truncated or augmented with additional text. Even when the bloc location is correct, the transcription on the image may not match exactly its corresponding entry in the lexicon because of some rare abbreviations or because of insertions of additional information inside the relevant transcription. All these factors and others lead to the conclusion that to be able to perform robust recognition, our models must be as flexible as possible by allowing skipping and inserting characters [3, 28]. Without this feature, any model, whatever robust at the character level it might be, will inevitably be mismatched with a transcription involving character omission/insertion. This is because in this case, one observation must be consumed each time the HMM makes a transition.

A convenient way to allow omission by an HMM is to add null transitions (transitions not consuming any observation) between states where an event is thought to be sometimes missing. For instance, to model a missing character in a word, every character HMM model can be augmented by a null transition between the first and last states (assuming that observations are emitted along transitions). To model insertion, we might add a noise model to take account of additional text. In this case, this noise model must be augmented by a null transition between its first and last states to allow the model to skip it in the more likely situation where no such "noise" text appears. This mix between noise models and null transitions is the key to getting as flexible HMM models as possible in highly adverse conditions. Note that as this noise data are quite rare in practice, it becomes clear that considering a noise model for every preceding character will ultimately lead to poor estimation of the noise models. One solution to overcome this problem is to tie the states of all the noise models, which is the same as considering one noise model regardless of the preceding character.

5.4.5.3 *Hybrid HMM/Neural Network*   An alternative approach to discrete or continuous HMM is the so-called hybrid HMM–NN [70] or hybrid HMM–MLP as usually the neural network is a multilayer perceptron (MLP), even though other neural network architectures have also been used. It consists of an HMM for which the distributions of the output probabilities are no longer estimated by a nonparametric discrete distribution or by a mixture of Gaussians, but directly by an MLP. This is achieved by first designing an MLP architecture for which the outputs are associated with the states of HMMs. Thus, the classes for the MLP are $s = 1, 2, \ldots, N$. The inputs to the MLP are the same observation frames as before, possibly augmented by neighboring frames, so as to take into account the context. It has been shown that under certain conditions (large training data, model topology sufficiently complex, etc.), the outputs of an MLP can be interpreted as estimates of a posteriori probabilities of the classes given the input [79]. Thus, the output of the neural network given the frame at time t is the probability of the state $s = j$ given o_t : $P(s = j|o_t)$, $j = 1, \ldots, N$. As HMMs rather need the state conditional likelihoods, the Bayes rule is used to perform the required transformation:

$$P(o_t|s = j) = \frac{P(s = j|o_t)P(o_t)}{P(s = j)}. \tag{5.48}$$

As $P(o_t)$ is constant during recognition, we can safely drop it and use directly the scaled likelihood:

$$\frac{P(o_t|s = j)}{P(o_t)} = \frac{P(s = j|o_t)}{P(s = j)}, \tag{5.49}$$

where $P(s = j)$ is readily obtained by

$$P(s = j) = \frac{\sum_y \sum_t P(s_t = j | O^y)}{\sum_y P(O^y)} = \frac{\sum_y \sum_t \alpha_t^y(j)\beta_t^y(j)}{\sum_y P(O^y)}. \qquad (5.50)$$

As MLP training requires the label for each input, a forced Viterbi alignment is carried out to map the feature vectors from a training input to its producing states. This is achieved by retrieving the best state sequence for every training observation sequence. Then the MLP training is run on the state-labeled feature vectors. This procedure is iterated until convergence is reached.

The main argument for using the MLP to obtain output probabilities is that the latter are trained discriminatively and that no assumption on their distribution is made as opposed to mixtures of Gaussians.

It is worth noting that discrimination brought by the neural net is local (at the state level) as the HMM and the neural network are trained alternatively and separately. Other approaches attempt to obtain discrimination at the world level by considering a global optimizing scheme [5, 6, 87].

Another way of using hybrid HMM–NN is to use the neural network to label HMM states (or transitions) with class labels [52]. In other words, instead of having the NN as the estimator of output probabilities, the winning class of the latter can be used as a discrete observation output by the state. A contextual approach based on this technique has been used in [31] for handwritten word recognition.

5.4.6   Summary of HMM-Based Recognition

This chapter has discussed an introduction to the theory of hidden Markov models and its use for text recognition. As shown in the previous discussions, HMMs offer many interesting features, which make them the technique of preferred choice for text recognition. HMMs are based on a sound theoretical probabilistic framework that makes them suitable for modeling highly variable data such as handwriting or degraded machine-printed text. HMM assumptions, although they lead to limitations in terms of modeling and accuracy, are behind the existence of efficient algorithms for parameter estimation and for recognition. To achieve optimal performance, many issues should be carefully addressed, such as the design of model topology, the choice between different kinds of models (discrete, continuous, and hybrid), the consideration of multiples feature alphabets, smoothing of unreliable parameters as well as coping with numerical limitations. Besides these considerations, overall robustness and efficiency heavily depend on preprocessing and feature extraction as well as on the recognition scheme and vocabulary representation.

5.5   HOLISTIC METHODS FOR HANDWRITTEN WORD RECOGNITION

As words are normally complex patterns and they contain great variability in handwriting styles, handwritten word segmentation is a difficult task. Handwritten word

two Hundred thirty Six-

One	Two	Three	Four	Five
Six	Seven	Eight	Nine	Ten
Eleven	Twelve	Thirteen	Fourteen	Fifteen
Sixteen	Seventeen	Eighteen	Nineteen	Twenty
Thirty	Forty	Fifty	Sixty	Seventy
Eighty	Ninety	Hundred	Thousand	Dollars
Dollar	and	Only		

FIGURE 5.23 Handwritten legal amount recognition involves the recognition of each word in the phrase matched against a static lexicon of 33 words.

recognition can be greatly aided by a lexicon of valid words, which is usually dependent on the application domain. For example, as shown in Figure 5.23 there are 33 different words that may appear in the so-called legal amounts on bank checks. Hence, the lexicon for this application is small and static (it is constant in all the instances). As another example, the lexicon used for street names in handwritten address reading generally comprises street name candidates generated from knowledge of the zip and the street number. As for each zip code we have different street names, this is an example of an application where the lexicon is small but dynamic (varying from one instance to the next). In some other applications, the size of the lexicon may be very large, for example, over 10,000 words for ordinary text. In any case, the nature of the lexicon is crucial to the design of the algorithm for handwritten word recognition in a particular application. A lexicon-driven word recognizer outputs, on inputting a word image, a reduced lexicon sorted by confidence measure [63].

5.5.1 Introduction to Holistic Methods

From the earliest days of research in handwritten word recognition, two approaches to this problem have been identified. The first approach, often called segmentation-based or analytical approach, treats a word as a collection of simpler subunits such as characters and proceeds by segmenting the word into these units, identifying the units, and building a word-level interpretation using a lexicon. The other approach treats the word as a single, indivisible entity and attempts to recognize the word as a whole. The latter approach is referred to as word-based or holistic approach. As opposed to the analytical approach, the holistic paradigm in handwritten word recognition treats the word as a single, indivisible object and attempts to recognize words based on features from the overall shape. The holistic paradigm was inspired in part

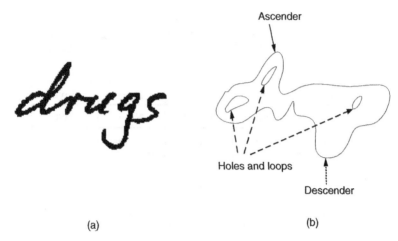

(a)   (b)

FIGURE 5.24 A word image (a) and its shape features (b) including "length," "ascenders," "descenders," "loops," and so on.

by psychological studies of human reading that indicate that humans use features of word shapes such as length, ascenders, and descenders in reading (see Fig. 5.24 for some of these features). A large body of evidence from psychological studies of reading points out that humans do not, in general, read words letter by letter. Therefore, a computational model of reading should include the holistic method.

Because analytical approaches decompose handwritten word recognition into the problem of identifying a sequence of smaller subunits of individual characters, the main problems they face are

- *Segmentation ambiguity*: deciding the boundaries of individual segments in the word image (see Fig. 5.25)
- *Variability of segment shape*: determining the identity of each segment (see Fig. 5.26)

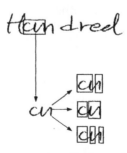

FIGURE 5.25 Ambiguities in segmentation: The letter(s) following "H" can be recognized as "w," "ui," "iu," or "iii."

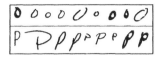

FIGURE 5.26 Large variability in shapes of handwritten characters ("O" and "P" in this example).

Holistic approaches circumvent these problems because they make no attempt to segment the word into subunits. Actually, holistic methods follow a two-step process: the first step performs feature extraction and the second step performs global recognition by comparing the representation of the unknown word with those of the references stored in the lexicon. This scheme leads to two important practical consequences: First, as letter segmentation is avoided and recognition is performed in a global way, these methods are usually considered to be tolerant to the dramatic deformations that affect unconstrained cursive scripts. Second, as they do not deal directly with letters but only with words, recognition is necessarily constrained to a specific lexicon of words. The second point is especially critical when training on word samples is required. In this case, the lexicon cannot be automatically updated from letter information. A training stage is thus mandatory to expand or modify the lexicon of possible words. This property makes this kind of method more suitable for applications where the lexicon is static (and not likely to change), like check recognition. For dynamic lexicons, the recognition system must have the reference models for all the words in the union of lexicons such that an unknown word can be compared with any subset of word references.

The term "holistic approach" has been used in the literature at least in two different senses: (1) an approach that matches words as a whole and (2) an approach that uses word shape features. It is important to distinguish holistic features from holistic approaches. A holistic approach may or may not use holistic word shape features. For example, it may use pixel direction distribution features. Conversely, a classifier may use word shape features in an approach that is not holistic to perform segmentation and/or character recognition. The term "global features" has been used by some researchers to refer to simpler aspects of word shapes that can be easily and reliably measured. Often, this refers to estimates of word length and counts of perceptual features such as ascenders and descenders (see Fig. 5.27).

If the size of the lexicon is small enough, the word shape or length of cursive words alone contains sufficient information to classify the image as one of the lexicon words with a very high confidence (see Fig. 5.28).

The process of constructing a lexicon in which each lexicon entry is represented by its holistic features, or statistics about holistic features (in the case of probabilistic methods), is sometimes referred to as "inverting the lexicon" [63]. Holistic methods described in the literature have used a variety of holistic features and representations. As an example, assume that the holistic features are [length, number of ascenders, number of descenders]. The lexicon shown in Figure 5.29 can be inverted based on these three features as follows:

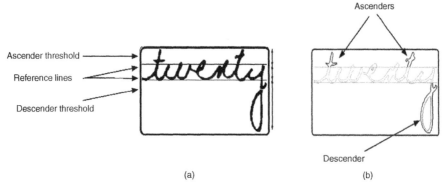

FIGURE 5.27 Ascenders and descenders are perceptual features. Features such as the proportion of pixels in the left and right segments, the number of extrema, the perimeter of the word are examples of holistic features.

FIGURE 5.28 If the size of the lexicon is small, the word shape or length of cursive words alone contains sufficient information to classify the image as one of the lexicon words.

beautiful [9, 4, 1]
Montreal [8, 3, 0]
Canada [6, 2, 0]
Beijing [7, 4, 2]

For training the word recognition system, when the lexicon is small and static, it becomes possible to collect a large number of training samples of each class (word). Training may then be performed in the traditional sense for estimating class-

FIGURE 5.29 A lexicon of four words that are inverted based on three holistic features. Each word is represented by a vector of three elements: [length, number of ascenders, number of descenders].

conditional densities of features from the training samples or storing prototypical feature vector exemplars for each class. Discriminative classifiers like neural networks and SVMs are also feasible for holistic word recognition of small static lexicons. In some methods (mainly in the online handwriting and printed domains), feature representation shown in the above example has been used for reduction of large lexicons. This feature representation has also been used internally by some analytical classifiers to rapidly discard dissimilar lexicon entries.

5.5.2   Overview of Holistic Methods

Holistic word recognition methods can be categorized from different perspectives [63]: the nature of application and lexicon, the level and data structure of feature representation, and so on.

1. *Application domain*: online or offline. Some holistic methods have been applied for online recognition of handwritten words, where words were written on an electronic tablet or with a light pen. Some other holistic methods deal with the problem of reading offline handwritten words. In such cases, handwritten words are typically scanned from a paper document and made available in the form of binary or gray-scale images to the recognition algorithm.

2. *Lexicon*: static or dynamic. In some applications like check recognition, the lexicon is static and fixed and each entry of the incoming words is compared with the same static lexicon. In the dynamic case, the lexicon varies from one instance to another instance, such as address recognition, where for each zip code there is a different lexicon for the names of the streets that exist in that zip code (region).

3. *Feature extraction*: low level, intermediate level, and high level. Local and low-level structural features such as stroke direction distributions have been applied successfully for holistic recognition of machine-printed words. Structural features at the intermediate level include edges, end points, concavities, diagonal and horizontal strokes, and they exhibit a greater abstraction from the image (pixel or trace) level. Perceptual features such as ascenders, descenders, loops, and length are easily perceived by the human eye, and we call them high-level features. There are evidences that these features are used in human reading. They are by far the most popular for holistic recognition of handwritten words.

4. *Feature representation*: vectors, assertions, sequences, or graphs. The representation scheme of features depends on the nature of features: whether they are low level, medium level, or high level. For example, feature vectors and matrix representations are commonly used to represent low level features. Counts and assertions, for example, the counts of ascenders and descenders, are the simplest representation of high-level features. Such simple features are also called "global features" and they are often used to discard dissimilar word

candidates from the lexicon. Sequences are also used for the representation of a whole word as a sequence of symbols representing a set of structural primitives, which correspond to intermediate- or high-level features or a combination of such features. With respect to intermediate-level features or higher level structural features such as edges, end points, and perceptual features, the presence or absence of each of these features is important. The presence of such features and their interrelationship can be represented by a graph, which denotes features as nodes and relationships as edges.

5. *Hybrid methods*. These refer to the methods that explicitly use a combination of other methodologies; for example, some methods adopt both analytical and holistic features. Although holistic and analytical approaches are commonly distinguished by the observation that the latter are segmentation-based, holistic and analytical paradigms comprise a continuum of approaches to word recognition. As noted by Casey and Lecolinet in [13], some form of segmentation is involved in all pattern recognition methods, for which holistic method is their feature extraction phase:

> The main difference lies in the level of abstraction of the segmented elements: features (that is to say low-level elements) in the case of holistic methods, versus pseudo-letters in the case of analytical methods.

5.5.3   Summary of Holistic Methods

The holistic paradigm in handwritten word recognition treats the word as a single, indivisible entity and attempts to recognize it using features of the word as a whole, and is inspired by psychological studies of human reading, which indicates that humans use features of word shapes such as length, ascenders, and descenders in reading. Holistic approaches circumvent the issues of segmentation ambiguity and character shape variability that are primary concerns for analytical approaches, and they may succeed on poorly written words where analytical methods fail to identify character content. Their treatment of lexicon words as distinct pattern classes has traditionally limited their application to recognition scenarios involving small, static lexicons. Nevertheless, the complexity of holistic recognition methods varies depending on the representation level of features and the efficiency of matching algorithm. The methods using simple global features and fast matching algorithms can be used for efficient lexicon reduction of large lexicons.

5.6   CHAPTER SUMMARY

This chapter first gives an overview of recognition methods for words and character strings, then goes into the details of oversegmentation, classification-based (character-model-based) recognition, HMM-based recognition, and ends with holistic word recognition. In classification-based recognition, we start with a probabilistic string classification model, under which the scores of character segmentation, character recognition, and linguistic context are unified. String-recognition-oriented classifier design techniques, including character-level training and string-level training techniques, are discussed. Path search and lexicon organization strategies, which both affect the time efficiency of string recognition, are also discussed in depth. In HMM-based recognition, we introduce the mathematical formulas of HMMs, describe the paradigms of applying HMMs to text recognition, and discuss the strategies for improving the recognition performance. Holistic recognition methods are reviewed from the perspectives of application domain, the nature of lexicon, the level and data structure of feature representation, and so on.

Although the classification-based recognition methods are primarily used for explicit segmentation, the HMM-based methods can be used for either explicit segmentation, implicit segmentation, or holistic recognition. Classification-based methods are particularly useful to applications where the linguistic context is weak, like numeral string recognition, or the number of character classes is large, like Chinese/Japanese character string recognition. They are, of course, applicable to other lexicon-driven recognition cases as well. In applications where the shapes of single characters are not discernable, like cursive word recognition, holistic recognition methods are very useful. Holistic methods, however, perform well only for small and static lexicons. For large lexicons, they can be used for lexicon reduction. In practice, no single word/string recognition method performs sufficiently, and it is often necessary to combine different methods to solve this difficult problem.

REFERENCES

1. N. Arica and F. T. Yarman-Vural. An overview of character recognition focused on off-line handwriting. *IEEE Transactions on Systems, Man and Cybernetics Part C*. **31**(2), 216–233, 2001.

2. L. R. Bahl, P. F. Brown, P. V. de Souza, and R. L. Mercer. Maximum mutual information estimation of hidden Markov model parameters for speech recognition. In *Proceedings of the IEEE International Conference on Acoustics, Speech and Signal Processing*, Tokyo, Japan, 1986, pp. 49–52.

3. L. R. Bahl and F. Jelinek. Decoding for channels with insertions, deletions, and substitutions with applications to speech recognition. *IEEE Transactions on Information Theory*, **21**(4), 404–411, 1975.

4. L. Bahl, F. Jelinek, and R. Mercer. A maximum likelihood approach to speech recognition. *IEEE Transactions on Pattern Analysis and Machine Intelligence*, **5**(2), 179–190, 1983.

5. Y. Bengio, R. De Mori, G. Flammia, and R. Kompe. Global optimization of a neural network-hidden Markov model hybrid. *IEEE Transactions on Neural Networks*, **3**(2), 252–259, 1992.

6. Y. Bengio, Y. LeCun, C. Nohl, and C. Burges. LeRec: a NN/HMM hybrid for online handwriting recognition. *Neural Computation*, **7**(5), 1289–1303, 1995.

7. A. Biem. Minimum classification error training for online handwritten word recognition. In *Proceedings of the 8th International Workshop on Frontiers in Handwriting Recognition.* Ontario, Canada, 2002, pp. 61–66.

8. A. Biem. Minimum classification error training for online handwriting recognition. *IEEE Transactions on Pattern Analysis and Machine Intelligence*, **28**(7), 1041–1051, 2006.

9. R. M. Bozinovic and S. N. Srihari. Off-line cursive script word recognition. *IEEE Transactions on Pattern Analysis and Machine Intelligence*, **11**(1), 68–83, 1989

10. T. M. Breuel. A system for the off-line recognition of handwritten text. In *Proceedings of the 12th International Conference on Pattern Recognition*. Jerusalem, Israel, 1994, Vol. 2, pp. 129–134.

11. J. Bromley and J. S. Denker. Improving rejection performance on handwritten digits by training with rubbish. *Neural Computation*, **5**, 367–370, 1993.

12. J. Cai and Z.-Q. Liu. Integration of structural and statistical information for unconstrained handwritten numeral recognition. *IEEE Transactions on Pattern Analysis and Machine Intelligence*, **21**(3), 263–270, 1999.

13. R. G. Casey and E. Lecolinet. A survey of methods and strategies in character segmentation. *IEEE Transactions on Pattern Analysis and Machine Intelligence*, **18**(7), 690–706, 1996.

14. F. Chang, C.-J. Chen, and C.-J. Lu. A linear-time component-labeling algorithm using contour tracing technique. *Computer Vision and Image Understanding*, **93**(2), 206–220, 2004.

15. C.-H. Chen. Lexicon-driven word recognition. In *Proceedings of the 3rd International Conferene on Document Analysis and Recognition*. Montreal, Canada, 1995, pp. 919–922.

16. W.-T. Chen and P. Gader. Word level discriminative training for handwritten word recognition. In *Proceedings of the 7th International Workshop on Frontiers of Handwriting Recognition*, Amsterdam, The Netherlands, 2000, pp. 393–402.

17. M. Y. Chen, A. Kundu, and S. N. Srihari. Variable duration hidden Markov model and morphological segmentation for handwritten word recognition. *IEEE Transactions on Image Processing*, **4**(12), 1675–1688, 1995.

18. D. Y. Chen, J. Mao, and K. M. Mohiuddin. An efficient algorithm for matching a lexicon with a segmentation graph. In *Proceedings of the 5th International Conference on Document Analysis and Recognition*. Bangalore, India, 1999, pp. 543–546.

19. Y. K. Chen and J. F. Wang. Segmentation of single or multiple touching handwritten numeral string using background and foreground analysis. *IEEE Transactions on Pattern Analysis and Machine Intelligence*, **22**, 1304–1317, 2000.

20. M. Cheriet, Y. S. Huang, and C. Y. Suen. Background region based algorithm for the segmentation of connected digits. In *Proceedings of the International Conference on Pattern Recognition*, The Hague, 1992, Vol. 2, pp. 619–622.

21. W. Chou. Discriminant-function-based minimum recognition error pattern-recognition approach to speech recognition. *Proceedings of IEEE*, **88**(8), 1201–1223, 2000.

22. C. K. Chow. On optimal recognition error and reject tradeoff. *IEEE Transactions on Information Theory*, **16**, 41–46, 1970.

23. A. P. Dempster, N. M. Laird, and D. B. Rubin. Maximum likelihood from incomplete data via the EM algorithm. *Journal of the Royal Statistical Society B*, **39**, 1–38, 1977.

24. A. Dengel, R. Hoch, F. Hönes, T. Jäger, M. Malburg, and A. Weigel. Techniques for improving OCR results. In H. Bunke and P. S. P. Wang, editors, *Handbook of Character Recognition and Document Image Analysis*. World Scientific, Singapore, 1997, pp. 227–254.

25. B. Dubuisson, M. Masson. A statistical decision rule with incomplete knowledge about classes. *Pattern Recognition*, **26**(1), 155–165, 1993.

26. A. J. Elms, S. Procter, and J. Illingworth. The advantage of using an HMM-based approach for faxed word recognition. *International Journal of Document Analysis and Recognition*. **1**(1), 18–36, 1998.

27. M. A. El-Yacoubi, M. Gilloux, and J.-M. Bertille. A statistical approach for phrase location and recognition within a text line: an application to street name recognition. *IEEE Transactions on Pattern Analysis and Machine Intelligence*, **24**(2), 172–188, 2002.

28. M. A. El-Yacoubi, M. Gilloux, R. Sabourin, and C.Y. Suen. An HMM based approach for offline unconstrained handwritten word modeling and recognition. *IEEE Transactions on Pattern Analysis and Machine Intelligence*, **21**(8), 752–760, 1999.

29. A. El-Yacoubi, R. Sabourin, M. Gilloux, and C.Y. Suen. Improved model architecture and training phase in an offline HMM-based word recognition system. In *Proceedings of the 13th International Conference on Pattern Recognition*. Brisbane, Australia, 1998, pp. 1521–1525.

30. A. El-Yacoubi, R. Sabourin, M. Gilloux, and C.Y. Suen. Off-line handwritten word recognition using hidden Markov models. In L. C. Jain and B. Lazzerini, editors, *Knowledge-Based Intelligent Techniques in Character Recognition*. CRC Press LLC, 1999, pp. 191–229.

31. C. Farouz, M. Gilloux, and J.-M. Bertille. Handwritten word recognition with contextual hidden Markov models. In S.-W. Lee, editor, *Advances in Handwriting Recognition*. World Scientific, Singapore, 1999, pp. 183–192.

32. P. D. Gader, M. Mohamed, and J.-H. Chiang. Handwritten word recognition with character and inter-character neural networks. *IEEE Transactions on Systems, Man and Cybernetics Part B*, **27**(1), 158–164, 1997.

33. S. Garcia-Salicetti, B. Dorizzi, P. Gallinari, A. Mellouk, and D. Fanchon. A hidden Markov model extension of a neural predictive system for on-line character recognition. In *Proceedings of the 3rd International Conference on Document Analysis and Recognition*, Montreal, Canada, 1995, Vol. 1, pp. 50–53.

34. M. Gilloux, M. Leroux, and J. M. Bertille. Strategies for cursive script recognition using hidden Markov models. *Machine Vision and Applications*, **8**(4), 197–205, 1995.

35. R. M. Gray. Vector quantization. *IEEE ASSP Magazine*, 4–29, 1984.

36. D. Guillevic and C. Y. Suen. HMM word recognition engine. In *Proceedings of the 4th International Conference on Document Analysis and Recognition*. Ulm, Germany, 1997, pp. 544–547.

37. R. M. Haralick and L. G. Shapiro. *Computer and Robot Vision*. Addison-Wesley, 1992, Vol. 1, pp. 28–48.

38. X. Huang, A. Acero, and H.-W. Hon. *Spoken Language Processing: A Guide to Theory, Algorithm and System Development*. Prentice-Hall, 2001.

39. X. D. Huang and M. A. Jack. Semi-continuous hidden Markov models for speech signals. *Computer Speech and Language*, **3**, 239–251, 1989.

40. H. Ikeda, N. Furukawa, M. Koga, H. Sako, and H. Fujisawa. A context-free grammar-based language model for string recognition. *International Journal of Computer Processing of Oriental Languages*, **15**(2), 149–163, 2002.

41. F. Jelinek. *Statistical Methods for Speech Recognition*. MIT Press, Cambridge, MA, 1998.

42. F. Jelinek and R. Mercer. Interpolated estimation of Markov source parameters from sparse data. In *Proceedings of the Workshop on Pattern Recognition in Practice*. Amsterdam, The Netherlands, 1980.

43. B.-H. Juang, W. Chou, and C.-H. Lee. Minimum classification error rate methods for speech recognition. *IEEE Transactions on Speech and Audio Processing*, **5**(3), 257–265, 1997.

44. B.-H. Juang and S. Katagiri. Discriminative learning for minimum error classification. *IEEE Transactions on Signal Processing*, **40**(12), 3043–3054, 1992.

45. A. Kaltenmeier, T. Caesar, J. M. Gloger, and E. Mandler. Sophisticated topology of hidden Markov models for cursive script recognition. In *Proceedings of the 2nd International Conference on Document Analysis and Recognition*. Tsukuba, Japan, 1993, pp. 139–142.

46. S. Katagiri, B.-H. Juang, and C.-H. Lee. Pattern recognition using a family of design algorithms based upon the generalized probabilistic descent method. *Proceedings of the IEEE*, **86**(11), 2345–2375, 1998.

47. G. Kim and V. Govindaraju. A lexicon driven approach to handwritten word recognition for real-time applications. *IEEE Transactions on Pattern Analysis and Machine Intelligence*, **19**(4), 366–379, 1997.

48. F. Kimura, Y. Miyake, and M. Sridhar. Handwritten ZIP code recognition using lexicon free word recognition algorithm. In *Proceedings of the 3rd International Conference on Document Analysis and Recognition*, Montreal, 1995, pp. 906–910.

49. F. Kimura, M. Sridhar, and Z. Chen. Improvements of a lexicon directed algorithm for recognition of unconstrained handwritten words. In *Proceedings of the 2nd International Conference on Document Analysis and Recognition*. Tsukuba, Japan, 1993, pp. 18–22.

50. F. Kimura, K. Takashina, S. Tsuruoka, and Y. Miyake. Modified quadratic discriminant functions and the application to Chinese character recognition. *IEEE Transactions on Pattern Analysis and Machine Intelligence*, **9**(1), 149–153, 1987.

51. A. L. Koerich, R. Sabourin, and C. Y. Suen. Large vocabulary off-line handwriting recognition: a survey. *Pattern Analysis and Applications*, **6**(2), 97–121, 2003.

52. P. Le Cerf, W. Ma, and D. Van Compernolle. Multilayer perceptrons as labelers for hidden Markov models. *IEEE Transactions on Speech and Audio Processing*, **2**(1), 185–193, 1994.

53. Y. LeCun and Y. Bengio. Word-level training of a handwritten word recognizer based on convolutional neural networks. In *Proceedings of the 12th International Conference on Pattern Recognition*, Jerusalem, Israel, 1994, Vol. 2, pp. 88–92.

54. A. Lifchitz and F. Maire. A fast lexically constrained Viterbi algorithm for on-line handwriting recognition. In *Proceedings of the 7th International Workshop on Frontiers of Handwriting Recognition*, Amsterdam, The Netherlands, pp. 313–322, 2000.

55. L. A. Liporace. Maximum likelihood estimation for multivariate observation of Markov sources. *IEEE Transactions on Information Theory*, **28**(5), 729–734, 1982.

56. C.-L. Liu, M. Koga, and H. Fujisawa. Lexicon-driven segmentation and recognition of handwritten character strings for Japanese address reading. *IEEE Transactions on Pattern Analysis and Machine Intelligence*, **24**(11), 1425–1437, 2002.

57. C.-L. Liu and K. Marukawa. Handwritten numeral string recognition: character-level training vs. string-level training. In *Proceedings of the 17th International Conference on Pattern Recognition*, Cambridge, UK, 2004, Vol. 1, pp. 405–408.

58. C.-L. Liu and M. Nakagawa. Precise candidate selection for large character set recognition by confidence evaluation. *IEEE Transactions on Pattern Analysis and Machine Intelligence*, **22**(6), 636–642, 2000.

59. C.-L. Liu, H. Sako, and H. Fujisawa. Discriminative learning quadratic discriminant function for handwriting recognition. *IEEE Transactions on Neural Networks*, **15**(2), 430–444, 2004.

60. C.-L. Liu, H. Sako, and H. Fujisawa. Effects of classifier structures and training regimes on integrated segmentation and recognition of handwritten numeral strings. *IEEE Transactions on Pattern Analysis and Machine Intelligence*, **26**(11), 1395–1407, 2004.

61. Z. Lu, Z. Chi, W. Siu, and P. Shi. A background-thinning-based approach for separating and recognizing connected handwriting digit strings. *Pattern Recognition*, **32**, 921–933, 1999.

62. Y. Lu and M. Sridhar. Character segmentation in handwritten words: an overview, *Pattern Recognition*, **29**(1), 77–96, 1996.

63. S. Madhvanath and V. Govindaraju. The role of holistic paradigms in handwritten word recognition. *IEEE Transactions on Pattern Analysis and Machine Intelligence*, **23**(2), 149–164, 2001.

64. J. Makhoul, S. Roucos, and H. Gish. Vector quantization in speech coding. *Proceedings of IEEE*, **73**, 1551–1588, 1985.

65. J. Makhoul, R. Schwartz, C. Lapre, and I. Bazzi. A script-independent methodology for optical character recognition. *Pattern Recognition*, **31**(9), 1285–1294, 1998.

66. S. Manke, M. Finke, and A. Waibel. A fast search technique for large vocabulary online handwriting recognition. In *Proceedings of the 5th International Workshop on Frontiers of Handwriting Recognition*, Colchester, UK, 1996, pp. 183–188.

67. U. V. Marti and H. Bunke. Using a statistical language model to improve the performance of an HMM-based cursive handwriting recognition system. *International Journal of Pattern Recognition and Artificial Intelligence*, **15**(1), 65–90, 2001.

68. C. L. Martin. Centered-object integrated segmentation and recognition of overlapping handprinted characters. *Neural Computation*, **5**, 419–429, 1993.

69. T. K. Moon. The expectation-maximization algorithm. *IEEE Signal Processing Magazine*, **13**(6), 47–60, 1996.

70. N. Morgan and H. Bourlard. Continuous speech recognition: an introduction to the hybrid HMM/connectionist approach. *IEEE Signal Processing Magazine*, **12**(3), 25–42, 1995.

71. H. Murase. Online recognition of free-format Japanese handwritings. *Proceedings of the 9th International Conference on Pattern Recognition*, Rome, Italy, 1988, pp. 1143–1147.

72. M. Nakagawa, B. Zhu, and M. Onuma. A model of on-line handwritten Japanese text recognition free from line direction and writing format constraints. *IEICE Transactions on Information and Systems*, **E-88D**(8), 1815–1822, 2005.

73. H. Ney and S. Ortmanns. Dynamic programing search for continuous speech recognition. *IEEE Signal Processing Magazine*, **16**(5), 64–83, 1999.

74. N. J. Nilsson. *Principles of Artificial Intelligence*. Springer, 1980.

75. R. Plamondon and S. N. Srihari. On-line and off-line handwriting recognition: a comprehensive survey. *IEEE Transactions on Pattern Analysis and Machine Intelligence*, **22**(1), 63–84, 2000.

76. L. R. Rabiner. A tutorial on hidden Markov models and selected applications in speech recognition. *Proceedings of IEEE*, **77**(2), 257–286, 1989.

77. L. Rabiner and B.-H. Juang. *Fundamentals of Speech Recognition*. Prentice-Hall, 1993.

78. E. H. Ratzlaff, K. S. Nathan, and H. Maruyama. Search issues in IBM large vocabulary unconstrained handwriting recognizer, In *Proceedings of the 5th International Workshop on Frontiers of Handwriting Recognition*, Colchester, UK, 1996, pp. 177–182.

79. M. D. Richard and R. P. Lippmann. Neural network classifiers estimate Bayesian a posteriori probabilities. *Neural Computation*, **3**(4), 461–483, 1991.

80. J. Sadri, C. Y. Suen, and T. D. Bui. Automatic segmentation of unconstrained handwritten numeral strings. In *Proceedings of the 9th International Workshop on Frontiers in Handwriting Recognition*, Tokyo, Japan, 2004, pp. 317–322.

81. Y. Saifullah and M. T. Manry. Classification-based segmentation of ZIP codes. *IEEE Transactions on Systems, Man and Cybernetics*, **23**(5), 1437–1443, 1993.

82. K. M. Sayre. Machine recognition of handwritten words: a project report. *Pattern Recognition*, **5**(3), 213–228, 1973.

83. T. Steinherz, E. Rivlin, and N. Intrator. Offline cursive script word recognition—a survey. *International Journal of Document Analysis and Recognition*, **2**(2), 90–110, 1999.

84. N. W. Strathy, C. Y. Suen, and A. Krzyzak. Segmentation of handwritten digits using contour features. In *Proceedings of the 2nd International Conference on Document Analysis and Recognition*, Tsukuba, Japan, 1993, pp. 577–580.

85. C. Y. Suen. *N*-gram statistics for natural language understanding and text processing. *IEEE Transactions on Pattern Analysis and Machine Intelligence*, **1**(2), 164–172, 1979.

86. K. Suzuki, I. Horiba, and N. Sugie. Linear-time connected-component labeling based on sequential local operations. *Computer Vision and Image Understanding*, **89**(1), 1–23, 2003.

87. Y. H. Tay, P. M. Lallican, M. Khalid, C. Viard-Gaudin, and S. Knerr. An analytical handwritten word recognition system with word-level discriminant training. In *Proceedings of the 6th International Conference on Document Analysis and Recognition*, Seattle, WA, 2001, pp. 726–730.

88. S. Tulyakov and V. Govindaraju. Probabilistic model for segmentation based word recognition with lexicon. *Proceedings of the 6th International Conference on Document Analysis and Recognition*, Seattle, WA, 2001, pp. 164–167.

89. A. Vinciarelli. A survey on off-line cursive word recognition. *Pattern Recognition*, **35**(7), 1433–1446, 2002.

90. A. Vinciarelli, S. Bengio, and H. Bunke. Offline recognition of unconstrained handwritten texts using HMMs and statistical language models. *IEEE Transactions on Pattern Analysis and Machine Intelligence*, **26**(6), 709–720, 2004.

91. P. H. Winston. *Artificial Intelligence*, 3rd edition. Addison-Wesley, 1992.

92. H. Xue and V. Govindaraju. Incorporating contextual character geometry in word recognition. In *Proceedings of the 8th International Workshop Frontiers in Handwriting Recognition*, Ontario, Canada, 2002, pp. 123–127.

93. S. J. Young, J. J. Odell, and P. C. Woodland. Tree-based state tying for high accuracy acoustic modeling. In *Proceedings of the ARPA Human Language Technology Workshop*, 1994, pp. 307–312.

CHAPTER 6

CASE STUDIES

A complete document analysis system involves multiple techniques in different processing steps. Even a character recognition or word recognition module (subsystem) needs
techniques traversing different chapters of this book. To exemplify how the techniques
are practically used and how they are performing, this chapter presents some concrete
examples of using multiple techniques to build practical recognition systems, and shows
experimental results on real image data. As described there are three cases: evolutionary
generation of pattern recognizers, offline handwritten Chinese character recognition,
date image segmentation and recognition on Canadian bank checks.

6.1 AUTOMATICALLY GENERATING PATTERN RECOGNIZERS WITH EVOLUTIONARY COMPUTATION

In this section, we describe the CellNet project. The CellNet project is an ongoing
attempt for creating an autonomous pattern recognizer—a system capable of self-
adapting to a pattern recognition task, in a given database with minimal human intervention. The CellNet system is an Evolutionary Algorithm that, beginning with a
set of preprogramed features, evolves binary classifiers capable of recognizing handwritten characters. To date, CellNet has been applied to several different sorts of
handwritten characters with minimal human adaptation, obtaining validation accuracies comparable to current technologies. We describe work originally published in
[17, 25].

The essential metaphor behind the CellNet system is that of hunters and prey. The
classifiers are hunters who must decide whether or not to accept images (prey). The

Character Recognition Systems: A Guide for Students and Practitioner, by M. Cheriet, N. Kharma,
C.-L. Liu and C. Y. Suen   Copyright © 2007 John Wiley & Sons, Inc.

hunters that correctly identify images (of a particular class) are declared more "fit," and hence are used to create new and ideally better hunters. These hunters evolve in cooperative coevolution, allowing hunters to merge, and operate as a single, more complex agent. Additionally, in later trials, a system of competitive coevolution was introduced. The images (prey) were allowed to apply a set of camouflage functions to themselves, in an attempt to fool the hunters. Prey are declared more "fit" by managing to fool as many hunters as possible, hence making the hunters' jobs more difficult. Through the use of cooperative and competitive coevolution, the CellNet system is able to overcome several initial difficulties in the context of handwritten character recognition, especially in combating overfitting and extending sparse training data. At the end of the chapter, we describe the potential of the system, and how it may be extended to other kinds of image recognition.

6.1.1 Motivation

The field of pattern recognition is vast. The shear volume of possible frameworks, conceptualizations, and systems is enormous. The amount of literature available to a practitioner in any subfield will span a wide berth of conceptual frameworks and highly specialized cases. A valuable tool in this environment is that of an autonomous pattern recognizer (APR). We envision a "black box" approach to pattern recognition in which the APR's operator need not be privy to the details of the mechanism used in order to generate a reliable recognition tool. A black box of this sort needs to accommodate many differing classes of input and output. Nearly no assumptions regarding the space of input can be made, nor any assumptions regarding clustering, features, and so on. The immediate trade-off found in this setting is the trade-off between the generality of the system and the so-called "curse of dimensionality." Any system that operates on a space of possible recognizers this large quickly confronts a computationally infeasible task: the space of all possible recognizers in any sufficiently general system is simply too huge. Hence, an APR needs to implement a means to "canalize development," a method of search through the space of possible recognizers that is limiting in terms of the subspace considered, but robust in terms of finding optima. In addition to these difficulties, such a system would need to be resistant to difficult problems such as over-fitting, the tendency of learning systems to "memorize data." In the context of adaptive systems, overfitting means that the system learns a very precise methodology for classifying a training set, which does not extend to classifying an independent validation set. An APR search technique needs to deal with the problem of over-fitting intrinsically in its methodology of subspace search, in addition to other global strategies.

6.1.2 Introduction

The aim of the CellNet project is to create a software system capable of automatically evolving complete pattern recognizers from arbitrary pattern sets, that is, to minimize the amount of expert input required for the creation of pattern recognizers given some set of target data. This is an ambitious goal, requiring much additional work. This chapter describes our initial successes in the context of handwritten character recognition. A pattern recognition system is almost always defined in terms of two functionalities: the description of patterns and the identification of those patterns. The first functionality is called feature extraction and the second, pattern classification. As there are many types of patterns in this world, ranging from the images of fruit flies to those of signatures and thumbprints, the focus of most research endeavors in pattern recognition has rightfully been directed toward the invention of features that can capture what is most distinctive about a pattern. This leads to two overlapping areas of research associated with features: feature selection and feature creation. Feature selection has to do with choosing a small set of features from a larger set, in order to maximize the rate of recognition of an associated classifier, and simultaneously reducing the (computational) cost of classification. Feature creation, on the contrary, has to do with the creation or growth of complex features from a finite set of simpler ones, also for the benefit of an associated classifier. CellNet blurs this distinction—beginning from a preprogramed set of functions, CellNet allows for the simultaneous selection and recombination of features into classifiers, evaluated at the highest level (largely accuracy). Hence, simple preprogramed features are both refined and combined into more complex features as a natural part of the functioning of the system.

In its current form, CellNet is capable of evolving classifiers for a given set of patterns. To achieve this it uses a specialized genetic operator: Merger. Merger is an operator, somewhat similar to that used in MessyGAs [8], designed to allow the algorithm to search increasingly larger feature spaces in an incremental manner. This is different from a normal GA, where the dimensionality of the search space is fixed at the start of the run, or, possibly, varies slightly and randomly through the recombination operator. CellNet is cast as a general system, capable of self-adapting to many handwritten alphabets, currently having been applied to both Arabic and Indian numbers (using the CEDAR database and a second database collected at CENPARMI, Montreal [1]). To achieve a truly autonomous pattern recognizer, however, several significant challenges need to be addressed.

1. Feature creation (or detection) mechanisms that are suited to large sets of classification problems. This is to eliminate the need for serious reconfiguration or worse still, reengineering, every time a different application domain is targeted.

2. Elimination of parameters, many of which need to be set by the user before the system is exploited. These parameters include: probability distributions types, probability values, population size, number of feature extractors, and so on.

3. Thorough testing of the evolved system against a diverse set of pattern databases, and in doing so, without subjecting the system to any significant amount of reconfiguration.

We view our work as a step toward the realization of points 2 and 3.

6.1.3 Hunters and Prey

As previously mentioned, the CellNet system is composed of both hunters (classifiers) and prey (images from a database). Here we describe both, with special attention on representation, as this is critical for an evolutionary algorithm.

6.1.3.1 Hunters The basis on which Hunters are built is a set of normalized feature functions. The functions (all applied to thinned figures) used by CellNet are the set parameterized histograms, central moments, Fourier descriptors, Zernike moments, normalized width (of a bounding box), normalized height, normalized length, number of terminators, number of intersections. However, for the sake of illustration, we shall assume that our examples use only two: F_1 and F_2. This assumption is made for ease of visual representation of a two-dimensional feature space and is easily generalized to higher-dimensional spaces.

A hunter is a binary classifier—a structure that accepts an image as input and outputs a classification. The fitness function that is used to drive the genetic algorithm determines which digit the agent classifies. For example, assuming our fitness function specifies that we are evolving hunters to recognize the digit "one," a hunter that returns "yes" given an image will implicitly be stating that the image is a "one," as opposed to "not-one," that is, any other digit. We will refer to these classes as the primary class and the nonprimary class. Hence, a hunter outputs a value from primary, nonprimary, uncertain when presented with an image. A hunter consists of cells, organized in a net. A cell is a logical statement—it consists of the index of a feature function along with bounds. Every cell is represented by the following format:

Feature function F_i	Bound b_1	Bound b_2

Provided with an image I, a cell returns true, if $b_1 < F_i(I) < b_2$, and false otherwise. A net is an overall structure that organizes cells into a larger tri-valued logical statement. That is, a net is a logical structure that combines the true/false values of its constituent cells to return a value from primary, nonprimary, uncertain. Chromosomes consist of trees that begin with a class bit—this bit determines whether or not the chromosome votes for "primary" or "nonprimary." Following the class bit is a tree of one or more cells, connected into a logical structure by and and not nodes. A chromosome may be represented as a string as follows:

Class bit C	[Not]	$Cell_1$	And	[Not]	$Cell_2$	And	$\cdots$

Hence, the latter part is a logical statement, which returns true or false. A chromosome will return C if the logical statement returns true, otherwise it will remain silent. Every chromosome is followed by two affinity bits, which we will describe later. A net is a collection of such chromosomes, connected via a vote mechanism. The vote mechanism collects input from each chromosome (although some chromosomes will remain silent), consisting of a series of zero or more values of "primary" or "nonprimary." The vote mechanism will tally the number of each, and output the majority,

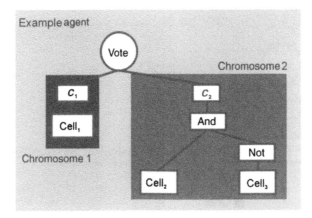

FIGURE 6.1   Tree diagram of an example agent.

or "uncertain" in the case of no input or a tie. For example, consider the following example agent specified by the two chromosomes, chromosome 1 and chromosome 2:

Chromosome 1	C_1	$Cell_1$			
Chromosome 2	C_2	$Cell_2$	And	Not	$Cell_3$

This example agent may be drawn as a tree as shown in Figure 6.1

Hence, a hunter is a net, that is, a hunter is an organized collection of one or more cells, which when presented with an image will return one of "primary," "nonprimary," or "uncertain." The complexity of a hunter is the number of cells it contains, regardless of organization.

Examples of Hunters of Complexity One. The following are some examples of hunters with complexity one and interpretations in a two-dimensional feature space. Assume the primary class is "one," and the nonprimary class is "not-one." Our first hunter, A_1, consist of a single cell in a single chromosome—it is illustrated in Figure 6.2 It is instructive to consider the feature space of all images on the basis of feature F_1 and F_2—every image maps to an (x,y) coordinate in this space, and hence may be drawn in a unit square. Agent A_1 may be viewed as a statement that partitions feature space into three disjoint sets—this is also illustrated in Figure 6.2 A second hunter, A_2, is illustrated in Figure 6.3 This agent's partition of the same feature space is also illustrated.

Merger. Thus far, we have given examples only of hunters with complexity one—this is the state of the CellNet system when initialized. What follows is a system of cooperative coevolution that generates agents of increasing complexity. Cooperative coevolution is achieved through the inclusion of a new genetic operator, augmenting the typical choices of crossover and mutation. This new operator, merger, serves to accept two hunters and produce a single new hunter of greater complexity. The com-

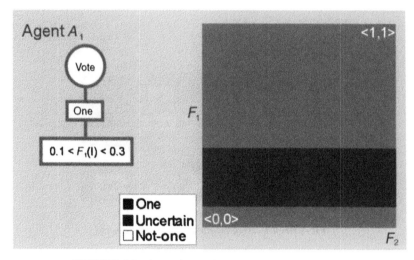

FIGURE 6.2 Agent A_1 and its partition of feature space.

plexity of the merged hunter will be the sum of the complexities of the parents. Merger operates at the level of chromosomes—when merging two hunters, chromosomes are paired randomly and merged either horizontally or vertically. Vertical merger simply places both chromosomes in parallel under the vote mechanism—they are now in direct competition to determine the outcome of the vote. Horizontal merger, on the contrary, combines the two chromosomes to produce a single and more complex chromosome, where the two original chromosomes are connected via a and or and-not connective. Hence, horizontal merger serves to refine a particular statement in the vote mechanism.

Table 6.1 shows illustrative examples for both vertical and horizontal merger. The decision of whether a merger will be horizontal or vertical is decided by comparing the affinity bits of the chromosomes in question. Consider the attempt to merge two

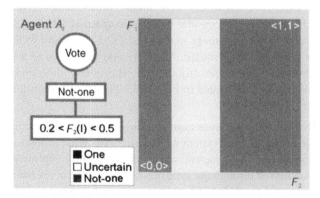

FIGURE 6.3 Agent A_2 and its partition of feature space.

TABLE 6.1 **Affinity-bits based control of the merger mechanism.**

A	B	Result (affinity bits–chromosome–conjunction–chromosome)
00	00	00 A and[not] B
00	01	01 A and[not] B
00	10	10 B and[not] A
00	11	Chromosomes laid out vertically
01	00	01 B and[not] A
01	01	Chromosomes laid out vertically
01	10	11 B and[not] A
01	11	Chromosomes laid out vertically
10	00	10 A and[not] B
10	01	11 A and[not] B
10	10	Chromosomes laid out vertically
10	11	Chromosomes laid out vertically
11	00	Chromosomes laid out vertically
11	01	Chromosomes laid out vertically
11	10	Chromosomes laid out vertically
11	11	Chromosomes laid out vertically

agents, A and B. Both A and B have chromosomes, each with a two-bit affinity. The affinity of a chromosome is the means through which it controls how it merges with another. Initially, all chromosomes in A and B are enumerated, and pairs are chosen, sequentially matching the chromosomes as they occur in the agent. Once chromosomes have been paired, they are merged according to the following mechanism: the conjunction is set as and or and-not on the basis of whether or not the vote classes agree.

The above (admittedly rather intricate) process was chosen for the amount of control that may be exploited genetically. For example, a chromosome may be declared complete by setting its affinity bits to "11." Or, it may "choose" to always be the first section of a chromosome (and hence, the section that defines which class is being voted for) by setting its affinity bits to "10."

Examples of Horizontal and Vertical Merger. The cooperative coevolution of hunters, as realized through Merger is a technical process, more easily explained visually. We reconsider agents A_1 and A_2 of merger, considering their children through the merger operator. Consider the horizontal merger of hunters A_1 and A_2—here, we produce agent A_3 by combining the chromosomes of A_1 and A_2 into one new one, linked via an and connective. As it is visible in Figure 6.4, horizontal merger may be viewed as the refinement of a partition created by two chromosomes.

In contrast, consider the vertical merger of these same two hunters, producing agent A_4—in this case, the chromosomes are combined directly under the vote mechanism. As shown in Figure 6.5, vertical merger may loosely be viewed as the union of the partitions generated by two chromosomes.

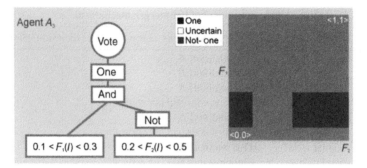

FIGURE 6.4 Agent A_3 and its partition of feature space.

6.1.3.2 Prey Initially, the prey in the CellNet system were simply images from a database. Hence, in initial experiments, prey had no genetic representation, as they were static structures. We shall refer to these in the future as *unaltered images*. In later experiments, we added a competitive component to our evolution of pattern recognizers—prey became agents, images that tried to disguise themselves using a set of camouflage functions. Hence, prey are altered images, rewarded genetically for "fooling" hunters. A prey consists of a simple genome—an image index and a series of bits.

Image index I	Bit b_1	Bit b_2	$\cdots$	Bit b_k

The image index points to a particular image in the database. The bits are boolean parameters, indicating whether a particular camouflage function is to be applied or not. Prey exist to be passed to hunters for classification—prior to this, however, all camouflage functions specified by the series of bits in a prey genome are applied—hence, a hunter views a transformed version of the original image specified by the image index. Camouflage functions used by the CellNet Co-Ev system consist of {salt–pepper, scaling, translation, and rotation}. These particular functions were chosen

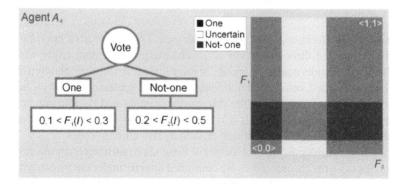

FIGURE 6.5 Agent A_4 and its partition of feature space.

as they were topologically invariant (save salt–pepper, unlikely to have a significant topological effect). Parameters for the functions were chosen such that simultaneous application of all to an image would still yield a human-readable image.

6.1.4 Genetic Algorithm

The CellNet system's primary driving force is a GA. It consists of a population of hunters and another of prey, initialized randomly, as the 0th generation. Following this begins an iterative procedure of refinement: Hunter agents are sorted according to an assigned fitness, according to which the next generation is prepared. The next generation is spawned from the "fittest" hunters of the first generation, utilizing some of the genetic operators: elitism, merger, crossover, and mutation. This process continues for a set number of generations. The reader interested in GAs is directed to Mitchell's text [38]. The GA functions to continuously update a population of hunters by selecting the best through a fitness function, then generating a new population via the genetic operators. The functioning of our genetic algorithm may be summarized as follows:

1. Initialize the initial population of randomly generated hunters;
2. evaluate the members of the population using the fitness function;
3. select the best percentage to go directly to the next generation (according to the rate of elitism);
4. select a set to be merged (according to the rate of merger);
5. select a set to be subjected to crossover (according to the rate of crossover);
6. apply mutation to all nonelite members according to the rate of mutation;
7. if the generation number is less than the maximum, go back to step 2.

In the case of competitive coevolution, not only did the population of hunters evolve but also the camouflaged images. Below, the steps are described in more detail. At the beginning of each generation, each hunter agent is evaluated for *fitness*. This is not necessarily the agent's accuracy (defined in Section 6.1.5.1), although for testing purposes raw accuracy is also computed. Rather, fitness is a function based loosely on accuracy and is meant to provide a measure that will drive the population's evolution. Explicit fitness functions depend on the trial in question and are shown in Section 6.1.5.

The first stage of the preparatory process is *elitism*: Here a percentage of the top-ranking agents are simply copied into the next generation without modification. This is done to provide additional stability in the genetic algorithm by ensuring that the maximal discovered fitness (although not necessarily accuracy) does not diminish. The population is then subjected to the *merger* operator: The mechanics of this operator has already been described in Section 6.1.3. Merger is an operator designed to select two fit agents and combine them to create a single agent with complexity equal to the sum of the parents'. The merged agents are then deposited in the new population as single units. Agents are selected for merger with a set probability, chosen using roulette-wheel selection (a.k.a. fitness-proportional selection). Next, the population is subjected to *crossover*. Crossover is always single point, although there are two variations on this—the first is a single point crossover in which indices are chosen randomly between the two agents (variable length) and the second in which one index is chosen for both agents (fixed length; it is assumed that parent agents will be of the same length in this case). In both cases, crossover is bitwise, where function limits are stored as integers. The final stage in the preparatory process, a *mutation* operator is applied. The mutation operator skims over all bits in all agents, making random changes with a set probability. The mutation operator applies to all hunter agents with the exception of those selected for elitism. The process is nearly identical for prey, whose genome consists of a simple series of bits. Crossover for a prey is single point and fixed length, occurring within the series of bits. Mutation flips a single bit in the series. This scheme was chosen as it appears to be the simplest example of a GA on a bit string and closest to a "standard" method.

6.1.5 Experiments

A series of experiments have been conducted using the CellNet system, attempting to automatically classify handwritten characters from two different databases. These experiments measured the ability of the CellNet system to find highly accurate classifiers under differing circumstances: different character sets, use of our novel merger operator, and use of competitive coevolution.

6.1.5.1 *Initial Experiments: Merger Versus the "Standard" GA* In the first series of experiments, a standard problem was chosen to evaluate the performance of CellNet in finding an effective pattern recognizer. The chosen problem was the ability of the system to distinguish between handwritten characters—specifically, to distinguish between the zero character and anything else. All examples were drawn from the CEDAR database. Data was drawn from two series of runs. The first set of runs consisted of the CellNet system evaluated without merger (hereby denoted the standard genetic algorithm or SGA trial). The second set of runs consisted of a similar run of the CellNet system, this time using the new merger operation (hereby denoted the merger-enabled or ME trial). In both the SGA and ME trials, a population of 500 agents was initialized randomly. Five hundred images were chosen as prey. In the ME and SGA trials, the prey were simple unaltered images. Fitness was assigned to each agent according to the function:

$$Fitness\,(A) = \frac{1}{|T|} \sum_{i \in T} \begin{cases} 1.0; & A \text{ correctly identifies image } i \\ 0.2; & A \text{ replies "uncertain" for image } i \\ 0.0; & A \text{ misidentifies image } i. \end{cases} \qquad (6.1)$$

This differs from raw accuracy, defined as

$$Accuracy_{train}(A) = \frac{1}{|T|} \sum_{i \in T} \begin{cases} 1.0; & A \text{ correctly identifies image } i \\ 0.2; & A \text{ replies "uncertain" for image } i \\ 0.0; & A \text{ misidentifies image } i \end{cases} \quad (6.2)$$

for the training sets (where T is the training set in question) and

$$Accuracy_{valid}(A) = \frac{1}{|V|} \sum_{i \in V} \begin{cases} 1.0; & A \text{ correctly identifies image } i \\ 0.0; & A \text{ does not correctly classify image } i \end{cases} \quad (6.3)$$

for the purposes of validation (where V is the set of images reserved for validation). Once fitness was assigned, the agents were sorted in descending order and prepared for the next generation. Two processes were used, one for the ME trials and one for the SGA trials. In the ME trials, we used a rate of elitism of 0.1, a rate of merger of 0.01, a rate of fixed-size crossover of 0.5, and a rate of mutation of 0.01. Agents in the ME trials were initialized with a complexity of one, and limited to a maximum complexity of 40 cells, meaning that all growth of complexity came from the merger operator through the generations. In the SGA trials, we used an elitism of 0.1, a fixed-size crossover of 0.5 and a mutation of 0.01. In the SGA trials, all agents were initialized with a complexity of 40 cells. For every 10 generations, a validation step was performed—accuracy was computed, instead using an independent verification set of 300 images. This independent verification had no feedback to the system— instead it was used to measure the progress of the system in a manner resistant to the effects of overfitting. A run in either trial was executed for 1000 generations. Each of the SGA and ME trials were repeated 15 times with little variance in results. Results of typical executions are shown in Figure 6.6

The maximum accuracy found on the validation set for the typical ME trial was 89.2%. For the SGA trial, the maximum validation accuracy found was 72.9%.

6.1.5.2 Autocomplexity Experiments
In the second set of experiments, the Autocomplexity (AC) experiments, the system was configured to again find a binary classifier in a manner similar to the ME trial described above. In the AC trials, however, no maximum complexity was enforced: agents were initialized with a complexity of one cell and could grow without bound. There were two sets of AC experiments: one using Latin digits (from the CEDAR database) and another using Indian digits from the CENPARMI database. These sets of experiments are called the Latin AC and the Indian AC or LAC and IAC experiments, respectively. For each run of the AC trial, the CellNet system was executed for 1000 generations, outputting data regarding its (raw) accuracy on the independent verification set and regarding the complexity of the agents produced. The initial prey population was initialized at 500 images, set to replace 3% from a total pool of 1000 every 5 generations. Again, prey consisted of simple, unaltered images. The LAC trial was run with a rate of elitism of 0.1, a rate of merger of 0.02, a rate of (variable-length) crossover of 0.4, a rate of mutation of

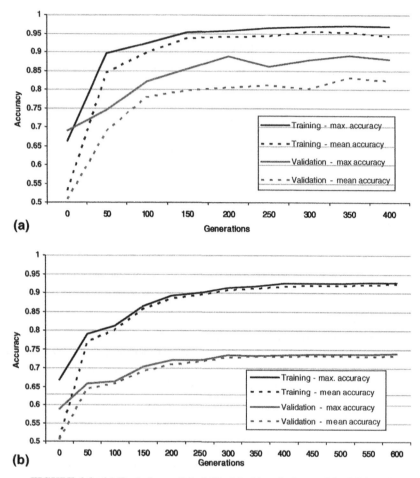

FIGURE 6.6   (a) Typical run of the ME trial; (b) typical run of the SGA trial.

0.01, and a complexity penalty (alpha) of 0.0005. The fitness function used was

$$Fitness\,(A) = \frac{1}{|T|}\left[\sum_{i\in T}\begin{cases}1.0; & A \text{ correctly identifies image } i\\ 0.2; & A \text{ replies ``uncertain'' on image } i\\ 0.0; & A \text{ misidentifies image } i\end{cases}\right] - \alpha\cdot|A|,$$

(6.4)

where T is t he (nonstatic) set of training images and $|A|$ is the complexity (number of cells) of A.

Results of the best (most accurate classifier found) LAC trial run is shown in Figure 6.7 The maximum validation accuracy found was 97.7%, found around generation 150. Following generation 250, maximum accuracy dropped slightly, settling around 94.8%, despite continued smooth increases in training accuracy. There

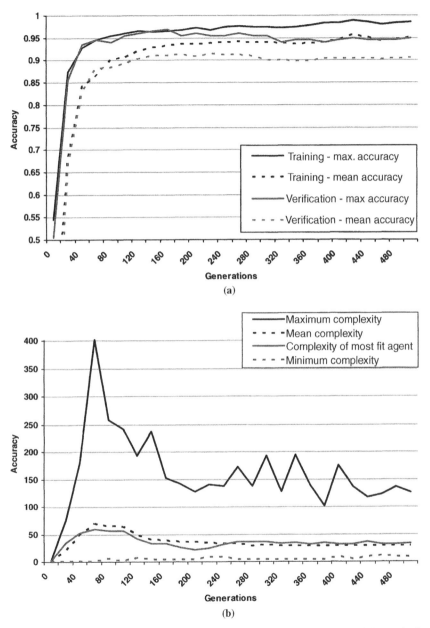

FIGURE 6.7   (a) Best run of the LAC trial accuracy; (b) best run of the LAC trial complexity.

was initially a large variance in the complexity measure, which are quickly stabilized. The majority of the most accurate agents displayed little variation from the mean. The mean maximum validation accuracy found between runs of the LAC trial was 0.954 (with a standard deviation of 0.0147). Complexity in all runs (for both the most accurate agent and the mean) settled at slightly higher than 30 cells, with very little variance between runs (approximately ±5 cells). Additional informal LAC trials were run involving Latin characters other than zero, with similar results. In the second set of experiments (the *IAC* trial), the system was configured identically to the LAC trial above, save that the prey was restructured to distinguish between "4" and not "4," using Indian handwritten characters from the CENPARMI database. The initial prey population was initialized at 400 images, set to replace 3% from a total pool of 600 every 5 generations. Figure 6.8 shows the results of the best (most accurate maximum classifier) run of 20.

The maximum validation accuracy found in the IAC trial was 97.4%, found around generation 380. Following generation 450, the accuracies followed a stable course, probably indicating convergence in the genetic algorithm. The IAC trial was run 20 times, each time with similar results. The mean maximum validation accuracy found between runs was 0.956 (with a standard deviation of 0.019), with typical mean complexities similar to those found in the best run above. A hunter with validation accuracy within 0.005 of the maximum validation accuracy found was generated in each trial prior to generation 500—convergence appears to have occurred in all cases prior to generation 1000.

6.1.5.3 Competitive Coevolutionary Experiments

In the final set of experiments, the competitive coevolutionary experiments, we allowed both hunters and prey to evolve. Both populations were initially spawned randomly. For each generation, each agent was evaluated against the entirety of the opposing population. Explicitly, let h be a member of the hunter population H, p a member of the prey population P. For each generation, each hunter h attempted to classify each prey p: let

$$ClassAttempt(h, p) = \begin{cases} 1; & h \text{ correctly classifies } p \\ 0.5; & h \text{ responds uncertain} \\ 0; & h \text{ incorrectly classifies } p. \end{cases} \tag{6.5}$$

Then the $accuracy_{train}$ of a hunter h was

$$Accuracy_{train}(h) = \frac{1}{P} \sum_{p \in P} classAttempt(h, p). \tag{6.6}$$

Fitness of a hunter was defined as

$$Fitness(h) = accuracy^2_{train}(h) - \alpha \cdot |h|, \tag{6.7}$$

where again, α is a system parameter designed to limit hunter complexity and $|h|$ is the number of cells in hunter h. In contrast, the fitness of a Prey p was defined as

$$Fitness(p) = \frac{1}{H} \sum_{h \in H} (1 - classAttempt(h, p)), \tag{6.8}$$

which is inversely proportional to the fitness for hunters.

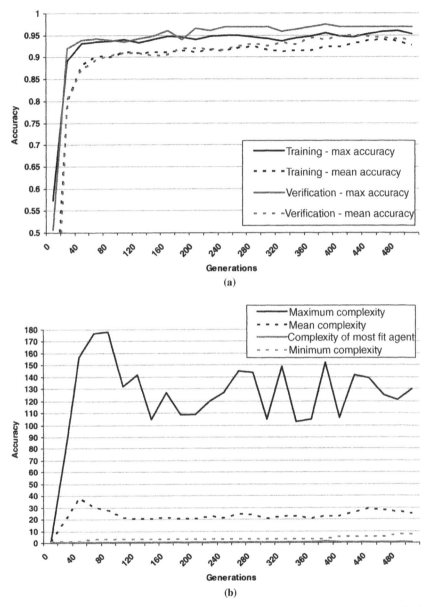

FIGURE 6.8   (a) Best run of the IAC trial accuracy; (b) best run of the IAC trial complexity.

In these experiments, the system's ability to recognize all Latin characters was tested. The system was configured to generate five binary hunters for each digit—these are labeled **h.x.y**, where x is the digit number and y an index from 0 to 4. Each hunter was trained using a base set of 250 training images and tested via an independent set of 150 validation images. Each run was executed for a maximum of

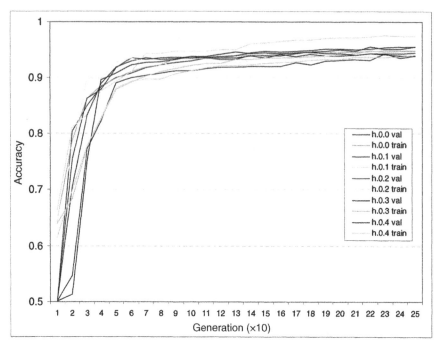

FIGURE 6.9 Maximum training (*light line*) and validation (*dark lines*) accuracies for the h.0 hunters.

250 generations, outputting data regarding validation accuracy each 10 generations. A typical run may be seen in the evolution of the h.0 hunters as illustrated in Figure 6.9 Training and validation accuracies are very close, although validation accuracy tends to achieve slightly higher levels—this behavior is typical of all digits. This is in contrast to previous experiments, where overfitting of approximately 2–3% was reported consistently. It is also noted that in initial generations of the runs, overfitting is common, as it can clearly be seen that the training plots are more accurate than the validation plots. This initial bonus, however, disappears by generation 60, where the validation plots overtake. However, also in contrast to previous experiments, complexity is vastly increased—in the case of the zero digit, mean complexity jumps from approximately 35 cells to approximately 65 cells, while the complexity of the most accurate agent jumps from 40 cells to seemingly random oscillations in the range of 50 cells to 350 cells. Figure 6.10 shows the complexities of the most accurate agents and the mean for the h.0 runs.

Table 6.2 shows the maximum training and validation accuracies for each binary hunter. The final columns compute the means for the validation and training accuracies for each class of hunter and compare the difference. It is shown that the mean difference between training and validation data is −0.006, implying slight *underfitting* of the classifiers to the training data.

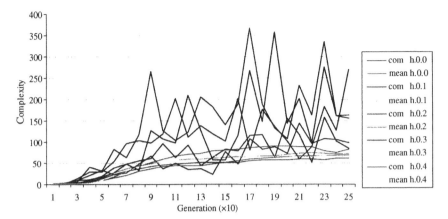

FIGURE 6.10   Complexity of most fit agents (*dark lines*) and mean complexity (*light lines*) for the h.0 runs.

Finally, a series of experiments was undertaken regarding the classifications of the evolved binary classifiers. The scope of these experiments was the determination of the relative independence of the errors made by the classifiers when classifying images. Hence, our goal was a measure of the variance found between the errors of the hunters for each particular digit. Each hunter evaluated a set of 300 previously unseen images—a note was made for each error. Each classifier for any particular digit then had an associated error list of images. These lists were contrasted, computing the total

TABLE 6.2   Maximum training and validation accuracies for the binary classifiers.

	h.dig.0		h.dig.1		h.dig.2		h.dig.3		h.dig.4		Mean		
	Train	Valid	Train	Valid	Train	Valid	Train	Valid	Train	Valid	Train	Valid	Diff
0	0.951	0.955	0.977	0.955	0.946	0.945	0.951	0.944	0.941	0.946	0.953	0.949	+0.004
1	0.992	0.981	0.984	0.971	0.992	0.982	0.987	0.977	0.990	0.984	0.981	0.979	+0.002
2	0.906	0.906	0.935	0.932	0.895	0.904	0.881	0.896	0.906	0.920	0.905	0.912	−0.007
3	0.922	0.910	0.894	0.919	0.894	0.910	0.895	0.908	0.906	0.926	0.902	0.915	−0.013
4	0.944	0.941	0.972	0.967	0.957	0.962	0.956	0.962	0.957	0.952	0.957	0.957	+0.000
5	0.890	0.919	0.935	0.937	0.899	0.919	0.919	0.922	0.894	0.914	0.907	0.922	−0.015
6	0.914	0.941	0.925	0.940	0.923	0.953	0.945	0.917	0.931	0.923	0.928	0.935	−0.007
7	0.937	0.934	0.937	0.954	0.954	0.954	0.946	0.940	0.961	0.939	0.947	0.944	+0.003
8	0.900	0.914	0.933	0.918	0.932	0.939	0.875	0.905	0.914	0.931	0.911	0.921	−0.010
9	0.882	0.911	0.938	0.944	0.915	0.917	0.918	0.926	0.924	0.939	0.915	0.927	−0.012
										Mean			−0.006

TABLE 6.3 Percentage agreement in errors made by classifiers by digit.

Digit	0	1	2	3	4	
#Errors	38	22	62	58	34	
Agreement	0.26	0.05	0.40	0.52	0.47	
Digit	5	6	7	8	9	Mean
#Errors	80	58	55	73	29	60.9
Agreement	0.19	0.19	0.35	0.25	0.25	0.29

number of errors (for all five hunters) and the percentage of the list shared by two or more hunters. These results are shown in Table 6.3 It is evident that there is much variance between the errors made by the various hunters.

6.1.6 Analysis

In the ME versus SGA trials, a clear bias is shown toward the ME system. The SGA system shows a slow gradual progression toward its optimal training accuracy and a poor progression toward a good validation accuracy. The latter is not surprising, as the SGA does not contain any measures to prevent overfitting to the training set, and the chosen features bear a high potential for precisely that. More interesting is the ME system's resilience to overfitting. We believe that this is a result of the structural construction of the agents, as early simple agents do not have the complexity necessary for "data memorization." The LAC and IAC trials showed a good progression toward a powerful pattern recognizer, without sacrificing much efficiency relative to the ME trial. Unlimited potential agent complexity proved a computational burden in the initial generations, but quickly decreased to a stable and relatively efficient maximum. The effects of overfitting can be seen here as well, but to a much lesser extent than in the SGA trial. A difference between training and verification accuracy of approximately 2% seems to be the norm, with overfitting nonexistent or negative in some runs. Variance existed between the runs within the LAC and IAC trials. There is nearly a 2% difference in validation accuracy between the typical and the best run. Indeed, this shows that several runs of the CellNet system are needed to guarantee a high-accuracy classifier.

Additionally, it shows that the CellNet system is capable of discovering several independent classifiers, based solely on the seed chosen for random parameter generation. This opens up the CellNet system for placement within a larger classifier-combining framework (e.g., bagging) utilizing agents found between independent runs of the system. The slowdown in convergence in the LAC and IAC trials relative to the ME trials is to be expected: the complexity of the ME trial was specifically chosen to be appropriate to the task at hand, while the agents in the AC trials had no such luxury. The AC trial was more computationally expensive in terms of time required for a single generation initially, but soon stabilized to a more reasonable range of agent complexities. The presence of large (100+ cells) and small (3 cells) agents throughout the trial is most likely the result of a preparatory process between generations, which was rather forgiving. With the augmentation of competitive coevolution, these issues are significantly improved. The creation of binary classifiers is accomplished for each Latin digit, showing little variance between trial runs. This contrasted against the use of several runs in the AC trials to find a good classifier. The inclusion of a genome for patterns and camouflage functions for diversification has resulted in an artificially difficult problem for classifiers, increasing overall performance, and lowering the system's propensity for "data memorization"—indeed, the typical coevolutionary run included *underfitting* rather than overfitting.

Finally, it has been demonstrated that although the reliability of the system's ability to generate classifiers has been improved, the error sets produced by the classifiers are largely independent. This matter is crucial for the creation of multiclassifiers (combinations of the binary classifiers to form a single multiple-class recognizer), a step that a practitioner may wish to take. The independence of the error rates of the classifiers implies that several hunters for each class may be used in a bagging or bootstrapping technique, methods that are expected to improve the accuracy of the overall multiclassifier. These results represent a significant step forward for the goal of an autonomous pattern recognizer. Competitive coevolution and camouflage is expected to aid in the problem of overfitting and reliability without expert tuning, and also in the generation of a larger and more diverse data set.

6.1.7   Future Directions

The idea of a truly autonomous pattern recognizer is enticing. A black box that could learn to recognize an arbitrary visual pattern in the best possible manner is an ambitious goal, and one not yet achieved by the CellNet system. However, work is currently under way involving several axis by which the CellNet system expositioned above may be improved. One very powerful aspect of evolutionary computing, little exploited thus far, is the ability of a "blind designer" to discover nonintuitive solutions that a human designer might overlook. Our current work centers on this goal—we are currently in the process of investigating a means to represent the "seed" patterns in an image independently of preprogram features. These seed patterns will form the basis for a means for transforming images, allowing simple patterns of trends to be exaggerated. Following this, we aim to classify images on the information provided by a set of descriptors describing the exaggerated images. It is our hope that this generalization of the CellNet system will allow it to self-adapt to a far wider set of databases of patterns without redesign of the initial preprogramed base.

Finally, a missing component in the creation of a truly autonomous system is the capacity of that system to select global parameters at run-time, as well as handling global strategies for the prevention of overfitting. Rather than specify parameters, such as elitism or crossover rate explicitly, the CellNet research group is involved in a redesign of the CellNet environment that will facilitate their inclusion intrinsically. This includes a divergence from the typical paradigm of genetic programing in favor of an immersive environment inspired by other experiments in simulated ecology. It is hoped that the natural competitive coevolution between camouflaged images and recognizing agents will provide a self-optimizing world in which global parameters are no longer necessary.

6.2 OFFLINE HANDWRITTEN CHINESE CHARACTER RECOGNITION

Chinese characters are used in daily communications by about one quarter of population of the world. Besides the characters in Chinese language, the Kanji characters in Japanese have similar shapes with original Chinese characters. Chinese characters are also divided into traditional characters and simplified characters, which are used in different areas of Chinese culture. Some traditional characters, which are not complex originally, have identical shapes with simplified ones since they were not simplified in the history. For recognition, we usually consider a character set of one type, say, traditional Chinese, simplified Chinese, or Japanese Kanji. Alphanumeric characters are often considered together because they are commonly mixed with Chinese text. Sometimes, traditional Chinese and simplified Chinese characters are also mixed in a text.

Compared to the recognition of other alphabets in the world, the recognition of Chinese characters has some special characteristics. First, the number of Chinese characters is very large. A standard of simplified Chinese, GB2312-80, has 3755 characters in the level-1 set and 3,008 characters in the level-2 set, 6763 in total. A new standard GB18030-2000 has 27,533 characters, including both traditional and simplified characters, as well as many other symbols. For ordinary people, the number of daily used characters is about 5000, but a commercial recognition system needs to accommodate nearly 10,000 characters. The large number of character classes poses a challenge to efficient classification. Second, many Chinese characters have complicated structures. To our knowledge, the number of strokes in Chinese characters ranges from 1 to 36. Many characters have hierarchical structures: a complicated character is composed of common substructures (called radicals) organized in 2D space. The structural complexity of Chinese characters makes the shape description difficult. On the contrary, the complicated structure provides rich information for shape identification. Third, there are many similar characters in Chinese. Some characters differ only in a small stroke or dot, such as those in Figure 6.11. Due to the similarity of shape between different characters, it is hard to achieve high recognition accuracy, especially in handwritten character recognition.

Since the first work of printed Chinese character recognition (PCCR) was published in 1966 [3], many research efforts have been contributed to both printed and

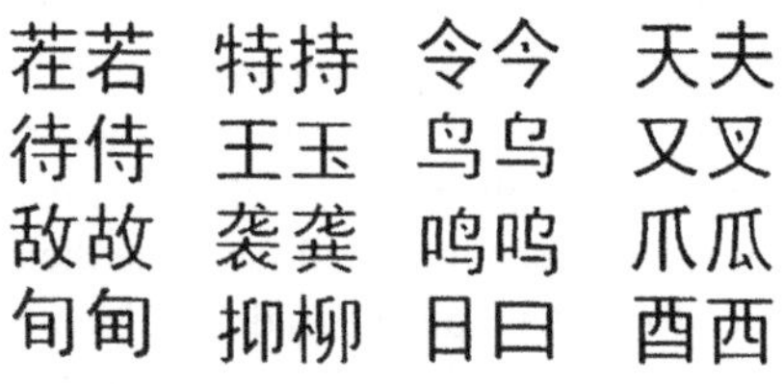

荏若　特持　令今　天夫
待侍　王玉　鸟乌　又叉
敌故　袭龚　鸣呜　爪瓜
旬甸　抑柳　日曰　酉西

FIGURE 6.11　Pairs of similar Chinese characters.

handwritten Chinese character recognition (HCCR). Many effective methods have been proposed, and the recognition performance has been constantly improved. From the 1990s, Japanese/Chinese OCR software packages have gained popularity by organizational and personal users for document entry, online handwritten character recognizers are widely applied to personal computers, PDAs (personal digital aides) and mobile phones for character entry, and offline handwritten character recognizers are applied to form processors, check readers, mail sorting machines, and so on.

Despite the many effective methods and successful results on Chinese character recognition reported in the literature, this section gives a case study of offline HCCR using representative methods of character normalization, feature extraction, and classification. Specifically, we will evaluate in experiments the performance of five normalization methods, three feature extraction methods, and three classification methods. The classification methods have been described in Chapter 4 of this book, three of the five normalization methods have been described in Chapter 2, and two of the three feature extraction methods have been described in Chapter 3. Those have not been described in previous chapters will be given in this section. We evaluate the recognition performance on two databases of handprinted characters, namely, ETL9B (Electro-Technical Laboratory, Japan) and CASIA (Institute of Automation, Chinese Academy of Sciences), with 3036 classes and 3755 classes, respectively.

In the rest of this section, we first give a brief review of previous works in HCCR, show the diagram of a Chinese character recognition system, then describe the methods used in our experiments, and finally, present the evaluation results.

6.2.1 Related Works

The first work of printed Chinese character recognition (PCCR) was reported in 1966 [3]. Research on online HCCR was started as early as PCCR [43], whereas offline HCCR was started in late 1970s, and has attracted high attention from the 1980s [39]. Since then, many effective methods have been proposed to solve this problem, and the recognition performance has advanced significantly [12, 48]. This study is mainly concerned with offline HCCR, but most methods of offline recognition are applicable to online recognition as well [32].

The approaches of HCCR can be roughly grouped into two categories: feature matching (statistical classification) and structure analysis. Based on feature vector representation of character patterns, feature matching approaches usually computed a simple distance measure (correlation matching), say, Euclidean or city block distance, between the test pattern and class prototypes. Currently, sophisticated classification techniques, including parametric and nonparametric statistical classifiers, artificial neural networks (ANNs), support vector machines (SVMs), and so on, can yield higher recognition accuracies. Nevertheless, the selection and extraction of features remain an important issue. Structure analysis is an inverse process of character generation: to extract the constituent strokes and radicals, and compute a structural distance measure between the test pattern and class models. Due to its resembling of human cognition and the potential of absorbing large deformation, this approach was pursued intensively in the 1980s and is still advancing [18]. However, due to the difficulty of stroke extraction and structural model building, it is not widely followed.

Statistical approaches have achieved great success in handprinted character recognition and are well commercialized due to some factors. First, feature extraction based on template matching and classification based on vector computation are easy to implement and computationally efficient. Second, effective shape normalization and feature extraction techniques, which improve the separability of patterns of different classes in feature space, have been proposed. Third, current machine learning methods enable classifier training with large set of samples for better discriminating shapes of different classes.

The methodology of Chinese character recognition has been largely affected by some important techniques: blurring [15], directional pattern matching [53, 52], nonlinear normalization [47, 51], modified quadratic discriminant function (MQDF) [21], and so on. These techniques and their variations or improved versions are still widely followed and adopted in most recognition systems. Blurring is actually a low-pass spatial filtering operation. It was proposed in the 1960s from the viewpoint of human vision and is effective to blur the stroke displacement of characters of the same class. Directional pattern matching, motivated from local receptive fields in vision, is the predecessor of current direction histogram features. Nonlinear normalization, which regulates stroke positions as well as image size, significantly outperforms the conventional linear normalization (resizing only). The MQDF is a nonlinear classifier suitable for high-dimensional features and large number of classes.

Recent advances in character normalization include curve-fitting-based nonlinear normalization [30], pseudo-two-dimensional (P2D) normalization [13, 34], and so on. The curve-fitting-based normalization performs as well as line-density-based nonlinear normalization at lower computational complexity. The P2D normalization, especially that based on line density projection interpolation [34], outperforms 1D normalization methods with the computational complexity increased only slightly. The advances in feature extraction include normalization-cooperated feature extraction (NCFE) [10] and continuous NCFE [31], gradient direction features on binary and on gray-scale images [29].

In the history, some extensions of direction feature, such as the peripheral direction contributivity (PDC) [9] and the reciprocal feature field [54], have reported higher accuracy in HCCR when a simple distance metric (correlation, Euclidean, or city block distance) was used for classification. These features, with very high dimensionality (over 1000), are actually highly redundant. As background features, they are sensitive to noise and connecting strokes. Extending the line element of direction feature to higher-order feature detectors (e.g., [42, 46]) helps discriminate similar characters, but the dimensionality also increases rapidly. The Gabor filter, also motivated from vision research, promises feature extraction in character recognition [50], but is computationally expensive compared to chaincode and gradient features, and at best, perform comparably with the gradient direction feature [35].

As to classification for large category set, discriminative prototype-based (learning vector quantization, LVQ) classifiers [28] and the MQDF have been demonstrated effective. The LVQ classifier yields much higher accuracy than the Euclidean distance (nearest mean) classifier, whereas the MQDF yields even higher accuracy at heavy burden of run-time complexity. Other popular discriminative classifiers, such as ANNs and SVMs, have not been widely applied to Chinese character recognition, because the straightforward training of ANNs and SVMs for a large category set is highly expensive. Using ANNs or SVMs to discriminate a subset of characters or training one-versus-all binary classifiers with subsets of samples have been tried by some researchers with success [4, 6, 23, 41].

The normalization methods evaluated in our experiments are linear normalization, line-density-based nonlinear normalization (NLN), moment-based normalization (MN), bi-moment normalization (BMN) [30], and modified centroid-boundary alignment (MCBA) [33]. We do not evaluate the P2D normalization methods here because the description of them (see [34]) is involving. For feature extraction, we focus on the direction features, which have been widely used with high success. We evaluate the chaincode direction feature, the NCFE, the continuous NCFE, the gradient direction features on binary and on gray-scale images. For classification, we compare the Euclidean distance classifier, the MQDF, and an LVQ classifier. In feature extraction, we will also investigate the effects of blurring and varying dimensionality. The MQDF and LVQ classifiers will be evaluated with various complexity of parameters.

6.2.2 System Overview

The diagram of an offline Chinese character recognition system is shown in Figure 6.12. The recognition process typically consists of three stages: preprocessing, feature extraction, and classification. Preprocessing is to reduce the noise in character image and, more importantly, to normalize the size and shape of character for improving the recognition accuracy. The feature extraction stage is usually followed by a dimensionality reduction procedure for lowering the computational complexity of classification and, possibly, improving the classification accuracy. For classification of a large category set, a single classifier cannot achieve both high accuracy and high speed. A complicated classifier, such as the MQDF, gives high accuracy with very high

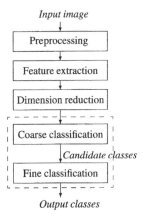

FIGURE 6.12 Diagram of offline Chinese character recognition system.

computational complexity. Using such a complicated classifier for fine classification, a low-complexity coarse classifier is used to select candidate classes for acceleration. The fine classifier only classifies the selected candidate classes, which are far less than the whole category set. The output of classification is a unique character class or a ranked list of multiple classes with their confidence scores.

6.2.3 Character Normalization

Normalization is to regulate the size, position, and shape of character images, so as to reduce the shape variation between the images of same class. Denote the input image and the normalized image by $f(x, y)$ and $g(x', y')$, respectively, normalization is implemented by coordinate mapping

$$\begin{cases} x' = x'(x, y) \\ y' = y'(x, y). \end{cases} \tag{6.9}$$

Most normalization methods use 1D coordinate mapping:

$$\begin{cases} x' = x'(x) \\ y' = y'(x). \end{cases} \tag{6.10}$$

Under 1D normalization, the pixels at the same row/column in the input image are mapped to the same row/column in the normalized image and hence, the shape restoration capability is limited. Nevertheless, 1D normalization methods are easy to implement, and if the 1D coordinate functions are designed appropriately, they lead to fairly high recognition accuracy.

Given coordinate mapping functions (6.9) or (6.10), the normalized image $g(x', y')$ is generated by pixel value and coordinate interpolation. In our implementation of 1D normalization, we map the coordinates forwardly from (binary) input image to normalized image, and use coordinate interpolation to generate the binary normalized image. For generating gray-scale normalized image, each pixel is viewed as a square of unit area. By coordinate mapping, the unit square of input image is mapped to a rectangle in the normalized plane, and each pixel (unit square) overlapping with the mapped rectangle is assigned a gray level proportional to the overlapping area [31].

In our experiments, the normalized image plane is set to a square of edge length L, which is not necessarily fully occupied. To alleviate the distortion of elongated characters, we partially preserve the aspect ratio of the input image by aspect ratio adaptive normalization (ARAN) using the aspect ratio mapping function

$$R_2 = \sqrt{\sin\left(\frac{\pi}{2} R_1\right)}, \tag{6.11}$$

where R_1 and R_2 are the aspect ratio of the input image and that of the normalized image, respectively (see Section 2.4.4). The normalized image with unequal width and height is centered on the square normalized plane.

The details of linear normalization, moment normalization, and nonlinear normalization have been described in Chapter 2 (Section 2.4.4). Linear normalization and nonlinear normalization methods align the bounding box of input character image to a specified rectangle of normalized plane. Moment normalization aligns the gravity center (centroid) of input image to the center of normalized plane and is a linear transformation of pixel coordinates. Nonlinear normalization is based on the histogram equalization of line density projections, and the coordinate transformation functions are not smooth. In the following, we describe two smooth nonlinear transformation methods based on functional curve fitting: bi-moment normalization [30] and modified centroid-boundary alignment (MCBA) [33].

Bi-moment normalization aligns the centroid of input character image to the center of normalized plane as moment normalization does. Unlike that moment normalization resets the horizontal and vertical boundaries of input image to be equally distant from the centroid, bi-moment normalization sets the boundaries unequally distant from the centroid to better account for the skewness of centroid. Specifically, the second-order central moment μ_{20} is split into two parts at the centroid:

$$\begin{cases} \mu_x^- = \sum_{x<x_c} \sum_y (x-x_c)^2 f(x,y) = \sum_{x<x_c} (x-x_c)^2 p_x(x), \\ \mu_x^+ = \sum_{x>x_c} \sum_y (x-x_c)^2 f(x,y) = \sum_{x>x_c} (x-x_c)^2 p_x(x), \end{cases} \tag{6.12}$$

where $p_x(x)$ is the horizontal projection profile of input image $f(x,y)$. The moment μ_{02}, calculated on the vertical projection profile, is similarly split into two parts μ_y^- and μ_y^+. The boundaries of input image are then reset to $[x_c - b\sqrt{\mu_x^-}, x_c + b\sqrt{\mu_x^+}]$ and $[y_c - b\sqrt{\mu_y^-}, y_c + b\sqrt{\mu_y^+}]$ (b is empirically set equal to 2). For the x axis, a quadratic function $u(x) = ax^2 + bx + c$ is used to align three points $(x_c - b\sqrt{\mu_x^-}, x_c, x_c + b\sqrt{\mu_x^+})$ to normalized coordinates $(0, 0.5, 1)$, and similarly, a quadratic function $v(y)$ is used for the y axis. Finally, the coordinate functions are

$$\begin{cases} x' = W_2 u(x), \\ y' = H_2 v(y), \end{cases} \tag{6.13}$$

where W_2 and H_2 are the width and the height of normalized image, respectively.

The quadratic functions can also be used to align the bounding box and centroid, that is, map $(0, x_c, W_1)$ and $(0, y_c, H_1)$ to $(0, 0.5, 1)$ (W_1 and H_1 are the width and the height of the bounding box of input image, respectively). We call this method centroid-boundary alignment (CBA). A modified CBA (MCBA) method [33] further adjusts the stroke density in central area by combining a sine function $x' = x + \eta_x \sin(2\pi x)$ with the quadratic functions

$$\begin{cases} x' = W_2[u(x) + \eta_x \sin(2\pi u(x))], \\ y' = H_2[v(y) + \eta_y \sin(2\pi v(y))]. \end{cases} \tag{6.14}$$

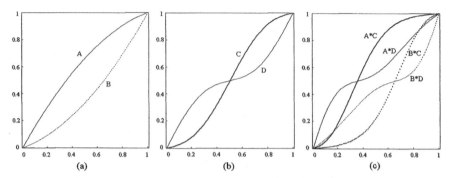

FIGURE 6.13 Curves of coordinate mapping: (a) Quadratic curve fitting for centroid alignment; (b) sine functions for adjusting inner density; (c) combination of quadratic and sine functions.

Figure 6.13(c) shows four curves of combined mapping functions. The curve "A*C" combines the effects of compressing the right side and stretching the inner, the curve "A*D" combines the effects of expanding the left side and compressing the inner, the curve "B*C" combines the effects of compressing the left side and stretching the inner, and the curve "B*D" combines the effect of expanding the right side and compressing the inner.

The amplitudes of sine waves, η_x and η_y, are estimated from the extent of the central area, which is defined by the centroids of half images divided by the global centroid (x_c, y_c). Denote the x coordinate of the centroid of the left half by x_1 and that of the right half by x_2. On centroid alignment with quadratic function, they are mapped to values $z_1 = ax_1^2 + bx_1$ and $z_2 = ax_2^2 + bx_2$. The extent of the central area is estimated by

$$s_x = z_2 - z_1 = ax_2^2 + bx_2 - ax_1^2 - bx_1. \tag{6.15}$$

The bounds of the central area is then reset to be equally distant from the aligned centroid: $0.5 - s_x/2$ and $0.5 + s_x/2$. The sine function aims to map these two bounds to coordinates 0.25 and 0.75 in the normalized plane. Inserting them into sine function $x' = x + \eta_x \sin(2\pi x)$, we obtain

$$\eta_x = \frac{s_x/2 - 0.25}{\sin(\pi s_x)}. \tag{6.16}$$

The amplitude η_y can be similarly estimated from the partial centroid coordinates y_1 and y_2 of half images divided at y_c.

The mapped coordinates x' and y' must be increasing with x or y such that the relative position of pixels is not reversed. This monotonicity is satisfied by

$$\frac{dx'}{dx} \geq 0,$$

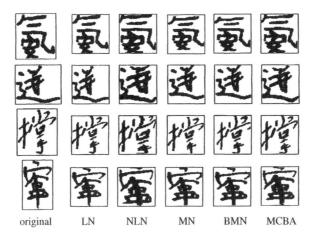

original   LN   NLN   MN   BMN   MCBA

FIGURE 6.14   Character image normalization by five methods. The leftmost image is the original and the other five are normalized ones.

which leads to

$$-\frac{1}{2\pi} \le \eta \le \frac{1}{2\pi}. \tag{6.17}$$

When the value of η computed by (6.16) is beyond this range, it is enforced to be a marginal value $-1/2\pi$ or $1/2\pi$, which corresponds to $s_x = 0.265$ or $s_x = 0.735$.

Figure 6.14 shows some examples of Chinese character normalization using five methods: linear (LN), line-density-based nonlinear (NLN), moment (MN), bi-moment (BMN), and MCBA.

6.2.4   Direction Feature Extraction

The implementation of direction feature is varying depending on the directional element decomposition, the sampling of feature values, the resolution of direction, and feature plane, and so on. Considering that the stroke segments of Chinese characters can be approximated into four orientations: horizontal, vertical, left-diagonal, and right-diagonal, early works usually decomposed the stroke (or contour) segments into these four orientations.

Feature extraction from stroke contour has been widely adopted because the contour length is nearly independent of the stroke-width variation. The local direction of contour, encoded as a chaincode, actually has eight directions (Fig. 3.3). Decomposing the contour pixels into eight *directions* instead of four *orientations* (a pair of opposite directions merged into one orientation) was shown to significantly improve the recognition accuracy [29]. This is because separating the two sides of stroke edge can better discriminate parallel strokes. The direction of stroke edge can also be measured by the gradient of image intensity, which applies to gray-scale images as well as binary images. The gradient feature has been applied to Chinese character recognition in 8-direction [36] and 12-direction [37].

Direction feature extraction is accomplished in three steps: image normalization, directional decomposition, and feature sampling. Conventionally, the contour/edge pixels of normalized image are assigned to a number of direction planes. The normalization-cooperated feature extraction (NCFE) strategy [10], instead, assigns the chaincodes of original image into direction planes. Though the normalized image is not generated by NCFE, the coordinates of edge pixels in original image are mapped to a standard plane, and the extracted feature is thus dependent on the normalization method.

Direction feature is also called direction histogram feature because at a pixel or a local region in normalized image, the strength values of N_d directions form a local histogram. Alternatively, we view the strength values of one direction as a directional image (direction plane).

In Chapter 3 (Section 3.1.3), we have described the blurring and sampling of orientation/direction planes, and the decomposition of chaincode and gradient directions. Both chaincode and gradient features are extracted from normalized character images. Whereas the chaincode feature applies to binary images only, the gradient feature applies to both binary and gray-scale images. In the following, we will describe the NCFE method that extracts chaincode direction feature from original character image incorporating coordinate mapping. The NCFE method has two versions: discrete and continuous, which generate discrete and continuous direction planes, respectively. We will then extend the chaincode feature, NCFE, and gradient feature from 8-direction to 12-direction and 16-direction.

To overcome the effect of contour shape distortion caused by character image normalization (stairs on contour are often generated), for chaincode feature extraction, the normalized binary image is smoothed using a connectivity-preserving smoothing algorithm [27]. The NCFE method characterizes the local contour direction of input image, which need not be smoothed. Though the gradient feature is also extracted from the normalized image, it does not need smoothing since the gradient is computed from a neighborhood and is nearly insensitive to contour stairs.

6.2.4.1 *Directional Decomposition for NCFE*

Directional decomposition results in a number of direction planes (with the same size as the normalize image plane), $f_i(x, y), i = 1, \ldots, N_d$. In binary images, a contour pixel is a black point with at least one of its 4-connected neighbors being white. If ignoring the order of tracing, the 8-direction chaincodes of contour pixels can be decided by raster scan (see Section 3.1.3).

In the NCFE method proposed by Hamanaka et al. [10], each contour pixel in the input character is assigned to its corresponding orientation (or direction) plane at the position decided by the coordinate mapping functions given by a normalization method. In an elaborated implementation [31], each chaincode in the original image is viewed as a line segment connecting two neighboring pixels, which is mapped to another line segment in a standard direction plane by coordinate mapping. In the direction plane, each pixel (unit square) crossed by the line segment in the main (x

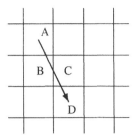

FIGURE 6.15   NCFE on continuous direction plane.

or y) direction is given a unit of direction contribution. To exploit the continuous nature of line segment, the strength of line direction falling in a pixel is proportional to the length of line segment falling in the unit square. This version of improvement is called continuous NCFE [31]. As in Fig. 6.15, where a line segment mapped from a chaincode overlaps with four unit squares A, B, C, and D. By discrete NCCF, the pixels A and C are assigned a direction unit, whereas by continuous NCCF, all the four pixels are assigned direction strengths proportional to the in-square line segment length.

Figure 6.16 shows the direction planes of three decomposition schemes: chaincode on normalized image, discrete NCFE, and gradient on binary normalized image. The direction planes of continuous NCFE and the gradient on gray-scale normalized image are not shown here because they are very similar to those of discrete NCFE and the gradient on binary image, respectively. We can see that the planes of chaincode directions (second row) are very similar to those of gradient directions (bottom row). The planes of NCFE, describing the local directions of the original image, show some

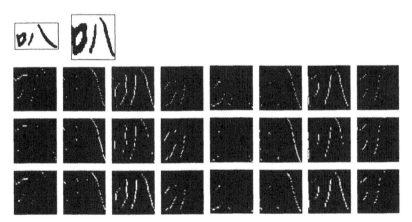

FIGURE 6.16   Original image and normalized image (top row), 8-direction chaincode planes of normalized image (second row), discrete NCFE (third row), and gradient on binary normalized image (bottom row).

difference. Comparing the original image and the normalized image, the orientation of the right-hand stroke, near left-diagonal orientation, deforms to near vertical. Consequently, the direction planes of left-diagonal orientation (2nd and 6th columns) of NCFE are stronger than those of chaincodes and gradient, whereas the planes of vertical orientation (3rd and 7th columns) of NCFE are weaker than those of chaincodes and gradient.

6.2.4.2 *Extension to More Directions* The extension of gradient decomposition into more than eight directions is straightforward: simply setting N_d standard directions with angle interval $360/N_d$ and typically, with one direction pointing to the east, then decompose each gradient vector into two components in standard directions and assign the component lengths to the corresponding direction planes. We set N_d equal to 12 first and then 16.

To decompose contour pixels into 16 directions, we follow the 16-direction extended chaincodes, which is defined by two consecutive chaincodes. In the weighted direction histogram feature of Kimura et al. [22], 16-direction chaincodes are downsampled by weighted average to form 8-direction planes.

Again, we can determine the 16-direction chaincode of contour pixels by raster scan. At a contour pixel (x, y), when its 4-connected neighbor $p_k = 0$ and the counterclockwise successor $p_{k+1} = 1$ or $p_{(k+2)\%2} = 1$, search the neighbors clockwise from p_k until a $p_j = 1$ is found. The two contour pixels, p_{k+1} or $p_{(k+2)\%2}$ and p_j, form a 16-direction chaincode. For example, in Figure 6.17, the center pixel has the east neighbor being 0, the north neighbor alone defining the 8-direction chaincode, and defining a 16-direction chaincode together with the southeast neighbor. The 16-direction chaincode can be indexed from a table of correspondence between the code and the difference of coordinates of two pixels forming the code, as shown in Figure 6.18. Each contour pixel has a unique 16-direction code.

For decomposing contour pixels into 12 directions, the difference of coordinates corresponding to a 16-direction chaincode is viewed as a vector (the dashed line in Fig. 6.17), which is decomposed into components in 12 standard directions as a gradient vector is done. In this sense, the 12-direction code of a contour pixel is not unique. For 12-direction chaincode feature extraction, a contour pixel is assigned

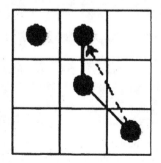

FIGURE 6.17 16-direction chaincode formed from two 8-direction chaincodes.

(-2,2)	(-1,2)	(0,2)	(1,2)	(2,2)
6	5	4	3	2
(-2,1)	(-1,1)	(0,1)	(1,1)	(2,1)
7	6	4	2	1
(-2,0)	(-1,0)	(0,0)	(1,0)	(2,0)
8	8		0	0
(-2,-1)	(-1,-1)	(0,-1)	(1,-1)	(2,-1)
9	10	12	14	15
(-2,-2)	(-1,-2)	(0,-2)	(1,-2)	(2,-2)
10	11	12	13	14

FIGURE 6.18   Difference of coordinates and the corresponding 16-direction chaincodes.

to two direction planes, with strength proportional to the component length. For (either discrete or continuous) NCFE, the two corresponding direction planes are assigned strengths proportional to the overlapping length of the line segment mapped by coordinate functions, as in Figure 6.15.

6.2.5   Classification Methods

Many classification methods have been described in Chapter 4. However, only a few classification methods have shown high success in large character set recognition. The modified quadratic discriminant function (MQDF) (Section 4.2.2.2) gives high classification accuracy at high complexity of storage/computation. It has the advantage that the parameters of each class are estimated on the training samples of one class only, so the training time is linear with the number of classes. Learning vector quantization (LVQ) classifiers (Section 4.3.6) yield good trade-off between the classification accuracy and the computational complexity. Though the training time of LVQ is proportional to the square of number of classes (the parameters of each class are estimated on the samples of all classes), it can be largely accelerated by hierarchical rival class search in training.

The complexity of parameters, namely, the number of eigenvectors per class for MQDF and the number of prototypes per class for LVQ, are variable and are usually set by trial-and-error or cross-validation. We will evaluate the classification performance of MQDF and LVQ classifiers with variable complexity, and compare with the performance of a simple classifier, the Euclidean distance (nearest mean) classifier.

6.2.6   Experiments

The normalization, feature extraction, and classification methods are evaluated on two databases of handprinted characters. The ETL9B database, collected by the Electro-Technical Laboratory (ETL) of Japan[1] contains the character images of 3036 classes

---

[1] ETL has been reorganized to the National Institute of Advanced Industrial Science and Technology (AIST) of Japan.

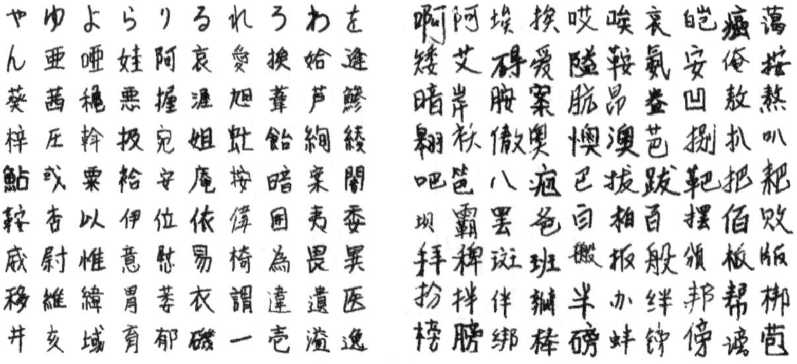

FIGURE 6.19 Some test images of ETL9B database (left) and CASIA database (right).

(71 hiragana, and 2965 Kanji characters in the JIS level-1 set), 200 samples per class. This database has been widely evaluated by the community [16, 22]. The CASIA database, collected by the Institute of Automation, Chinese Academy of Sciences, in early 1990s, contains the handwritten images of 3755 Chinese characters (the level-1 set in GB2312-80 standard), 300 samples per class.

In the ETL9B database, we use the first 20 and last 20 samples of each class for testing and the remaining samples for training classifiers. In the CASIA database, we use the first 250 samples of each class for training and the remaining 50 samples per class for testing. Some test images of two databases are shown in Figure 6.19.

In our experiments, a character pattern is represented by a feature vector. The feature vector undergoes two transformations: variable transformation and dimensionality reduction. Variable transformation [7, 49] is also called as Box-Cox transformation [11]. By this transformation, each feature variable x is replaced by $x^\alpha, 0 < \alpha < 1$. For causal variables, this is helpful to make the probability density of features closer to Gaussian. We set $\alpha = 0.5$, which was shown to perform satisfactorily. After variable transformation, the feature vector is projected onto a low-dimensional linear subspace learned by Fisher linear discriminant analysis (LDA) [7] (see also Section 3.1.10.2 in Chapter 3). We set the reduced dimensionality to 160 for all feature types.

For classification by MQDF, we use 40 principal eigenvectors for each class. The minor eigenvalues are forced to be a class-independent constant, which is proportional to the average feature variance σ^2, with the multiplier β $(0 < \beta \le 1)$ selected by fivefold holdout validation on the training data set. This is to say, the class means, principal eigenvectors and eigenvalues are estimated on 4/5 of training samples, several values of β are tested on the remaining 1/5 of training samples to select the optimal value of β that yields the highest accuracy. On fixing β, all the parameters are then reestimated on the whole training set.

The classification of MQDF is speeded up by selecting 100 candidate classes using Euclidean distance. The MQDF is then computed on the candidate classes only. Candidate selection is further accelerated by clustering the class means into groups. The input feature vector is first compared to cluster centers and then compared to the class means contained in a number of nearest clusters. We set the total number of clusters to 220 for the ETL9B database and 250 for the CASIA database.

To compare the performance of various normalization and feature extraction methods, we use the Euclidean distance and MQDF for classification. Afterwards, we evaluate MQDF and LVQ with variable parameter complexity on a selected feature type.

First, we evaluate the effect of blurring in feature extraction on two normalization methods: linear normalization and nonlinear normalization, and compare the 4-orientation and 8-direction versions of chaincode feature. We then justify the advantage of continuous NCFE over discrete NCFE, and the advantage of gradient feature on gray-scale image over that on binary image. We evaluate three types of direction features (chaincode, continuous NCFE, gray-scale-based gradient) with variable sampling resolution, and based on the optimal sampling scheme, we compare the performance of five normalization methods. Last, on a selected normalization-feature combination, we evaluate MQDF with variable number of principal eigenvectors and LVQ with variable number of prototypes.

6.2.6.1 *Effects of Blurring*   We compare the recognition performance of chaincode feature sampled by zoning and by blurring (spatial filtering), and compare the 4-orientation and 8-direction versions. In this experiment, the character image is normalized by two methods: linear normalization and line-density-based nonlinear normalization (Tsukumo and Tanaka [47]). The size of normalized plane (and orientation/direction planes) is set to 64×64 pixels. Each orientation/direction plane is sampled 8×8 feature values by zoning or blurring. Thus, the original dimensionality of feature vector is 256 for 4-orientation and 512 for 8-direction. In either case, the dimensionality is reduced to 160 by LDA.

Table 6.4 shows the test accuracies using Euclidean distance and MQDF classifiers on ETL9B database and Table 6.5 shows the accuracies on CASIA database. In each table, the left half shows the accuracies of Euclidean distance and the right half of MQDF. "4-zone" and "4-blur" indicate 4-orientation features sampled by zoning and blurring, respectively, and "8-zone" and "8-blur" indicate 8-direction features sampled by zoning and blurring, respectively.

On both ETL9B and CASIA databases, comparing the accuracies of either Euclidean distance or MQDF, we can see that the accuracies of 8-direction feature are evidently higher than those of 4-orientation feature. Comparing the features by

TABLE 6.4   **Accuracies (%) of 4-orientation and 8-direction chaincode feature by zoning and by blurring on ETL9B database.**

	Euclidean				MQDF			
Norm.	4-zone	4-blur	8-zone	8-blur	4-zone	4-blur	8-zone	8-blur
Linear	89.15	90.83	90.72	92.65	95.57	96.95	96.20	97.44
Nonlin	95.81	96.60	96.16	97.06	98.18	98.80	98.42	98.95

TABLE 6.5   Accuracies (%) of 4-orientation and 8-direction chaincode feature by zoning and by blurring on CASIA database.

Norm.	Euclidean				MQDF			
	4-zone	4-blur	8-zone	8-blur	4-zone	4-blur	8-zone	8-blur
Linear	81.60	84.46	83.97	87.08	92.42	94.78	93.52	95.55
Nonlin	91.62	93.06	92.47	93.93	96.20	97.23	96.75	97.63

zoning and by blurring, it is evident that blurring yields higher recognition accuracies than zoning. Thus, this experiment justifies the superiority of 8-direction feature over 4-orientation feature and the superiority of feature blurring over zoning. In the following, we experiment with 8-direction features sampled by blurring only.

6.2.6.2 Justification of NCFE and Gradient

We compare the recognition performance of discrete NCFE and continuous NCFE, and the performance of gradient features on binary normalized image and on gray-scale normalized image (the input character image is binary). As in the previous experiment, two normalization methods (linear and nonlinear) are undertaken, the size of normalized plane is 64×64, and 8×8 feature values are sampled from each direction plane by blurring. The 512-dimensional feature vector is reduced to 160 by LDA.

Tables 6.6 and 6.7 show the test accuracies on ETL9B database and CASIA database, respectively. On both two databases, comparing the accuracies of Euclidean distance or MQDF, it is true that the continuous NCFE yields higher recognition accuracy than the discrete NCFE, and the gradient feature on gray-scale normalized image yields higher accuracy than that on binary normalized image. The difference of accuracy, however, is not so prominent as that between feature zoning and feature blurring.

6.2.6.3 Effects of Feature Sampling Resolution

In this experiment, we compare the performance of three direction features (chaincode, continuous NCFE, and gradient feature on gray-scale image) with varying direction resolutions with a common normalization method (line-density-based nonlinear normalization). The

TABLE 6.6   Accuracies (%) of 8-direction NCFE (discrete and continuous, denoted by *ncf-d* and *ncf-c*) and gradient feature (binary and gray, denoted by *grd-b* and *grd-g*) on ETL9B database.

Norm.	Euclidean				MQDF			
	ncf-d	ncf-c	grd-b	grd-g	ncf-d	ncf-c	grd-b	grd-g
Linear	92.35	92.62	92.98	93.26	97.40	97.51	97.49	97.75
Nonlin	97.27	97.39	97.26	97.42	99.08	99.13	99.02	99.10

OFFLINE HANDWRITTEN CHINESE CHARACTER RECOGNITION

TABLE 6.7 **Accuracies (%) of 8-direction NCFE (discrete and continuous, denoted by**
ncf-d **and** *ncf-c*) **and gradient feature (binary and gray, denoted by** *grd-b* **and** *grd-g*) **on**
CASIA database.

	Euclidean					MQDF			
Norm.	ncf-d	ncf-c	grd-b	grd-g		ncf-d	ncf-c	grd-b	grd-g
Linear	86.73	87.22	87.66	88.45		95.73	95.92	95.72	96.30
Nonlin	94.46	94.65	94.18	94.66		97.97	98.08	97.70	97.99

direction resolution of features is set equal to 8, 12, and 16. For each direction resolution, three schemes of sampling mesh are tested. For 8-direction features, the mesh of sampling is set to 7×7 (M1), 8×8 (M2), and 9×9 (M3); for 12-direction, 6×6 (M1), 7×7 (M2), and 8×8 (M3); and for 16-direction, 5×5 (M1), 6×6 (M2), and 7×7 (M3). We control the size of normalized image (direction planes) to be around 64×64, and the dimensionality (before reduction) to be less than 800. The settings of sampling mesh are summarized in Table 6.8.

On classifier training and testing using different direction resolutions and sampling schemes, the test accuracies on ETL9B database are listed in Table 6.9, and the accuracies on CASIA database are listed in Table 6.10. In the tables, the chaincode direction feature is denoted by *chn*, continuous NCFE is denoted by *ncf-c*, and the gradient feature on gray scale image by *grd-g*.

We can see that on either database, using either classifier (Euclidean or MQDF), the accuracies of 12-direction and 16-direction features are mostly higher than those of 8-direction features. This indicates that increasing the resolution of direction decomposition is beneficial. The 16-direction feature, however, does not outperform the 12-direction feature. To select a sampling mesh, let us focus on the results of 12-direction features. We can see that by Euclidean distance classification, the accuracies of M1 (6×6) and M2 (7×7) are mostly higher than those of M3 (8×8), whereas by MQDF, the accuracies of M2 and M3 are higher than those of M1. Considering that M2 and M3 perform comparably while M2 has lower complexity, we take the sampling mesh M2 with 12-direction features for following experiments. The original dimensionality of direction features is now $12 \times 7 \times 7 = 588$, which is reduced to 160 by LDA.

TABLE 6.8 **Settings of sampling mesh for 8-direction, 12-direction, and 16-direction**
features.

Mesh	M1		M2		M3	
	Zones	Dim	Zones	Dim	Zones	Dim
8-dir	7×7	392	8×8	512	9×9	648
12-dir	6×6	432	7×7	588	8×8	768
16-dir	5×5	400	6×6	576	7×7	784

TABLE 6.9 Accuracies (%) of 8-direction, 12-direction, and 16-direction features on ETL9B database.

	Euclidean			MQDF		
chn	M1	M2	M3	M1	M2	M3
8-dir	97.09	97.06	96.98	98.92	98.95	98.91
12-dir	97.57	97.44	97.46	98.98	99.00	**99.03**
16-dir	97.48	**97.60**	97.48	98.80	99.00	99.00
ncf-c	M1	M2	M3	M1	M2	M3
8-dir	97.39	97.39	97.29	99.07	99.13	99.11
12-dir	97.95	97.94	97.87	99.18	**99.23**	99.22
16-dir	97.95	**97.96**	97.89	99.02	99.15	99.21
grd-g	M1	M2	M3	M1	M2	M3
8-dir	97.41	97.42	97.34	99.07	99.10	99.11
12-dir	97.73	97.70	97.69	99.06	99.14	99.14
16-dir	97.71	**97.81**	97.75	98.92	99.06	**99.15**

6.2.6.4 _Comparison of Normalization Methods_   On fixing the direction resolution (12-direction) and sampling mesh (7×7), we combine the three types of direction features with five normalization methods: linear (LN), nonlinear (NLN), moment (MN), bi-moment (BMN), and MCBA. The accuracies on the test sets of two databases are listed in Tables 6.11 and 6.12, respectively. It is evident that the

TABLE 6.10 Accuracies (%) of 8-direction, 12-direction, and 16-direction features on CASIA database.

	Euclidean			MQDF		
chn	M1	M2	M3	M1	M2	M3
8-dir	94.05	93.93	93.78	97.55	97.63	97.63
12-dir	94.66	94.56	94.47	97.66	97.74	**97.79**
16-dir	94.56	**94.72**	94.63	97.28	97.65	97.71
ncf-c	M1	M2	M3	M1	M2	M3
8-dir	94.69	94.65	94.51	98.06	98.08	98.00
12-dir	95.56	95.52	95.45	98.14	98.25	**98.27**
16-dir	95.47	**95.58**	95.48	97.83	98.08	98.18
grd-g	M1	M2	M3	M1	M2	M3
8-dir	94.69	94.66	94.59	97.95	97.99	98.03
12-dir	95.06	95.10	95.02	97.91	98.00	**98.05**
16-dir	95.02	95.15	**95.20**	97.58	97.91	98.02

TABLE 6.11   Accuracies (%) of various normalization methods on ETL9B database.

Norm.	Euclidean			MQDF		
	chn	ncf-c	grd-g	chn	ncf-c	grd-g
LN	93.64	94.06	94.03	97.62	97.91	97.89
NLN	97.44	97.94	97.70	99.00	**99.23**	99.14
MN	97.65	97.93	97.88	99.05	99.17	99.18
BMN	**97.67**	**97.96**	**97.91**	**99.08**	99.19	**99.20**
MCBA	97.48	97.81	97.73	99.00	99.16	99.14

linear normalization (LN) is inferior to the other four normalization methods, whereas the difference of accuracies among the latter four normalization methods is smaller.

Comparing the normalization methods except LN, we can see that on both two databases, BMN outperforms MN and MCBA. MN and MCBA perform comparably on ETL9B database, but on CASIA database, MCBA outperforms MN. The performance of NLN depends on the feature extraction method. With chaincode feature and gradient feature, the accuracies of NLN are comparable to those of BMN and MCBA. However, on both databases, the combination of NLN with continuous NCFE gives the highest accuracy. This is because with nonsmooth coordinate mapping, NLN yields larger shape distortion than the other normalization methods. The chaincode feature and gradient feature, extracted from the normalized image, are sensitive to the shape distortion. NCFE combines the contour direction feature of the original image and the coordinate transformation of nonlinear normalization, thus yields the best recognition performance.

6.2.6.5 *Comparison of Classifiers*   The inferior classification performance of Euclidean distance to MQDF has been manifested in the above experiments. Now, we compare the performance of MQDF and LVQ classifiers with variable parameter complexity on a selected normalization-feature combination, namely, 12-direction continuous NCFE with bi-moment normalization (BMN). We experiment on the CASIA database only. As shown in Table 6.12, the accuracies of Euclidean distance and MQDF are 95.46 and 98.07%, respectively.

TABLE 6.12   Accuracies (%) of various normalization methods on CASIA database.

Norm.	Euclidean			MQDF		
	chn	ncf-c	grd-g	chn	ncf-c	grd-g
LN	88.62	89.54	89.69	95.89	96.51	96.46
NLN	94.56	94.52	95.10	**97.74**	**98.25**	98.00
MN	94.39	95.11	95.10	97.50	97.96	97.94
BMN	**94.70**	**95.46**	**95.44**	97.65	98.07	**98.08**
MCBA	94.52	95.27	95.17	97.69	98.13	98.04

TABLE 6.13 Accuracies (%) of MQDF with variable number k of principal eigenvectors and LVQ with variable number p of prototypes on CASIA database.

MQDF (k)	1	2	3	4	5	10	15
Accuracy (%)	96.44	96.92	97.13	97.3	97.40	97.75	97.87
MQDF (k)	20	25	30	35	40	45	50
Accuracy (%)	97.95	97.97	98.02	98.05	98.07	**98.09**	98.08
LVQ (p)	1	2	3	4	5		
Accuracy (%)	96.82	97.15	97.21	**97.22**	97.12		

The number of principal eigenvectors per class for MQDF classifier is set equal to $k = 1, 2, 3, 4, 5, 10, 15, 20, 25, 30, 40, 45, 50$, and the number of prototypes per class for LVQ classifier is set equal to $p = 1, 2, 3, 4, 5$. The class prototypes are initialized to be the cluster centers of the training samples of each class and are adjusted to optimize the minimum classification error (MCE) criterion (see details in Section 4.3.6) on all the training samples.

The test accuracies of MQDF with variable number of principal eigenvectors and LVQ classifier with variable number of prototypes are shown in Table 6.13. We can see that as the number of principal eigenvectors increases, the classification accuracy of MQDF increases gradually and almost saturates when $k \geq 40$. As for the LVQ classifier, the accuracy is maximized when each class has $p = 3$ prototypes. Using more prototypes, the training samples are separated better, but the generalization performance does not improve. When $p = 1$, the initial prototypes are class means, which gives the test accuracy 95.46% (Euclidean distance). By prototype optimization in LVQ, the test accuracy is improved to 96.82%.

Comparing the performance of MQDF and LVQ classifiers, it is evident that the highest accuracy of LVQ is lower than that of MQDF. However, the MQDF needs a large number of parameters to achieve high accuracies. For M-class classification with k principal eigenvectors per class, the total number of vectors to be stored in the parameter database is $M \cdot (k + 1)$ (the class means are stored as well). For LVQ classifier, the total number of vectors is $M \cdot p$. The computational complexity in recognition is linearly proportional to the total number of parameters. Thus, the run-time complexity of MQDF with k principal eigenvectors per class is comparable to that of LVQ classifier with $k + 1$ prototypes per class. When we compare two classifiers with similar complexity, say $k = 1$ versus $p = 2, \ldots, k = 4$ versus $p = 5$, we can see that when the parameter complexity is low ($k \leq 3$), the LVQ classifier gives higher classification accuracy than the MQDF. So, we can say that the LVQ classifier provides better trade-off between the classification accuracy and the run-time complexity.

That the MQDF and LVQ classifiers have different trade-off between accuracy and run-time complexity is due to the different training algorithms. The parameters of MQDF are estimated generatively: each class is estimated independently; whereas the parameters of LVQ classifier are trained discriminatively: all parameters are adjusted to separate the training samples of different classes. Discriminative classifiers give higher classification accuracy at low complexity of parameters, but are time consuming in training, and needs large number of training samples for good generalization. In the case of training on CASIA database on personal computer with 3.0 GHz CPU, the training time of MQDF is about 1 h (computation mainly for diagonalization of covariance matrices), whereas the training time of LVQ with acceleration by two-level rival prototype search is over 4 h.

6.2.7   Concluding Remarks

We introduced an offline handwritten Chinese character recognition (HCCR) system and evaluated the performance of different methods in experiments. For feature extraction, we justified the superiority of feature blurring over zoning, the advantage of continuous NCFE over discrete NCFE, and the advantage of gradient feature on gray-scale image over that on binary image. In comparing direction features with variable sampling resolutions, our results show that the 12-direction feature has better trade-off between accuracy and complexity. The comparison of normalization methods shows that bi-moment normalization (BMN) performs very well, and line-density-based nonlinear normalization (NLN) performs well when combined with NCFE. The comparison of classifiers shows that the LVQ classifier gives better trade-off between accuracy and run-time complexity than the MQDF. The MQDF yields high accuracies at very high run-time complexity.

The current recognition methods provide fairly high accuracies on handprinted characters, say, over 99.0% on the ETL9B database and over 98.0% on the CASIA databases. The recognition of unconstrained handwritten Chinese characters has not been studied extensively because a public database of such samples is not available. We can expect that on unconstrained characters, the accuracies will be much lower, and so, efforts are needed to improve the current methods and design new methods.

Better performance can be expected by using P2D normalization methods [34]. Despite the superiority of direction features, even better features should be automatically selected from a large number of candidate features. Thus, feature selection techniques will play an important role in future research of character recognition. For classification, many discriminative classifiers have not been exploited for large category set. As a generative classifier, the MQDF is promising for HCCR, particularly for handprinted characters. Recently, even higher accuracies have been achieved by discriminative learning of feature transformation and quadratic classifier parameters [36, 37]. Discriminative classifier design for large category problems will be an important direction in the future.

6.3   SEGMENTATION AND RECOGNITION OF HANDWRITTEN DATES ON CANADIAN BANK CHEQUES

This case study describes a system developed to recognize date information handwritten on Canadian bank checks. A segmentation-based strategy is adopted in this system. In order to achieve high performance in terms of efficiency and reliability, a knowledge-based module is proposed for the date segmentation and a cursive month word recognition module is implemented based on a combination of classifiers. The interaction between the segmentation and recognition stages is properly established by using multihypotheses generation and evaluation modules. As a result, promising performance is obtained on a test set from a real-life standard check database.

6.3.1   Introduction

Research on OCR began in the 1950s, and it is one of the oldest research areas in the field of pattern recognition. Nowadays, many commercial systems are available for reliably processing cleanly machine-printed text documents with simple layouts, and some successful systems have also been developed to recognize handwritten texts, particularly, isolated handprinted characters and words. However, the analysis of documents with complex layouts, recognition of degraded machine-printed texts, and the recognition of unconstrained handwritten texts demand further improvements through research. In this case study, our research aims at developing an automatic recognition system for unconstrained handwritten dates on bank checks, which is a very challenging topic in OCR.

The ability to recognize the date information handwritten on bank checks is very important in application environments (e.g., in Canada) where checks cannot be processed prior to the dates shown. At the same time, date information also appears on many other kinds of forms. Therefore, there is a great demand to develop reliable automatic date processing systems.

The main challenge in developing an effective date processing system stems from the high degree of variability and uncertainty in the data. As shown in Figure 6.20, people usually write the date zones on check in such free styles that little a priori knowledge and few reliable rules can be applied to define the layout of a date image. For example, the date fields can contain either only numerals or a mixture of alphabetic letters (for *month*) and numerals (for *day* and *year*), punctuations, suffixes, and the article "Le" may also appear. (The dates can be written in French or in English in Canada, and a "Le" may be written at the beginning of a French date zone.)

Perhaps because of this high degree of variability, there has been no published work on this topic until the work on the date fields of machine-printed checks was reported in 1996 [14]. This reference also considered date processing to be the most difficult target in check processing, given that it has the worst segmentation and

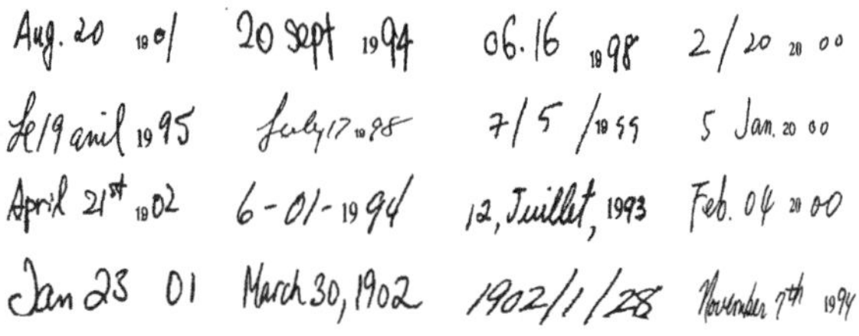

FIGURE 6.20   Sample dates handwritten on standard Canadian bank checks.

recognition performance. In 2001, a date processing system for recognizing handwritten date images on Brazilian checks was presented in [40]. A segmentation-free method was used in this system, that is, an HMM (Hidden Markov Model) based approach was developed to perform segmentation in combination with the recognition process.

The system addressed in this case study (which is an extension of a previous work [5]) is the only published work on processing of date zones on Canadian bank checks. In our system, date images are recognized by a segmentation-based method, that is, a date image is first segmented into *day*, *month*, and *year*, the category of *month* (alphabetic or numeric) is identified, and then an appropriate recognizer is applied for each field. In the following, the main modules of the whole system will be discussed, together with some experimental results.

6.3.2 System Architecture

Since the date image contains fields that may belong to different categories (alphabetic or numeric), it is difficult to process the entire date image at the same time efficiently. Therefore, a segmentation-based strategy is employed in our system.

The main procedures in the system consist of segmenting the date image into fields through the detection of the separator or transition between the fields, identifying the nature of each field, and applying an appropriate recognizer for each.

Figure 6.21 illustrates the basic modules in our date processing system. In addition to the two main modules related to segmentation and recognition, a preprocessing module has been designed to deal with simple noisy images and to detect and process possible appearances of "Le." In the postprocessing stage, in order to further improve the reliability and performance of the system, a two-level verification module is designed to accept valid and reliable recognition results, and to reject the others.

In order to improve the performance and efficiency of the system, a knowledge-based segmentation module is used to solve most segmentation cases in the segmentation stage. Ambiguous cases are handled by a multihypotheses generation module at this stage, for a final decision to be made when more contextual information and syntactic and semantic knowledge are available, that is, multihypotheses evaluation is made in the recognition stage.

6.3.3 Date Image Segmentation

Figure 6.22 shows an overview of the date segmentation module. As described above, the date image segmentation module divides the entire image into three subimages corresponding to *day*, *month*, and *year*, respectively, and also makes a decision on how *month* is written so that it can be processed using an appropriate recognizer. Since there is no predefined position for each field, and no uniform or even obvious spacing between the fields, it is difficult to implement the segmentation process with a high success rate by using simple structural features.

The first step of our segmentation module is to separate *year* from *day&month* based on structural features and the characteristics of the *Year* field, and then the

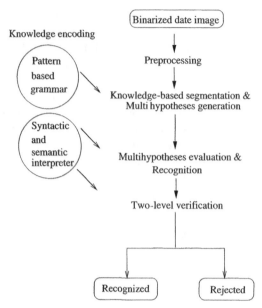

FIGURE 6.21   Diagram of date processing system.

knowledge-based segmentation module and the multihypotheses generation and evaluation modules are applied in the *day&month* segmentation stage.

6.3.3.1   *Year Detection*   Based on our database analyses (CENPARMI Check database and CENPARMI_IRIS Check database), the writing styles of date zones on Canadian bank checks can be grouped into two categories, which are defined as standard format and free format in this case study. The standard format is used when "2" and "0" or "1" and "9" are printed as isolated numerals on the date zone indicating the century and the free format is adopted when the machine-printed "20" or "19" does not appear on a date zone. Since about 80% of date zones are of the standard format in our databases, *year* with the standard format is first detected in the *year* detection module. If the detection is not successful, *year* with the free format is detected.

The detection of *year* for the standard format is based on the detection of the machine-printed "20" or "19." For detecting *year* with the free format, we first assume: (i) *year* is located at one end of the date zone; (ii) *year* belongs to one of the two patterns: 20** (or 19**) and **, where * is a numeral; and (iii) a separator is used by writers to separate a *year* field from a *day&month* field. Based on these assumptions, we use a candidate_then_confirmation strategy to detect the *year* with the free format. *Year* candidates are first detected from the two ends of the date zone, and then one of the candidates is confirmed as the *year* by using the recognition results from a digit recognizer and by using the assumption that a *year* field contains either four numerals starting with "20" (or "19") or two numerals. Here *year* candidates are obtained by detecting the separator (a punctuation or a big gap) between *year* and *day&month*

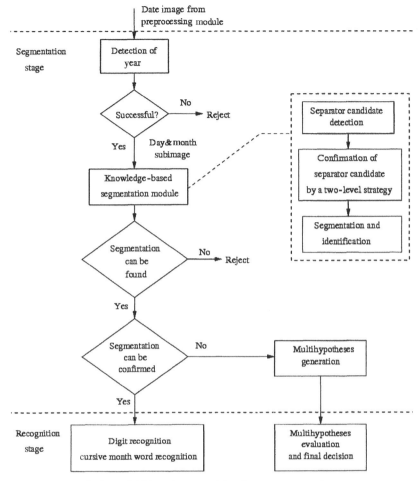

FIGURE 6.22   Diagram of date image segmentation.

fields, and these separator candidates are detected from structural features [5, 45]. In the confirmation stage, the confidence value for a *year* candidate with the pattern 20∗∗ (or 19∗∗) is considered to be higher than that of a *year* candidate with the pattern ∗∗.

6.3.3.2 *Knowledge-Based Day&Month Segmentation*   The tasks of this knowledge-based segmentation module include (i) detecting the separator between *day* and *month*; and (ii) segmenting the *day&month* field into *day* and *month*, and identifying the category of the *month*.

For the separator detection, the separators can be punctuations, such as slash "/," hyphen "-," comma "," and period "." or big gaps, and a candidate_then_confirmation strategy is used here to detect them. Separator candidates are first detected by shape and spatial features [5, 45]. While some of the candidates with high confidence values of the features can be confirmed immediately, others should be evaluated by considering more information.

Based on our database analyses, some relationship between the type of separator and the writing style of *day&month* has been found, for example, slashes or hyphens usually appear when both *day* and *month* are written in numerals. So some separator candidates are easily confirmed or rejected using a set of rules if the knowledge about the writing style can be obtained [45]. Furthermore, these rules can be designed in the training stage based on human knowledge and syntactic constraints, and in general they can be encoded in a pattern-based grammar. Here pattern means image pattern, and usually a *day&month* subimage can appear in any one of the three patterns: *NSN*, *NSA*, and *ASN*, where *S* denotes the separator, *N* denotes a numeric string (*day* or *month* field), and *A* denotes an alphabetic string (*month* field). The pattern-based grammar used to detect the separator can be expressed as

Given image Pattern:
If $<$ *condition* $>$ then $<$ *action* $>$
 $<$ *condition* $> \Rightarrow$ the separator candidate (S) in the image Pattern is P, $P \in \Sigma$
  $<$ *action* $> \Rightarrow$ Confidence Adjustment|
      Add New Feature|
      Adjust Segmentation Method|...

Here Σ represents the set of separator types. Some explanations about this pattern-based grammar are given in Table 6.14, where *day&month* image "patterns" are given as the entries. When we try to confirm a separator candidate and the "condition" is that the separator candidate is a "/" or "−" or ..., we can take the corresponding "action."

For this knowledge-based method, two approaches have been developed in our system to determine the writing styles. The first method adopts a *distance_to_numeral* measure to represent the likelihood of a subimage in *day&month* field being numeric, and it is based on feedbacks of a digit recognizer and structural features. A system of combining multiple multilayer perceptron (MLP) networks is the second method to realize this writing style analysis task. More details about these two methods can

TABLE 6.14 Examples of condition-action rules.

Pattern	Condition	Action
NSN	−	"−" candidate is confirmed.
NSA	/	New "long" features are checked.
ASA	NULL	Separator candidate is not confirmed.
...	...	...

be found in [45] and [26], respectively. The effectiveness of these two methods has been proven in the experiments. However, their results may be inconclusive in some ambiguous segmentation cases, where the multihypotheses generation and evaluation are introduced.

After the separator detection step, *day&month* segmentation and identification can be conducted. Based on the separators detected, a set of rules have been developed to segment the *day&month* field into *day* and *month* fields and to determine whether the month field is written in numeric or alphabetic form [5, 45].

6.3.3.3 *Multihypotheses Generation*   In the knowledge-based *day&month* segmentation module, only the separators with high confidence values and the writing styles with high confidence values to indicate "common styles" would be confirmed. Here "common styles" are determined based on database analyses, including that the gap separator usually occurs at the transition between numeric and alphabetic fields, the subimages on both sides of slash or hyphen are often numeric, and a period separator is usually used in *ASN* pattern. Otherwise, the multihypotheses generation module is activated. This module produces and places multiple hypotheses in a multihypotheses list, where each hypothesis consists of a possible segmentation of *day&month* field.

6.3.3.4 *Multihypotheses Evaluation*   Each possible segmentation in the multihypotheses list includes a separator candidate and segments on both sides of the separator candidate. For each such hypothesis, the multihypotheses evaluation module estimates its confidence values for the three writing styles or types (*NSN*, *NSA*, or *ASN*) as the following weighted sums:

$$
\begin{aligned}
Confidence_{Type1} = \ &w_1 * DigitConfidence_{Left} \\
&+ w_1 * DigitConfidence_{Right} \\
&+ w_3 * SeparatorConfidence \\
Confidence_{Type2} = \ &w_1 * DigitConfidence_{Left} \\
&+ w_2 * WordConfidence_{Right} \\
&+ w_3 * SeparatorConfidence \\
Confidence_{Type3} = \ &w_2 * WordConfidence_{Left} \\
&+ w_1 * DigitConfidence_{Right} \\
&+ w_3 * SeparatorConfidence
\end{aligned}
$$

$DigitConfidence_{Left}$ and $DigitConfidence_{Right}$ are confidence values from a digit recognizer for the left and right sides of the separator candidate, respectively. $WordConfidence_{Left}$ and $WordConfidence_{Right}$ are confidence values from a cursive month word recognizer. $SeparatorConfidence$ is the confidence value of the separator candidate, which is derived from the segmentation stage. The weights w_1, w_2, and w_3 are determined in the training stage. For example, $w_1 = 0.8$, and $w_2 = 1$ because the distribution of confidence values from the digit recognizer is different from that of the word recognizer, and these weights are set to make the confidence values of the digit and word recognizers comparable.

For each hypothesis in the list, usually the Type with the maximum $Confidence_{Type}$ value is recorded to be compared with the corresponding information from other hypotheses in the list. In the application, some semantic and syntactic constraints can be used to improve the performance of this multihypotheses evaluation module. First, based on semantic constraints, if the recognition result from a Type is not a valid date, the corresponding $Confidence_{Type}$ would be reduced by a small value (α) before $Confidence_{Type}$ values are compared in Type selection. In addition, as we discussed above, "common styles" have been determined based on database analyses. These "common styles" can be used as syntactic constraints to modify the Type selection procedure, that is, if the interpretation of the first choice is not a "common style" and the difference between the confidence values of the top two choices is very small, the second choice that has the second largest $Confidence_{Type}$ value should be the final selection if the interpretation of this choice is a "common style" and is a valid date.

6.3.4 Date Image Recognition

In the date recognition stage, if one segmentation hypothesis can be confirmed in the segmentation stage, an appropriate recognizer (digit recognizer or cursive month word recognizer) is invoked for each of *day*, *month* and *year* fields. Otherwise, the multi-hypotheses evaluation module that makes use of the results from the digit and word recognizers is invoked. The digit recognizer used in our date processing system was originally developed for processing the courtesy amount written on bank checks [44], and a 74% recognition rate (without rejection) was reported for processing these courtesy amounts. For the recognition of cursive month words, a new combination method with an effective conditional topology has been implemented in our date processing system. More discussions about this cursive month word recognizer are given below.

6.3.4.1 Introduction to Cursive Month Word Recognizers The recognition of month words on bank checks poses some new problems in addition to the challenges caused by the high degree of variability in unconstrained handwriting. First, improper binarization and preprocessing can have a great impact on the quality of month word images. Second, although a limited lexicon is involved, many similarities exist among the month word classes, and this can give rise to problems in feature extraction and classification. Furthermore, month words on bank checks can be written in French or in English in Canada, which increases the complexity of the problem. In addition, most month words have very short abbreviated forms, which make it more difficult to differentiate between them.

According to our literature review, only two systems have been reported for month word recognition. In [40], a system was developed for handwritten month word recognition on Brazilian bank checks by using an explicit segmentation-based HMM classifier. Another system mentioned in [19] is the previous work of this case study, which deals with English month word recognition by combining an MLP

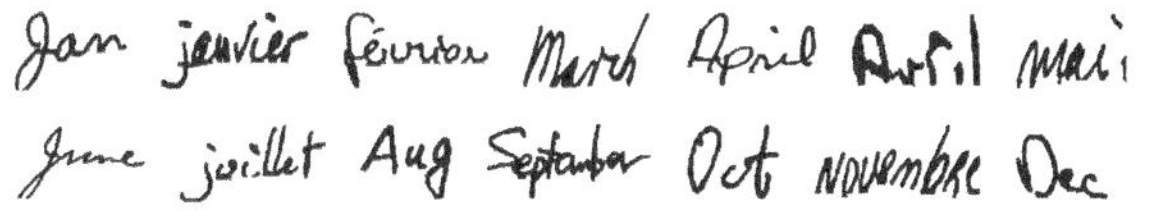

FIGURE 6.23 Month word samples from CENPARMI_IRIS check database.

classifier and an HMM classifier. Based on the previous work, some improvements and modifications have been made in our current recognition system to recognize both French and English month words. An effective conditional combination topology is presented to combine two MLP and one HMM classifiers, and a proposed modified Product fusion rule gives the best recognition rate so far.

6.3.4.2 *Writing Style Analysis for Cursive Month Word Recognition*   In this section, the writing styles of month words are analyzed based on CENPARMI bank check databases. Some samples extracted from the databases are shown in Figure 6.23, including both English and French month words.

Altogether, 33 English/French month word classes have been observed in the databases including full and abbreviated forms. Based on analyses of the databases, many similarities have been found among the word classes:

- The most similar classes are: "September" and "Septembre", "October" and "Octobre", "November" and "Novembre", and "December" and "Décembre". Since each pair of classes are very similar in shape and they represent the same month, they are assigned to the same class. So only 29 classes will be considered.
- Some other similarities among the words are due to their shapes, for example, "Jan," "June" and "Juin," and "Mar," "Mai," and "Mars." They can also contain similar subimages, for example, "September," "November" and "December." These similarities can affect the performance of the recognition systems, and they have been considered in the feature extraction and individual classifier design.

From analyses of the databases, we also find that about one third of the month word classes consist of only three letters, mostly from abbreviations. In addition, French month words have been found to have freer writing styles, for example, both capital and lowercase letters can be used at the beginning of words, and mixtures of capital and lowercase letters appear more frequently.

These writing style analyses show some of the challenges for month word recognition. Some solutions to this problem will be given in the following sections using three individual classifiers and a combination system.

6.3.4.3 *Individual Classifiers*   In this section, three individual month word classifiers will be discussed, including two MLP classifiers and one HMM classifier.

(1) *MLP classifiers*
Two MLP classifiers have been developed in CENPARMI for month word recognition. MLPA and MLPB were originally designed for the recognition of the legal amount [20] and the courtesy amount [44], respectively, and both of them have been modified for month word recognition. The two MLPs use a holistic approach. Their architectures and features can be summarized below. More details can be found from the references cited.

- MLPA is a combination of two networks using different features. The combination adopts the "fusion" scheme, that is, MLPA is implemented by combining the two networks at an architectural level, and uses the outputs of the neurons in two hidden layers as new input features [20]. There are two feature sets designed for the two networks. The feature set 1 consists of mesh features, direction features, projection features, and distance features, whereas the feature set 2 consists of gradient features.
- MLPB is a combination of three MLPs. The combination scheme is a "hybrid" strategy that combines the output values of the MLPs. The MLPs are implemented by using three different sets of input features. Feature set 1 consists of pixel distance features (PDF), whereas feature sets 2 and 3 consist of size-normalized image pixels from different preprocessed images [44].

(2) *HMM classifier*
A segmentation based grapheme level HMM has been developed using an analytical feature extraction scheme for month word recognition [19]. This model has the potential to solve the over- or under-segmentation problem in cursive script recognition. The features used to obtain pseudotemporal sequences for the HMM come from shape feature, direction code distribution feature, curvature feature, and moment feature.

(3) *Performances*
The training set for implementing the individual month word classifiers consists of 6201 month word samples, and the test set consists of 2063 month word samples. Both the training set and the test set were extracted from CENPARMI databases (CENPARMI check database and CENPARMI_IRIS check database).

The performances of the two MLP classifiers and the HMM classifier on the test set are shown in Table 6.15. In the table, the 12 outputs mean mapping of the month words into the 12 months of a year. These results indicate that the two MLP classifiers produce comparable recognition rates, whereas the HMM classifier produces the worst results. In addition, performance improvements can be noticed from these results when the number of output classes is changed from 29 to 12. This is a result of the fact that some month words of different classes, but representing the same month, can have very similar shapes (e.g., "Mar," "Mars," "March," etc). Samples of these month words are sometimes recognized as words of different classes but belonging to the same month, which become correct classifications when the output classes are reduced to 12.

TABLE 6.15   **Performances of individual classifiers.**

Classifier	Recognition rate (%)	
	29 outputs	12 outputs
MLPA	76.44	78.87
MLPB	75.42	77.27
HMM	66.89	69.70

In order to enhance the recognition performance, combinations of these classifiers are considered in the following section.

6.3.4.4 *Combinations*   Combination systems are expected to produce better recognition results. Before discussing the combination topology and combination rule for month word classification, further analyses of the performances of the three individual classifiers are presented while focusing on the correlation of the recognition results among these classifiers. Some useful information has been obtained from these analyses.

(1) *Correlation of recognition results among individual classifiers*
The correlation of recognition results among the three individual classifiers is shown in Table 6.16. It was obtained from an additional training set (2063 samples extracted from CENPARMI check databases) separate from the training set for training the individual classifiers. This additional training set is also used in the experiments for developing the combination topology and combination rules to be discussed later.

In Table 6.16, "correct" means all the classifiers are correct on the sample, "error" means all the classifiers are incorrect, and "C/E" means both correct and wrong results are produced by the classifiers. The errors can be further grouped into two classes ErrorI and ErrorII, which mean the samples are assigned different and the same incorrect classes by the classifiers, respectively. From these results, we can infer the following:

• Even though combining these three classifiers may improve the recognition rate considerably due to the independence in errors, a combination would have a lower bound of 1.26% (26/2063) for the error rate without rejections, because all the three classifiers produce the same 26 errors.

TABLE 6.16   **Correlation of recognition results among classifiers.**

Classifiers	Correct	Error	C/E	ErrorI	ErrorII
MLPA and MLPB	1297	326	440	213	113
MLPA and HMM	1162	282	619	217	65
MLPB and HMM	1167	295	601	218	77
MLPA, MLPB, and HMM	1032	199	832	173	26

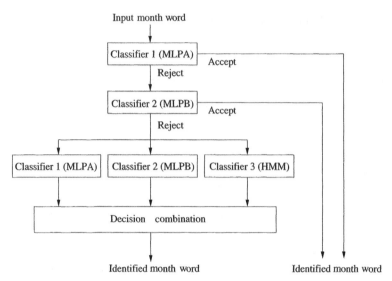

Input month word

Classifier 1 (MLPA)   Accept

Reject

Classifier 2 (MLPB)   Accept

Reject

| Classifier 1 (MLPA) | Classifier 2 (MLPB) | Classifier 3 (HMM) |

Decision   combination

Identified month word   Identified month word

FIGURE 6.24   Conditional combination topology for month word recognition.

- If two classifiers are to be combined, the combination of MLPA and MLPB makes the same errors more frequently than the other combinations, which can be observed from the ErrorIIs produced. Although the performances of both MLPA and MLPB are much better than that of the HMM classifier, they are homogeneous classifiers and tend to produce the same errors, which is not useful in combination. The idea that more distinct classifiers can better complement each other has been shown here, and it is very important for designing combination systems. In addition, three classifiers would be less likely to make the same errors than two of them as observed from the ErrorIIs.

(2) *Combination topology*
Combination topologies can be broadly classified as parallel, serial, hybrid, and conditional. Based on considerations of both speed and accuracy, as well as experimental findings, a conditional combination topology is proposed here to combine the three classifiers to improve the performance of month word recognition as shown in Figure 6.24.

In this architecture, MLPA and MLPB are first applied using serial strategy in the first stage; for samples rejected by both MLPA and MLPB, the decisions of all the three classifiers are combined in the second stage. The rejection conditions of both MLPA and MLPB have been set very strictly in the first stage (total error rate introduced in the first stage is 0.6% on the test set). A rejection is made in MLPA when the top confidence value is less than a threshold (0.99), and a rejection is made in MLPB when the difference between the top two confidence values is less than a threshold (0.4). This stage proved to be very efficient and effective based on the results of our experiments. About 33% of the samples can be recognized in the first stage. More difficult samples are rejected and sent to the three classifiers in parallel, and decisions from these classifiers are combined to get better results. Since the samples have already been processed by MLPA and MLPB by this stage, additional processing is required only from the HMM classifier. Combination rules used in the second stage will be discussed below.

(3) *Combination rule*
Many strategies have been designed and implemented by various researchers to combine parallel classifiers with measurement outputs. For example, the sum, product, maximum, minimum, median, and voting rules have been studied extensively [24]. The main advantages of these rules are their simplicity and that they do not require any training. Three of these widely used combination rules have been applied in our month word recognition module. These are the majority vote on decisions, ave (sum) and product. Since detailed descriptions of these rules can be found in the literature [24], here we only describe the weighted product rule used in our system. Let x be an input sample, and $D_k(x)$, where $k = 1, 2, 3$, be the decision on x by the three classifiers MLPA, MLPB, and HMM. $C_{k,w_j}(x)$, where $k = 1, 2, 3$ and $j = 1, 2, \ldots, 29$ represent the confidence values assigned to class w_j by classifier k. $D(x)$ and $C_{w_j}(x)$ are the final combined result and confidence, respectively. The weighted product rule can be described as follows:

$$C_{w_j}(x) = \prod_{k=1}^{3}(C_{k,w_j}(x))^{Weight_k}, j = 1, 2, \ldots, 29,$$
$$D(x) = w_j, \text{ if } C_{w_j}(x) = \max_{i=1}^{29}(C_{w_i}(x)).$$

Here the weighting factors $Weight_k$s are determined in the training stage, and they are chosen to make the confidence values of the three classifiers comparable because the distributions of the confidence values for the three classifiers are different.

Experiments have been conducted to compare these combination rules based on the conditional topology implemented. The results are shown in Table 6.17 without rejections.

Among these three combination rules, weighted product rule is superior to the other two. However, based on error analyses, a problem of product rule is found regarding the effect of very low output (confidence) values, which is also reported in [2] as veto effect. A modified product rule is proposed in this case study to reduce this effect, which will be discussed below.

TABLE 6.17 **Performance of different combination rules on the test set.**

	Recognition rate (%)	
Combination	29 outputs	12 outputs
Majority vote	76.59	79.01
sum	79.69	81.39
Weighted product	84.97	86.67

(4) *Modified product rule*

The veto effect in the product rule is caused by small classifier measurement output values dominating the product, that is, by giving close to zero values. The basic idea is to modify the output of a classifier if it falls below a specified threshold. A modified product (Mproduct) has been presented in [2]. They suggested that if the outputs of more than half classifiers are larger than a threshold for a class, then the output of a classifier which is less than the threshold should be modified by setting the output to the threshold. Better results have been obtained in their experiments. However, they did not give the method for selecting the threshold value. In our experiments, we find that the performance is sensitive to this threshold and it is hard to make a decision except using heuristics.

Based on the above method, a new modified product rule is proposed in our combination system. This method can be described as follows:

If two of $C_{k,w_j}(x)$, $k = 1, 2, 3$ on class w_j are larger than a high threshold (thr1)
and the other $C_{m,w_j}(x)$ is less than a low threshold (thr2), $j = 1, 2, \ldots, 29$
  $C_{m,w_j}(x) = $ thr2.
else
  $C_{k,w_j}(x)$, where $k = 1, 2, 3$ and $j = 1, 2, \ldots, 29$ remain unaltered.

After the above recomputation of $C_{k,w_j}(x)$, the same steps as shown in the previous weighted product rule are adopted. Here the modification of using two different thresholds has been made. The method of choosing these two thresholds in our system is as follows:

- thr1 is larger than thr2, but it is not a real "high" threshold. We set thr1 to a value such that when a classifier measurement output is larger than this value, this output is very often among the top three choices of all the classifiers in the system.
- thr2 is not a fixed value. The measurement output value of the fifth choice of the classifier considered is used as the threshold in our system. (If the value is zero, then thr2 is set to the small value of 0.05 instead.)

In this modified product rule, the thresholds are related to the measurement outputs in the system. It is a reasonable process that provides an automatic setting of the thresholds. Based on this new modified product rule, better results have been obtained for our month word recognition as shown in Table 6.18.

TABLE 6.18   Combination results of modified product rule on the test set.

	Recognition rate (%)	
Combination	29 outputs	12 outputs
Modified product	85.36	87.06

This is the best result produced by our month word recognition system. It is difficult to compare this result with the other systems [19, 40] due to the use of different databases. In [40], a 91% recognition rate for 12 Brazilian month word classes has been reported on a small test set (402 samples). In another system [19], 21 English classes are recognized, and an 87.3% recognition rate was obtained on a test set of 2152 samples. Because both French and English month words are processed in our system, a total of 29 classes are involved, giving rise to greater complexities. The recognition rate of 85.36% for 29 classes in our system is very promising.

Comparing the combination results obtained here with those of the individual classifiers, significant performance improvements have been observed. In addition, some experiments have been conducted to compare the effect of combining three or two classifiers. Combining the MLPA and HMM classifiers using the weighted product rule produces the best results among the combinations of two classifiers, with recognition rates of 83.18% and 85.26% for 29 and 12 outputs, respectively, which is not as good as those of combining the three classifiers.

6.3.5 Experimental Results

Several experiments have been designed to test the performance of our date processing system. Since the performance of the cursive month word recognition has been discussed above, the performances of only the date segmentation module and the entire system will be given below.

6.3.5.1 Date Image Segmentation The test set is derived from the CENPARMI_IRIS check database, and it contains 3399 date images, of which 1219 samples are written in English, and the other 2180 samples are written in French. The segmentation results based on different rejection thresholds are given in Table 6.19 for both English and French samples. Some discussions on these results are given below.

- *The Rejection.* Rejection rate 1 is obtained by trying to recognize every date sample, and a rejection is made when at least one of the three fields corresponding to *year*, *month*, and *day* cannot be found, for example, *day&month* is written or binarized as one component or the *year* field cannot be found. If strong noise such

TABLE 6.19   Performances of date segmentation system for the English and French sets.

English	Correct(%)	Rejection(%)	Error(%)
Rejection rate 1	90.40	1.81	7.79
Rejection rate 2	74.57	21.16	4.27
French	Correct(%)	Rejection(%)	Error(%)
Rejection rate 1	82.94	4.22	12.84
Rejection rate 2	65.69	26.97	7.34

as a big blob of ink (due to improper binarization) or too many components have been detected in the preprocessing stage, a rejection is also made. For rejection rate 2, the recognition result should be a valid date and the average confidence value of the three fields should exceed a threshold; otherwise, a rejection is made.

- *The Performance.* The performance for the English set is better than that for the French set. Several reasons can account for this difference. First, based on the database analysis, it was found that more French checks have the free format. With this format, more variations exist and detecting the handwritten *year* is more difficult than detecting the machine-printed "20" or "19." Therefore, more errors and rejections occur. In addition, the article "Le" sometimes used at the beginning of French date zones, together with freer writing styles on French checks, also increase the difficulty of segmentation.

- *Error Analysis.* Based on our experiments in the training stage, the errors made can be categorized into two classes. The first class contains date images of very poor quality, and it includes: (i) date images having touching fields; (ii) strong noise introduced by improper binarization; and (iii) incomplete dates with one of the three fields missing. Our current segmentation module cannot process this type of date image, so a rejection is often made when rejection rate 2 is imposed. Based on the experiments, about 40% of the errors belong to this type when rejection rate 1 is adopted. The second class of errors consist of errors generated by all segmentation modules in the system.

- *The Efficiency.* The date segmentation is divided into two stages in order to improve the performance and efficiency of the system. The knowledge-based segmentation module is used in the first stage to solve most segmentation cases. Ambiguous cases are handled by multihypotheses generation and evaluation modules in a later stage. Experimental results show that 74.19% of the date images in English and 71.41% of the date images in French are processed in the knowledge-based segmentation module, and the others are processed by the multihypotheses generation and evaluation modules.

6.3.5.2 *Overall Performances*

The overall performances are given in Table 6.20, where the rejections are made under the same conditions as in Table 6.19.

TABLE 6.20 Performances of date processing system for the English and French sets.

English	Correct (%)	Rejection (%)	Error (%)
Rejection rate 1	62.34	1.81	35.85
Rejection rate 2	61.69	21.16	17.15
French	Correct (%)	Rejection (%)	Error (%)
Rejection rate 1	57.75	4.22	38.03
Rejection rate 2	53.67	26.97	19.36

The errors of the date processing system mainly come from segmentation, month word misrecognition and/or numeral misrecognition. Currently a more effective verification module in the postprocessing stage is being considered to reduce the error rates and improve the recognition rates. Also, new checks have been designed and used in practice to facilitate recognition and reading. They include boxes for entering the day, month, and year on the check.

6.3.6 Concluding Remarks

This case study proposes a system for automatically recognizing the date information handwritten on Canadian bank checks. In the system, the date segmentation is implemented at different levels using knowledge obtained from different sources. Simple segmentation cases can be efficiently solved by using the knowledge-based segmentation module, which makes use of some contextual information provided by writing style analyses. For ambiguous segmentation cases, the multihypotheses generation and evaluation modules are invoked to make the final decision based on the recognition results and semantic and syntactic constraints. In addition, a new cursive month word recognizer has been implemented based on a combination of classifiers. The complete system has produced promising performance on a test set from a real-life standard check database.

REFERENCES

1. Y. Al-Ohali, M. Cheriet, and C. Suen. Databases for recognition of handwritten Arabic cheques. In *Proceedings of the 7th International Workshop on Frontiers of Handwriting Recognition*, Amsterdam, The Netherlands, 2002, pp. 601–606.

2. F. M. Alkoot and J. Kittler. Improving the performance of the product fusion strategy. In *Proceedings of the 15th International Conference on Pattern Recognition*, Barcelona, Spain, September 2000, Vol. 2, pp. 164–167.

3. R. Casey and G. Nagy. Recognition of printed Chinese characters. *IEEE Transactions on Electronic Computers.* **15**(1), 91–101, 1966.

4. J. X. Dong, A. Krzyzak, C. Y. Suen. An improved handwritten Chinese character recognition system using support vector machine. *Pattern Recognition Letters.* **26**(12), 1849–1856, 2005.

5. R. Fan, L. Lam, and C. Y. Suen. Processing of date information on cheques. In A. C. Downton and S. Impedovo, editors, *Progress in Handwriting Recognition*, World Scientific, Singapore, 1997, pp. 473–479.

6. H. C. Fu and Y. Y. Xu. Multilinguistic handwritten character recognition by Bayesian decision-based neural networks. *IEEE Transactions on Signal Processing.* **46**(10), 2781–2789, 1998.

7. K. Fukunaga. *Introduction to Statistical Pattern Recognition*, 2nd edition. Academic Press, 1990.

8. D. E. Goldberg, B. Korb, and K. Deb. Messy genetic algorithms: motivation, analysis, and first results. *Complex Systems.* **3**(5), 493–530, 1989

9. N. Hagita, S. Naito, and I. Masuda. Handprinted Chinese characters recognition by peripheral direction contributivity feature. *Transactions on IEICE Japan.* **J66-D**(10), 1185–1192, 1983 (in Japanese).

10. M. Hamanaka, K. Yamada, and J. Tsukumo. Normalization-cooperated feature extraction method for handprinted Kanji character recognition. In *Proceedings of the 3rd International Workshop on Frontiers of Handwriting Recognition*, Buffalo, NY, 1993, pp. 343–348.

11. R. V. D. Heiden and F. C. A. Gren. The box-cox metric for nearest neighbor classification improvement. *Pattern Recognition.* **30**(2), 273–279, 1997.

12. T. H. Hildebrandt and W. Liu. Optical recognition of Chinese characters: advances since 1980. *Pattern Recognition.* **26**(2), 205–225, 1993.

13. T. Horiuchi, R. Haruki, H. Yamada, and K. Yamamoto. Two-dimensional extension of nonlinear normalization method using line density for character recognition. In *Proceedings of the 4th International Conference on Document Analysis and Recognition*, Ulm, Germany, 1997, pp. 511–514.

14. G. F. Houle, D. B. Aragon, R. W. Smith, M. Shridhar, and D. Kimura. A multi-layered corroboration-based check reader. In *Proceedings of IAPR Workshop on Document Analysis Systems*, Malvern, USA, October 1996, pp. 495–546.

15. T. Iijima, H. Genchi, and K. Mori. A theoretical study of the pattern identification by matching method. In *Proceedings of the First USA-JAPAN Computer Conference*, October 1972, pp. 42–48.

16. N. Kato, M. Suzuki, S. Omachi, H. Aso, and Y. Nemoto. A handwritten character recognition system using directional element feature and asymmetric Mahalanobis distance. *IEEE Transactions on Pattern Analanalysis and Machine Intelligence.* **21**(3), 258–262, 1999.

17. N. Kharma, T. Kowaliw, E. Clement, C. Jensen, A. Youssef, and J. Yao. Project CellNet: Evolving an autonomous pattern recognizer. *International Journal on Pattern Recognition and Artificial Intelligence.* **18**(6), 1039–1056, 2004.

18. I.-J. Kim and J. H. Kim. Statistical character structure modeling and its application to handwritten Chinese character recognition. *IEEE Transactions on Pattern Analysis and Machine Intelligence*; **25**(11), 1422–1436, 2003.

19. J. Kim, K. Kim, C. P. Nadal, and C. Y. Suen. A methodology of combining HMM and MLP classifiers for cursive word recognition. In *Proceedings of the 15th International Conference on Pattern Recognition*, Barcelona, Spain, September 2000, Vol. 2, pp. 319–322.

20. J. Kim, K. Kim, and C. Y. Suen. Hybrid schemes of homogeneous and heterogeneous classifiers for cursive word recognition. In *Proceedings of the 7th International Workshop on Frontiers in Handwriting Recognition*, Amsterdam, the Netherlands, September 2000, pp. 433–442.

21. F. Kimura, K. Takashina, S. Tsuruoka, and Y. Miyake. Modified quadratic discriminant functions and the application to Chinese character recognition. *IEEE Transactions on Pattern Analysis and Machine Intelligence*; **9**(1), 149–153, 1987.

22. F. Kimura, T. Wakabayashi, S. Tsuruoka, and Y. Miyake. Improvement of handwritten Japanese character recognition using weighted direction code histogram. *Pattern Recognition.* **30**(8), 1329–1337, 1997.

23. Y. Kimura, T. Wakahara, and A. Tomono. Combination of statistical and neural classifiers for a high-accuracy recognition of large character sets. *Transactions of IEICE Japan.* **J83-D-II**(10), 1986–1994, 2000 (in Japanese).

24. J. Kittler, M. Hatef, R. P. W. Duin, and J. Matas. On combining classifiers. *IEEE Transactions on Pattern Analysis and Machine Intelligence*; **20**(3), 226–239, 1998.

25. T. Kowaliw, N. Kharma, C. Jensen, H. Mognieh, and J. Yao. Using competitive co-evolution to evolve better pattern recognizers. *International Journal Computational Intelligence and Applications*. **5**(3), 305–320, 2005.

26. L. Lam, Q. Xu, and C. Y. Suen. Differentiation between alphabetic and numeric data using ensembles of neural networks. In *Proceedings of the 16th International Conference on Pattern Recognition*, Quebec City, Canada, August 2002, Vol. 4, pp. 40–43.

27. C.-L. Liu, Y-J. Liu, and R-W. Dai. Preprocessing and statistical/structural feature extraction for handwritten numeral recognition. In A. C. Downton and S. Impedovo, editors, *Progress of Handwriting Recognition*, World Scientific, Singapore, 1997, pp. 161–168.

28. C.-L. Liu and M. Nakagawa. Evaluation of prototype learning algorithms for nearest neighbor classifier in application to handwritten character recognition. *Pattern Recognition*. **34**(3), 601–615, 2001.

29. C.-L. Liu, K. Nakashima, H. Sako, and H. Fujisawa. Handwritten digit recognition: Benchmarking of state-of-the-art techniques. *Pattern Recognition*. **36**(10), 2271–2285, 2003.

30. C.-L. Liu, H. Sako, and H. Fujisawa. Handwritten Chinese character recognition: alternatives to nonlinear normalization. In *Proceedings of the 7th International Conference on Document Analysis and Recognition*, Edinburgh, Scotland, 2003, pp. 524–528.

31. C.-L. Liu, K. Nakashima, H. Sako, and H. Fujisawa. Handwritten digit recognition: investigation of normalization and feature extraction techniques. *Pattern Recognition*. **37**(2), 265–279, 2004.

32. C.-L. Liu, S. Jaeger, and M. Nakagawa. Online recognition of Chinese characters: the state-of-the-art. *IEEE Transactions on Pattern Analysis Machanic Intelligence*; **26**(2), 198–213, 2004.

33. C.-L. Liu and K. Marukawa. Global shape normalization for handwritten Chinese character recognition: a new method. In *Proceedings of the 9th International Workshop on Frontiers of Handwriting Recognition*, Tokyo, Japan, 2004, pp. 300–305.

34. C.-L. Liu and K. Marukawa. Pseudo two-dimensional shape normalization methods for handwritten Chinese character recognition. *Pattern Recognition*. **38**(12), 2242–2255, 2005.

35. C.-L. Liu, M. Koga, and H. Fujisawa. Gabor feature extraction for character recognition: comparison with gradient feature. In *Proceedings of the 8th International Conference on Document Analysis and Recognition*, Seoul, Korea, 2005, pp. 121–125.

36. C.-L. Liu. High accuracy handwritten Chinese character recognition using quadratic classifiers with discriminative feature extraction. In *Proceedings of the 18th International Conference on Pattern Recognition*, Hong Kong, 2006, Vol. 2, pp. 942–945.

37. H. Liu and X. Ding. Handwritten character recognition using gradient feature and quadratic classifier with multiple discrimination schemes. In *Proceedings of the 8th International Conference on Document Analysis and Recognition*, Seoul, Korea, 2005, pp. 19–23.

38. M. Mitchell. *An Introduction to Genetic Algorithms*. MIT Press, 1998.

39. S. Mori, K. Yamamoto, and M. Yasuda. Research on machine recognition of handprinted characters. *IEEE Transactions on Pattern Analysis Machanic Intelligence*; **6**(4), 386–405, 1984.

40. M. Morita, A. El Yacoubi, R. Sabourin, F. Bortolozzi, and C. Y. Suen. Handwritten month word recognition on Brazilian bank cheques. In *Proceedings of the 6th International Conference on Document Analysis and Recognition*, Seattle, USA, September 2001, pp. 972–976.

41. K. Saruta, N. Kato, M. Abe, Y. Nemoto. High accuracy recognition of ETL9B using exclusive learning neural network-II (ELNET-II). *IEICE Transactions of Information and Systems*. **79-D**(5), 516–521, 1996.

42. M. Shi, Y. Fujisawa, T. Wakabayashi, and F. Kimura. Handwritten numeral recognition using gradient and curvature of gray scale image. *Pattern Recognition*. **35**(10), 2051–2059, 2002.

43. W. Stalling. Approaches to Chinese character recognition. *Pattern Recognition*. **8**, 87–98, 1976.

44. N. W. Strathy. Handwriting recognition for cheque processing. In *Proceedings of the 2nd International Conference on Multimodal Interface*, Hong Kong, January 1999, pp. 47–50.

45. C. Y. Suen, Q. Xu, and L. Lam. Automatic recognition of handwritten data on cheques—fact or fiction? *Pattern Recognition Letters*. **20**(11–13), 1287–1295, 1999.

46. L.-N. Teow and K.-F. Loe. Robust vision-based features and classification schemes for off-line handwritten digit recognition. *Pattern Recognition*. **35**(11), 2355–2364, 2002.

47. J. Tsukumo and H. Tanaka. Classification of handprinted Chinese characters using non-linear normalization and correlation methods. In *Proceedings of the 9th International Conference on Pattern Recognition*, Rome, Italy, 1988, pp. 168–171.

48. M. Umeda. Advances in recognition methods for handwritten Kanji characters. *IEICE Trans. Information and Systems*. **E29**(5), 401–410, 1996.

49. T. Wakabayashi, S. Tsuruoka, F. Kimura, and Y. Miyake. On the size and variable transformation of feature vector for handwritten character recognition. *Transactions of IEICE Japan*. **J76-D-II**(12), 2495–2503, 1993 (in Japanese).

50. X. Wang, X. Ding, and C. Liu. Gabor filter-base feature extraction for character recognition. *Pattern Recognition*. **38**(3), 369–379, 2005.

51. H. Yamada, K. Yamamoto, and T. Saito. A nonlinear normalization method for hanprinted Kanji character recognition—line density equalization. *Pattern Recognition*. **23**(9), 1023–1029, 1990.

52. Y. Yamashita, K. Higuchi, Y. Yamada, and Y. Haga. Classification of handprinted Kanji characters by the structured segment matching method. *Pattern Recognition Letters*. **1**, 475–479, 1983.

53. M. Yasuda and H. Fujisawa. An improvement of correlation method for character recognition. *Transactions of IEICE Japan*. **J62-D**(3), 217–224, 1979 (in Japanese).

54. M. Yasuda, K. Yamamoto, H. Yamada, and T. Saito. An improved correlation method for handprinted Chinese character recognition in a reciprocal feature field. *Transactions of IEICE Japan*. **J68-D**(3), 353–360, 1985 (in Japanese).

INDEX

Printed and bound by CPI Group (UK) Ltd, Croydon, CR0 4YY

27/10/2024

14580254-0002